Springer Series in Advanced Manufacturing

Series Editor

Professor D.T. Pham
Intelligent Systems Laboratory
WDA Centre of Enterprise in Manufacturing Engineering
University of Wales Cardiff
PO Box 688
Newport Road
Cardiff
CF2 3ET
UK

Other titles in this series

Assembly Line Design
B. Rekiek and A. Delchambre

Advances in Design
H.A. ElMaraghy and W.H. ElMaraghy (Eds.)

Effective Resource Management in Manufacturing Systems: Optimization Algorithms in Production Planning
M. Caramia and P. Dell 'Olmo

Condition Monitoring and Control for Intelligent Manufacturing
L. Wang and R.X. Gao (Eds.)

Optimal Production Planning for PCB Assembly
W. Ho and P. Ji

Trends in Supply Chain Design and Management: Technologies and Methodologies
H. Jung, F. F. Chen and B. Jeong (Eds.)

Process Planning and Scheduling for Distributed Manufacturing
L. Wang and W. Shen (Eds.)

Collaborative Product Design and Manufacturing Methodologies and Applications
W.D. Li, S.K. Ong, A.Y.C. Nee and C.s McMahon (Eds.)

Decision Making in the Manufacturing Environment
R. Venkata Rao

Reverse Engineering: An Industrial Perspective
V. Raja and K.J. Fernandes (Eds.)

Frontiers in Computing Technologies for Manufacturing Applications
Y. Shimizu, Z. Zhong and R. Batres

Automated Nanohandling by Microrobots
S. Fatikow

A Distributed Coordination Approach to Reconfigurable Process Control
N.N. Chokshi and D.C. McFarlane

ERP Systems and Organisational Change
B. Grabot, A. Mayère and I. Bazet (Eds.)

ANEMONA
V. Botti and A. Giret

Theory and Design of CNC Systems
S.-H. Suh, S.-K. Kang, D.-H. Chung and I. Stroud

Machining Dynamics
K. Cheng

Hoda A. ElMaraghy
Editor

Changeable and Reconfigurable Manufacturing Systems

Hoda A. ElMaraghy, PhD, PEng, FSME, FCSME
Intelligent Manufacturing Systems (IMS) Center
University of Windsor
204 Odette Building
401 Sunset Avenue
Windsor, Ontario, N9B 3P4
Canada

ISBN: 978-1-84882-066-1 e-ISBN: 978-1-84882-067-8

DOI 10.1007/978-1-84882-067-8

Springer Series in Advanced Manufacturing ISSN 1860-5168

A catalogue record for this book is available from the British Library

Library of Congress Control Number: 2008940430

Cover design: eStudioCalamar S.L., Girona, Spain

Printed on acid-free paper

9 8 7 6 5 4 3 2 1

springer.com

Preface

"The only thing we know about the future is that it will be different."
Peter Drucker (1909–2005)

Change has become a constant in today's manufacturing environment. While change is inevitable, it is important to take advantage of it and make it happen efficiently through good designs and by developing effective change enablers. The advantages of ***change ability*** are well known, and have been demonstrated by many examples as early as the invention of the movable type printing machines.

Globalization, unpredictable markets, increased products customization and the quest for competitive advantages are but a few of the many challenges facing manufacturing enterprises now and in the future. Frequent changes in products, production technologies and manufacturing systems are evident today along with their significant implementation cost. One key strategy for success is to satisfy the market need for products variations and customization, utilizing the new technologies, while reducing the resulting variations in their manufacturing and associated cost. This trend is on the rise in view of the paradigm shifts witnessed in manufacturing systems and their increased flexibility and responsiveness to cope with the evolution of both products and systems.

A host of external and internal change drivers exist that affect the manufacturing enterprises at various levels from strategic planning for re-positioning the business, down to the actual production facilities to achieve a high degree of adaptability. The drivers relate to business strategy modification, market volatility and products/production variations. The changing manufacturing environment, characterized by aggressive competition on a global scale, scarce resources and rapid changes in process technology, requires careful attention in order to prolong the life of manufacturing systems by making them easily adaptable and facilitating the integration of new technologies and new functions. Changes can most often be anticipated but some go beyond the design range. This requires providing innovative change enablers and adaptation mechanisms to achieve modularity, scalability and compatibility. While changes may not always be anticipated, the behavior of their enablers should be pre-planned for all scenarios to ensure cost effective adaptability.

Changeability is defined as the characteristics to economically accomplish early and foresighted adjustments of the factory's structures and processes on all levels, in response to change impulses.

Several manufacturing systems paradigms have emerged as a result of these changes including agile, adaptable, flexible and reconfigurable manufacturing. The ability to cope with change is the common denominator among all these paradigms, each of which presents a set of technological solutions to enable changes to occur efficiently and profitably. Flexible manufacturing for example changes the system behavior without changing its configuration, while reconfigurable manufacturing would change the system behavior by changing its configuration.

There are two types of change enablers: hard or physical enablers and soft or logical enablers. The "physical/hard" change enablers include the physical attributes that facilitate change. These characteristics are not only limited to the machinery but they also apply to the factories infrastructures, physical plant and buildings. Hardware changes also require major changes at the "logical/soft" enablers level, such as the software systems used to control individual machines, complete cells, and systems as well as to process plan individual operations and to plan and control the whole production. The logical enabling technologies extend beyond the factory walls to the strategic planning levels, logistics and supply chains. In addition, manufacturing changes are not limited to the technical systems; they include the business organization and employees that should also be planned and managed effectively.

The role of changeability enablers can be well illustrated, as mentioned, by the example of the invention of the movable type printing machine. In the early days, books were either copied out by hand on scrolls and paper or printed from hand-carved wooden blocks, each block is used to print a whole page, a part of a page or even individual letters. This took a long time, and even a short book could take months to complete. The woodwork was extremely time-consuming, the carved letters or blocks were very fragile and the susceptibility of wood to ink gave such blocks a limited lifespan. Moreover, the same hand-carved letters did not look the same. Johannes Gutenberg (1397–1468) is generally credited with the invention of practical movable type. He made metal moulds, by the use of dies, into which he could pour hot liquid metal, in order to produce separate letters having the same shape as those written by hand. These letters were consistent, more readable and more durable than wooden blocks. They could be arranged and re-arranged many times to create different pages from the same set of letters. The Koreans (in 1234, over 200 years ahead of Gutenberg's feat) and the Chinese (between 1041 to 1048) have independently invented movable type. However, it was not until Gutenberg introduced around 1450 the use of the enabling printing press technology (used in his times by the wine industry) to press the arranged type letters against paper that this invention took off. The press enabled sharp impressions to be made on both sides of a sheet of paper and allowed many repetitions as well as letters re-use.

Movable print is a perfect example of early applications of standardization, modularity, compatibility, inter-changeability, scalability, flexibility and reconfigurability. Regardless of earlier introductions of the movable print, it was Gutenberg's combination of the printing press; movable type, paper and ink that helped the invention evolve into an innovative and practical process. By combining these elements into a production system, he made the rapid printing of written materials feasible, which lead to an information explosion in Renaissance Europe. The print invention is regarded by many as the invention of the millennium, thanks to Gutenberg, who provided the ***change ability*** and ***technological enablers*** to make it a success, which lead to mass printing practices that changed our world.

In this book, the technological enablers of changeability are particularly emphasized. Many important perspectives on change in manufacturing and its different facets are provided. The book presents the new concept of Changeability as an umbrella framework that encompasses many paradigms such as agility, adaptability, flexibility and reconfigurability, which are in turn enablers of change. It establishes the relationship among these paradigms and presents a hierarchical classification that puts them in context at all levels of a manufacturing enterprise. It provides the definitions and classification of key terms in this new field. The book places great emphasis on the required change enablers. It contains original contributions and results from senior international experts, experienced practitioners and accomplished researchers in the field of manufacturing. It presents cutting edge technologies, the latest thinking and research results as well as future directions to help manufacturers stay competitive. In addition, most chapters contain either industrial applications or case studies to clearly demonstrate the applicability of these important concepts and their impact.

The book is organized in 5 parts and 22 chapters by authors from Canada, Europe, Japan and Asia. It offers balanced and comprehensive treatment of the subjects as well as in depth analysis of many related issues. **Part I** introduces manufacturing changeability, its definitions, characteristics, enablers and strategies, presents models and enablers for changing and evolving products and their systems, and discusses the concept of focused flexibility in production systems. **Part II** deals with the physical technological change enablers for machine tools and robots configuration and re-configuration and control, including new unified dynamic and control models, and highlights the important, but less discussed, changeable and reconfigurable assembly systems. **Part III** focuses on the logical change enablers. It presents new unified dynamic and control models for reconfigurable robots as well as reconfigurable control systems. It introduces novel methods for reconfiguring process plans, new perspectives on adaptive as well as change ready production and manufacturing planning and control systems, and models for capacity planning and its complexity. **Part IV** discusses the topic of managing and justifying change in manufacturing including the effect of changeability on the design of products and systems, the use and programming of CNC machine tools, quality and maintenance strategies for reconfigurable and changeable manufacturing and the economic and strategic justification of these systems. **Part V** sheds light on some important future directions such as the cognitive factory, the migration manufacturing new concept for automotive body production and an architectural view of changeable factory buildings.

The book will serve as a comprehensive reference in this subject for industrial professionals, managers, engineers, specialists, consultants, researchers and academics in manufacturing, industrial and mechanical engineering; and general readers who are scientifically bent and interested to learn about the new and emerging manufacturing paradigms and their potential impact on the work place and future jobs. It can also be used as a primary or supplementary textbook for both post-graduate and senior under-graduate courses in Manufacturing Paradigms, Advanced Manufacturing Systems, Flexible/Reconfigurable Manufacturing, Integrated Manufacturing, and Management of Technology.

I hope you will enjoy reading this book, and would like to leave you with a final thought best expressed by the following interesting quote:

"I do not know whether it becomes better if it changes.
But it must change if it should become better."
German Philosopher,
Georg Christoph Lichtenberg (1742–1799)

Windsor, Ontario, Canada *Hoda A. ElMaraghy*
July 2008

Acknowledgments

The contributions of all authors and their cooperation throughout the preparation of their manuscripts have been instrumental in shaping this book and are sincerely acknowledged. The important work of many colleagues in the Working Group on Changeability of the International Academy of Production Engineering Research (CIRP), lead by Professor Hans-Peter Wiendahl, provided the inspiration and impetus for this book. Several of these international experts have contributed valuable chapters in the book.

The research support I received from the Canada Research Chairs program since 2002 has made it possible for me to conduct a comprehensive research program in the field of manufacturing systems. It enabled me to supervise and train many researchers including Post Doctoral fellows, Ph.D. candidates, Master's students and research engineers, and disseminate the resulting research outcomes in the last six years in 100 referred journal and conference papers. Sample outputs of this research appear in 9 co-authored chapters in the book. Many insightful discussions, input and critique from members of my research group, too many to list here, have been very constructive, and for which I am grateful.

The expert assistance of Miss Zaina Batal, the Administrative and Research Assistant at the Intelligent Manufacturing Systems (IMS) Center at the University of Windsor, in the compilation, checking, verification and coordination of all contributions and helping me complete this project in a timely manner is greatly appreciated. Finally, the support and guidance of the Springer editorial staff, Mr. Anthony Doyle and Mr. Simon Rees, have been very useful throughout the book proposal and manuscript preparation stages.

Windsor, Ontario, Canada *Hoda A. ElMaraghy*
July 2008

Contents

Contributors

Abele, E., Prof. Dr.-Ing.
Professor
Institute of Production Management, Technology and Machine Tools (PTW)
Technische Universität Darmstadt
Petersenstraße 30
64287 Darmstadt, Germany

AlGeddawy, T.N., M.Sc.,
Ph.D. Candidate
Intelligent Manufacturing Systems (IMS) Center
University of Windsor
204 Odette Bldg., 401 Sunset Ave.
Windsor, ON N9B 3P4, Canada

Azab, A., Ph.D.
Assistant Professor
Industrial and Manufacturing Systems Engineering Department
Intelligent Manufacturing Systems (IMS) Center
University of Windsor
204 Odette Bldg., 401 Sunset Ave.
Windsor, ON N9B 3P4, Canada

Beetz, M., Ph.D.
Professor
Computer Science Department
Technical University of Munich
Boltzmannstraße 15
85748 Garching, Germany

Bender, K., Prof. Dr.-Ing.
Professor
Institute of Information Technology in Mechanical Engineering (itm)
Technical University of Munich
Boltzmannstraße 15,
85748 Garching, Germany

Cao, Y., Ph.D.
Assistant Professor
School of Engineering
University of British Columbia
3333 University Way
Kelowna, BC V1V 1V7, Canada

Deif, A., Ph.D.
Assistant Professor
Industrial Systems Engineering
Faculty of Engineering
University of Regina
Regina, SASK, S4S 0A2, Canada

Djuric, A., Ph.D.
Researcher
Intelligent Manufacturing Systems (IMS) Center
University of Windsor
204 Odette Bldg., 401 Sunset Ave.
Windsor, ON N9B 3P4, Canada

ElMaraghy, H.A., Ph.D., P.Eng., FCIRP, FSME, FCSME
Canada Research Chair in Manufacturing Systems
Professor,
Department of Industrial & Manufacturing Systems Engineering
Director, Intelligent Manufacturing Systems (IMS) Center
University of Windsor
204 Odette Bldg., 401 Sunset Ave.
Windsor, ON N9B 3P4, Canada

ElMaraghy, W.H., Ph.D., P.Eng., FCIRP, FCSME, FASME
Professor and Head,
Department of Industrial & Manufacturing Systems Engineering
Director, Intelligent Manufacturing Systems (IMS) Center
University of Windsor
204 Odette Bldg., 401 Sunset Ave.
Windsor, ON N9B 3P4, Canada

Engelhard, M., Dipl.-Tech.-Math.
Researcher
Institute of Product Development
Technical University of Munich
Boltzmannstraße 15
85748 Garching, Germany

Ertelt, C., Dipl.-Ing.
Researcher
Institute of Product Development
Technical University of Munich
Boltzmannstraße 15
85748 Garching, Germany

Friedrich, M., Dipl.-Ing.
Researcher
Institute of Information Technology in Mechanical Engineering (itm)
Technical University of Munich
Boltzmannstraße 15
85748 Garching, Germany

Fujishima, M., Ph.D.
Director and General Manager
Mori Seiki Co. Ltd.
362 Idono-cho
Yamato-Koriyama City, Nara 639-1183, Japan

Hedrick, R., MA.Sc.
Software Specialist, CAM Multi-Tasking
MasterCam Canada Inc.
LaSalle, ON N9J 3H6, Canada

Herle, S., S l.Ing.
Lecturer
Faculty of Automation and Computer Science
Department of Automation
Technical University of Cluj-Napoca, Romania

Ismail., M., M.Sc., MBA.
Ph.D. Candidate
Intelligent Manufacturing Systems (IMS) Center
University of Windsor
204 Odette Bldg., 401 Sunset Ave.
Windsor, ON N9B 3P4, Canada

Kircher, C., Dipl.-Ing.
Researcher
Institute for Control Engineering of Machine Tools and Manufacturing Units (ISW)
University of Stuttgart
Seidenstraße 36
70174 Stuttgart, Germany

Kupke, D., Dipl.-Ing. Dipl.-Wirt. Ing.
Production Management
Chair for Production Engineering
Laboratory for Machine Tools and Production Engineering (WZL)
University of Aachen
Steinbachstraße 53B
52074 Aachen, Germany

Kuzgunkaya, O., Ph.D.
Assistant Professor
Department of Mechanical and Industrial Engineering
Concordia University, EV4.139
1455 de Maisonneuve Blvd. West
Montreal, Quebec, H3G 1M8, Canada

Labib, A., Ph.D.
Professor
Department of Strategy and Business Systems (SBS)
Portsmouth Business School (PBS)
University of Portsmouth
Richmond Building, Portland Street
Portsmouth, PO1 3DE U.K.

Lau, C., Dipl.-Wi.-Ing.
Researcher
Institute for Machine Tools and Industrial Management (*iwb*)
Technische Universität München
Beim Glaspalast 5
86153 Augsburg, Germany

Lenders, M., Dipl.-Ing.
Chief Engineer, Department Manager Innovation Management
Chair for Production Engineering
Laboratory for Machine Tools and Production Engineering (WZL)
University of Aachen
Steinbachstraße 53B
52074 Aachen, Germany

Lotter, B., Prof. Dr.-Ing.
Assembly Consultant
Kirchberg 8
75038 Oberderdingen, Germany

Meichsner, T., Dr.-Ing.
Executive Director
Wilhelm Karmann GmbH
Karmannstraße 1
49084 Osnabrück, Germany

Meselhy, K.T., M.Sc.
Ph.D. Candidate
Intelligent Manufacturing Systems (IMS) Center
University of Windsor
204 Odette Bldg., 401 Sunset Ave.
Windsor, ON N9B 3P4, Canada

Mori, M., Ph.D.
President
Mori Seiki Co. Ltd.
362 Idono-cho
Yamato-Koriyama City, Nara 639-1183, Japan

Nussbaum, C., Dipl.-Ing.
Innovation Management
Chair for Production Engineering
Laboratory for Machine Tools and Production Engineering (WZL)
University of Aachen
Steinbachstraße 53B
52074 Aachen, Germany

Ostgathe, M., Dipl.-Ing.
Researcher
Institute for Machine Tools and Industrial Management (*iwb*)
Technical University of Munich
Boltzmannstraße 15
85748 Garching, Germany

Pritschow, G., Prof. Dr.-Ing. Dr.h.c.mult. Dr. E.h.
Professor
Institute for Control Engineering of Machine Tools & Manufacturing Units (ISW)
University of Stuttgart
Seidenstraße 36
70174 Stuttgart, Germany

Reichardt, J.
Prof. J. Reichardt Architekten BDA
Im Walpurgistal 10
45136 Essen, Germany

Reinhart, G., Prof. Dr.-Ing.
Head
Institute for Machine Tools and Industrial Management (*iwb*)
Technical University of Munich
Boltzmannstraße 15
85748 Garching, Germany

Ruehr, T., Dipl.-Inf.
Researcher
Computer Science Department, Chair IX
Technical University of Munich
Boltzmannstraße 15
85748 Garching, Germany

Samy, S.N., M.Sc.
Ph.D. Candidate
Intelligent Manufacturing Systems (IMS) Center
University of Windsor
204 Odette Bldg., 401 Sunset Ave.
Windsor, ON N9B 3P4, Canada

Schuh, G., Prof. Dr.-Ing. Dipl.-Wirt. Ing.
Professor & Innovation Management Chair for Production Engineering
Laboratory for Machine Tools and Production Engineering (WZL)
University of Aachen
Steinbachstraße 53B
52074 Aachen, Germany

Seyfarth, M., Dipl.-Ing.
Researcher
Institute for Control Engineering of Machine Tools & Manufacturing Units (ISW)
University of Stuttgart
Seidenstraße 36
Stuttgart, Germany

Shea, K., Prof. PhD.
University Professor
Virtual Product Development
Institute of Product Development
Mechanical Engineering Department
Technical University of Munich
Boltzmannstraße 15
85748 Garching, Germany

Terkaj, W., Ing.
Ph.D Candidate
Department of Mechanical Engineering
Politecnico di Milano
Via La Masa 1
20156 Milano, Italy

Tolio, T., Ph.D.
Professor
Department of Mechanical Engineering
Politecnico Di Milano
Via La Masa 1
20133 Milan, Italy

Urbanic, R.J., Ph.D.
Assistant Professor
Department of Industrial & Manufacturing Systems Engineering (IMSE)
University of Windsor
401 Sunset Ave.
Windsor, ON N9B 3P4, Canada

Valente, A., Ph.D.
Mechanical Engineer
Division of Manufacturing and Production Systems
Politecnico di Milano
Piazza Leonardo da Vinci, 32
20133 Milano, Italy

Vogl, W., Dipl.-Ing.
Researcher
Institute for Machine Tools and Industrial Management (*iwb*)
Technical University of Munich
Boltzmannstraße 15
85748 Garching, Germany

Wiendahl, H.H., Dr.-Ing.
Specialty Advisor
Fraunhofer Institute for Manufacturing Engineering and Automation IPA in Stuttgart (IPA FHG)
Fraunhofer University
Auftragsmanagement und Logistik
Nobelstrasse 12
70569 Stuttgart, Germany

Wiendahl, H.P., Prof. em. Dr.-Ing E.h. Dr. sc h.c. ETH Dr.-Ing.
Professor Emeritus
Institute of Production Systems and Logistics (IFA)
University of Hannover
An der Universität 2
30823 Garbsen, Germany

Wiesbeck, M., Dipl.-Wi.-Ing.
Researcher
Institute for Machine Tools and Industrial Management (*iwb*)
Technical University of Munich
Boltzmannstraße 15
85748 Garching, Germany

Wörn, A., M.Sc.
Scientific Assistant
Institute of Production Management, Technology and Machine Tools (PTW)
Technische Universität Darmstadt
Petersenstraße 30
64287 Darmstadt, Germany

Wurst, K-H., Dr.-Ing.
Researcher
Institute for Control Engineering of Machine Tools & Manufacturing Units (ISW)
University of Stuttgart
Seidenstraße 36
70174 Stuttgart, Germany

Yuniarto, M.N., Ph.D.
Division of Manufacturing Engineering
Department of Mechanical Engineering
ITS-Surabaya, 60111, Indonesia

Zaeh, M.F., Prof. Dr.-Ing.
Head
Institute for Machine Tools and Industrial Management (*iwb*)
Technical University of Munich
Boltzmannstraße 15
85748 Garching, Germany

Part I
Definitions and Strategies

Chapter 1
Changeability – An Introduction

H. ElMaraghy[1] **and H.-P. Wiendahl**[2]

Abstract Manufacturing has been experiencing dynamically changing environment that presents industrialists and academics with formidable challenges to adapt to these changes effectively and economically while maintaining a high level of responsiveness, agility and competitiveness. Advances in manufacturing technologies, equipment, systems and organizational strategies are helping manufacturers meet these challenges. The ability to change and effectively manage this change is a fundamental pre-requisite for surviving and prospering in this turbulent environment. Changeability is presented as an umbrella concept that encompasses many change enablers at various levels of an industrial company throughout the life cycle of the manufacturing system. In this introduction, the scope of manufacturing changeability is outlined, the objects of change are defined, the change enablers are introduced and discussed and the change management strategy is highlighted.

Keywords changeability objects, changeability enablers, changeability strategy

1.1 Motivation

Manufacturing systems have evolved over the years in response to many external drivers including the introduction of new manufacturing technologies and materials, the constant evolution of new products and the increased emphasis on quality as well as the escalating global competition and pressing need for responsiveness, agility and adaptability. Internally the desire to reduce waste and increase efficiency and productivity while creating/preserving/increasing high value jobs with meaningful human involvement is an equal challenge.

---

[1] Intelligent Manufacturing Systems (IMS) Center, University of Windsor, Canada

[2] University of Hannover, Germany

Several manufacturing systems paradigms have appeared as a result over the years in view of these drivers and against a back drop of volatility in market demands, changing customer's preferences and need for more products differentiation and customization. In addition, global and distributed supply chains and added pressures of labor issues, competition from developing countries and currency fluctuations increase the pressure.

Recently introduced manufacturing systems paradigms, such as Flexible and Reconfigurable manufacturing, are responding to these needs in different ways. They can be viewed as enablers of change and transformation at different levels.

Flexible manufacturing allows changing individual operations, processes, parts routing and production schedules. This corresponds to variations in products within a pre-defined scope of a parts family. It also allows adjusting production capacity within the limits of the existing system. Therefore, FMS offers generalized flexibility that is built-in a priory and permits changes and adaptation of processes and production volumes, within the pre-defined boundaries, without physically changing the manufacturing system itself.

Reconfigurable manufacturing allows changeable functionality and scalable capacity (Koren, 2006) by physically changing the components of the system through adding, removing or modifying machine modules, machines, cells, material handling units and/or complete lines. Hence, RMS responds to changes by offering focused flexibility on demand by physically reconfiguring the manufacturing system. Hardware reconfiguration also requires major changes in the software used to control individual machines, complete cells, and systems as well as to plan and control the individual processes and overall production.

Both flexible and reconfigurable manufacturing will be briefly reviewed in this chapter due to their importance as enablers of change.

The significant reduction in product development time and faster introduction of new products models and variations were not paralleled in the field of design and development of manufacturing systems. These manufacturing systems must be designed to satisfy certain products requirements and constraints that vary over time, and the rate of such changes has accelerated recently.

In the context of manufacturing systems, one can envisage a life cycle (ElMaraghy 2005), as outlined in Fig. 1.1, which starts with the initial system design and synthesis according to the specified objectives and constraints followed by modeling, analysis and simulation, then the final design is realized, implemented and used in production. The manufacturing system undergoes re-design and reconfiguration, throughout its operation and as new requirements emerge and changes are required, aimed meeting the requirements of the changed environment. Therefore, both soft/logical and hard/physical reconfiguration (ElMaraghy, 2005) and flexibility are in fact enablers of change and can extend the utility, usability, and life of manufacturing systems.

One result of the dynamics of markets is the mutation of the product life cycle characteristic and the increasing divergence of the life cycles of the associated manufacturing processes and equipment, as shown in Fig. 1.2 (Wirth, 2004).

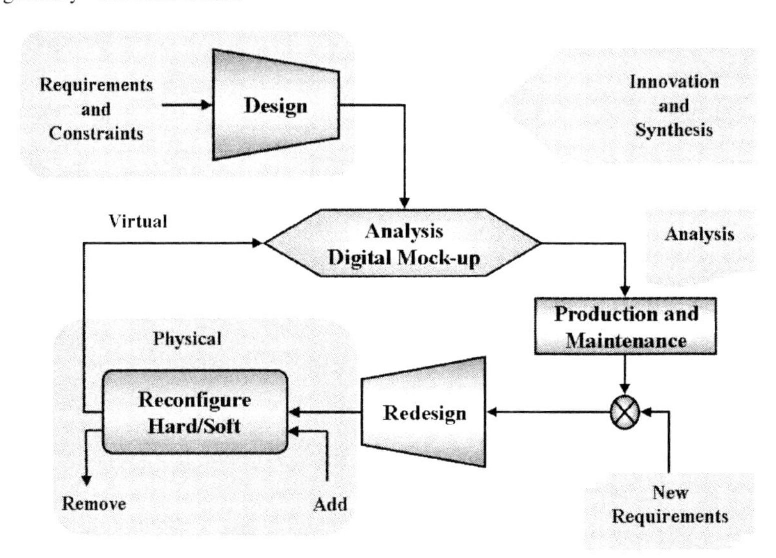

Fig. 1.1 Manufacturing systems life cycle (ElMaraghy, 2005)

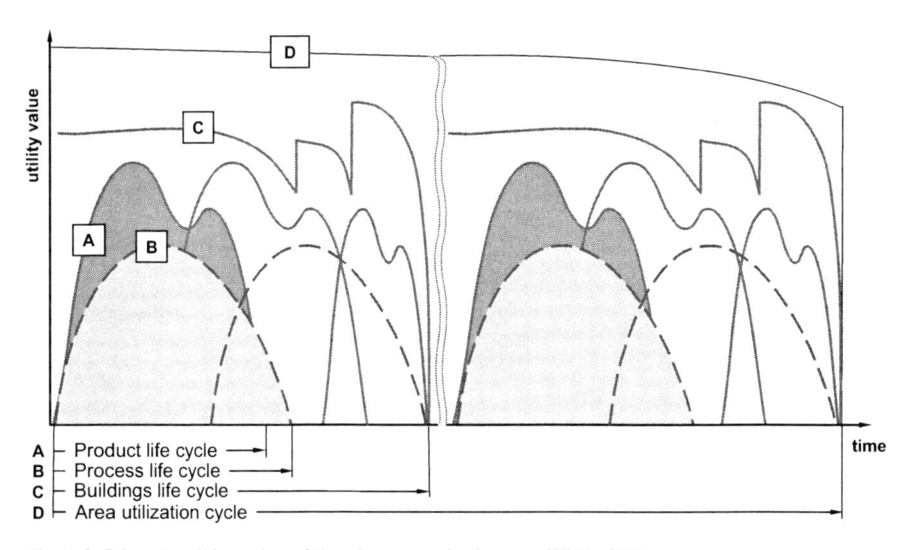

Fig. 1.2 Diverging life cycles of the elements of a factory (Wirth, 2004)

In the past, a steady production volume increase was observed after the release of a product, followed by quite long and stable periods then a ramp down. At present, production volumes climb much faster to the first peak, then go down and reach a second peak after further promotion activities. A face-lift of the product often follows shortly thereafter, and then a sudden reduction of the production volume occurs, mainly due to introducing a new product.

Furthermore, it has been observed that the life cycle of a product used to be approximately in the same range of the life cycle of the machine tools and technological processes. With the ever-decreasing market life of products, the installed processes and the equipment are now expected to produce the next product generation(s). The factory buildings have to be adaptable to two and sometime three product generations as well, and even the site must follow new requirements, for example regarding logistics and environmental regulations.

Increasing outsourcing, manufacturing at different sites and the manifold cooperation taking place within global networks, along with the new nature of products, systems and factories life cycles and the exploding number of product models and variants, also increase the complexity of production processes. Therefore, the operation of global supply chains becomes a reality for an increasing number of manufacturing companies (Wiendahl, 2006).

This development leads to another fundamental change in the role of manufacturing; in the past, product development and order handling were regarded as primary processes whereas order fulfillment and distribution were seen more as auxiliary functions. However, at present, reliable delivery of customized products in globally distributed markets has the highest priority. This priority increasingly determines the development of products, processes and production facilities, including logistics, under the following guidelines:

- The design of business processes and supply chains has to be carried out primarily taking into consideration the globally distributed market needs.
- Adaptable, and if necessary temporary, production units located close to the market are required instead of centralized factories with a high vertical range of manufacture.
- Production logistics is governed by the supply and distribution logistics.
- Different order types must be mastered by the planning and control systems in the same factory.
- Product structures have to be adapted to the changing requirements of an internationally distributed production.
- Production and assembly methods must consider both local and global conditions.

Hence, the prerequisites for successful participation in dynamic and global production networks require that the production processes, resources, plants structures, manufacturing systems layouts as well as their logistical and organizational concepts be adaptable quickly and with low effort. This ability is necessary for production companies to withstand the continuous changes and the turbulent manufacturing environment facing them, and can be described as '*changeability*'.

1.2 Evolution of Factories

To meet these challenges, factories have undergone several major steps of evolution as depicted in Fig. 1.3 along with their main features (Wiendahl, 2001). The *functional factory* model, with highly flexible resources and know how bundling for specific technologies, was quite adaptable to both product and volume changes but resulted in long delivery time and high inventory. It was suited for stable and well predicted markets. Logistical control was maintained by the 'push' principle; Orders were planned and scheduled according to the desired date and then released for manufacturing. Orders were monitored by feedback data and schedule deviations were corrected by prioritizing.

With increasing orientation towards customers' need for fast delivery, the *segmented factory* model provided an answer by structuring the factory into manufacturing areas, buffers for semi-finished goods and an assembly area arranged according to the processes flow. The manufacturing and assembly activities themselves were organized in cells, fractals or segments. Manufacturing produced order anonymous part families, whereas assembly finished customer specific final products upon request within short time periods, sometimes in days or even hours. The buffer in between was filled on demand using the 'pull' principle and not based on forecast: Whenever the level of an item falls below a predefined stock level a new lot is produced within a fixed time.

In the meantime, different goals such as the decrease of a previously unknown high entrepreneurial complexity are becoming increasingly important. Due to the tendency to focus on main competences and the consistent outsourcing of procurement, production, distribution up to development processes, the reduction of costs has been the goal of these concepts. These *strategic supply chains* are especially found in the automobile industry. The OEM (Original Equipment Manufacturer)

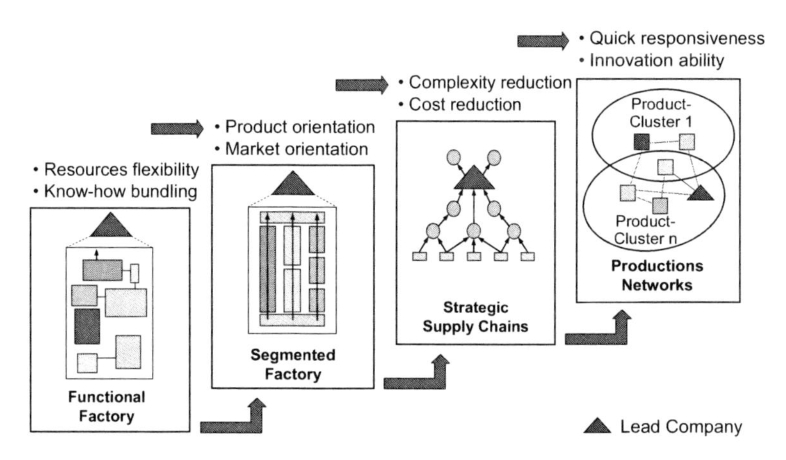

Fig. 1.3 Evolution of factories

acts here as a lead company, which establishes and runs the system of interlinked companies.

At present, we see evolving *production networks* with a temporary cooperation mostly lasting for the product life of a product family. Here, the partners do not have a hierarchical relationship, although for the customer only the lead company is visible. Changeable and reconfigurable production processes, resources, structures and layouts as well as logistic and organizational concepts are becoming prerequisites for the successful participation of a firm in production networks.

The next generation factories are described as adaptive, transformable, high performing and intelligent. The EU Manufacture strategy underlines this vision in the proposal for the 7th framework (www.manufuture.org).

In addition to the design of the production structure, production enterprises have to fundamentally re-consider their internal processes. This is frequently done by installing a 'production system', the origin of which is the Toyota Production System (Ohno, 1988). One outcome is the concept of lean production, the central idea of which is to avoid waste in every process with respect to time, space, movement, energy, material, etc. (Womack, 2003).

1.3 Deriving the Objects of Changeability

As already stated, production firms have to understand what their main change drivers are and to define and take necessary and appropriate actions at the right time. One main aspect is to define the objects that have to be changeable and their appropriate degree of changeability. Figure 1.4 summarizes the main steps to reach this goal (see also (Shi, 2003) and (Dashchenko, 2006)).

The impulse for a change is triggered by *change drivers*, whose first category is the demand volatility measured by volume fluctuation over time. Variety is the scope of the products' variants, both in basic models as well as in variants within the models with respect to size, material and additional features. A major change driver is a new company strategy, e.g. a decision to enter a new market, to sell or buy a product line, or to start a strategic turn around program, etc.

The impacts of the change drivers may be external or internal. The external focus targets the product and its added value for the customers, for instance a product with lower life cycle cost or faster delivery. In addition, the production volume trend during the life cycle and the number of different products may be considered. The internal focus typically arises if the performance of the firm is not satisfactory, mainly with respect to loss of profit caused by badly organized business processes. This targets the production with its processes, facilities and the work organization.

Out of these two fields of change the aspired *change strategy* arises: Should a change simply fulfill the immediate need on the operational level? These changes are just necessary to survive; they are more defensive and are typically performed within the given structures and procedures, such as the installation or replacement

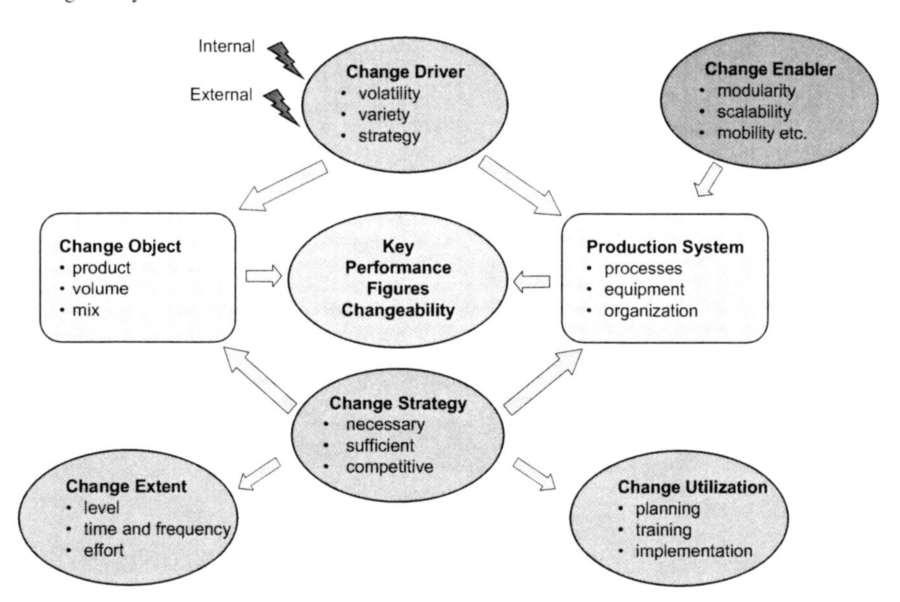

Fig. 1.4 Steps to define change objects

of machines. Or, should the planned change be more tactical and fulfill sufficiently the needs of the foreseeable future? These changes are more proactive and occur typically in business processes like order fulfillment or service. Or, is a strategic investment in changeability finally chosen with the aim of gaining a superior competitive advantage (Gerwin, 1993). This could mean a complete re-arrangement of the manufacturing procedures like introducing the six-sigma or lean production concept.

The selected change potential obviously very much determines the *change extent*. First, the level of the factory on which the changeability has to be ensured, must be determined. Secondly, the expected change frequency and the time allowed for each change has to be estimated and thirdly, the necessary effort in equipment, manpower, knowledge and time, are typically measured as the cost of change. Typically, changeability beyond the immediate necessity requires additional investment. Therefore, the effort has to be comparable with the expected benefit.

In order to ensure the production system's ability to adapt itself according to the aforementioned change drivers it needs *change enablers* like modularity, scalability, mobility and others, which will be explained in more detail in Sect. 1.8.

Changes on levels above a single workstation are complex. Therefore, they have to be planned, workers have to be trained and the implementation has to be fast without loss of product quality. This aspect takes into account that an installed changeability has to be kept up.

Finally a performance measurement system has to be utilized in order to check the impact of the implemented changeability measures with respect to the output performance of the factory. Typical performance indicators are delivery time, due date performance, turn around rate, inventory, days of supply and overhead cost.

1.4 Elements of Changeable Manufacturing

The challenge here is how to handle such a broad topic and apply it to the real production world. From the introductory paragraphs, it became obvious that the scope of changeability has to be widened from the manufacturing system that makes various work pieces to encompass the whole factory that produces different products in various variants. It should be noted that the terms "flexibility and reconfiguration" are generally specific to certain factory levels.

Therefore, *changeability* has been proposed as an umbrella concept that encompasses many aspects of change on many levels within the manufacturing enterprise (Wiendahl et al., 2007).

Changeability in this context is defined as *the characteristics to accomplish early and foresighted adjustments of the factory's structures and processes on all levels, due to change impulses, economically.* In the following sections, the term changeability will be interpreted according to the factory level.

Figure 1.5 depicts the elements of such a changeable manufacturingg and their specific properties. The product is placed in the center to signify the importance of its interaction with both the physical and logical levels of changeable manufacturing. Although its design is not included in this scope of changeable manufacturing; it is

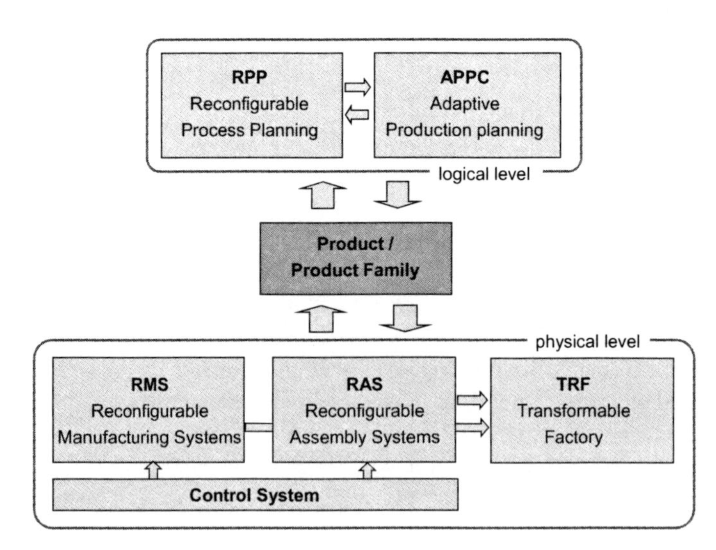

Fig. 1.5 Scope of changeable manufacturing

acknowledged that it will also be influenced by the changing requirements of the physical and logical manufacturing domains.

On the physical level, the manufacturing and assembly systems (have to be reconfigurable (RMS and RAS) and the factory with its technical infrastructure including buildings should be transformable (TRF). Major modules at the logical level are necessary to operate a factory and must also be changeable in order to support the changes at the physical levels. This calls for a process planning system that is able to react to changes in the products design or the physical manufacturing resources and, hence, is called "Reconfigurable Process Planning (RPP)" (ElMaraghy, 2006 and 2007 and Azab and ElMaraghy, 2007). The production planning and control also has to react to changes in production volume, product mix or reconfigured process plans. Therefore, it is called "Adaptive Production Planning and Control (APPC)". A specific additional component is a control loop to monitor external or internal change drivers and to trigger change activities on the physical and/or logical levels. Finally, an evaluation procedure is necessary to justify the additional expenses due to the changeability of the physical and logical objects.

1.5 Factory Levels

The main application of flexibility has traditionally been mainly on the manufacturing and assembly levels. This scope has to be extended to the whole factory, as previously justified. Six levels of factory structuring have been proposed by Westkämper (2006) and Nyhuis (2005), based on Wiendahl (2002), as shown in Fig. 1.6 (left).

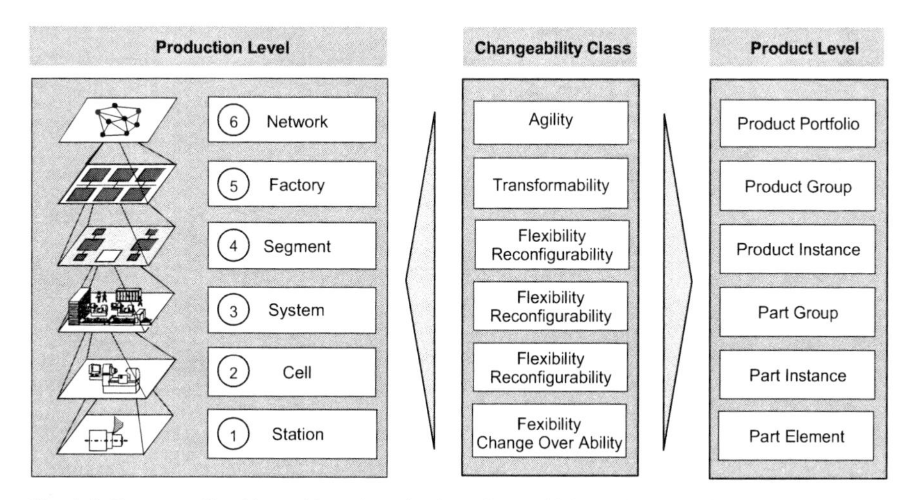

Fig. 1.6 Corresponding hierarchies of production, changeability and product levels

The common ground for this structure is given by the underlying *processes,* which are performed by either machines and/or workers. This level is within the domain of technology and ergonomics and is beyond the scope of this chapter.

The main concern on the level above the processes is with the single *workstations* (level 1) and their value adding operations on a work piece element, often called feature.

Often several manufacturing resources are arranged into *cells* that typically perform most of the necessary operations to finish a part or an assembly including quality assurance (level 2). The operations are usually executed partly by machines and partly by workers.

If several cells are more or less automatically interlinked, the terms manufacturing *system* and assembly *system* are commonly used (level 3). They manufacture the variants of a part, often called part family.

The next level up refers to *segments* in which whole products are typically manufactured in a ready-to-ship state (level 4). Segments commonly include facilities for manufacturing, assembly, buffers, quality measurement devices, packing, etc.

A *factory* describes a production unit at a given site with more than one production segment and may often serve as a node of a production network or a supply chain (level 5).

The highest structuring level is the production *network* (level 6), which can be interpreted from the resource perspective as geographically separated production units linked by material and information flows along the supply chain. This level will not be treated in this chapter because the arrangement of the network is more of a strategic task carried out because of the expansion or restructuring of the whole enterprise.

1.6 Changeability Classes

If the five structuring levels of the factory, which will be discussed further, are mapped to the associated products level, a hierarchy emerges that allows the definition of five types of changeability (Wiendahl, 2002). Figure 1.6 provides an overview of this hierarchical structure where any given level encompasses those below it.

The hierarchy of product levels starts at the top with the products portfolio that a company offers to the market. Then the product or a product family follows. The product is usually structured into sub-products or assembly groups that contain work pieces, which consist of features. Five classes of changeability evolve from this (Wiendahl, 2004).

- *Changeover ability* designates the operative ability of a single machine or workstation to perform particular operations on a known work piece or subassembly at any desired moment with minimal effort and delay.

- *Flexibility* describes the operative ability of a manufacturing or assembly system to switch with minimal effort and delay within a pre-defined family of work pieces or sub-assemblies by logically re-programming, re-routing and re-scheduling of the same system.
- *Reconfigurability* refers to the tactical ability of an entire production and logistics area to switch with reasonably little time and effort to new – although similar – members of a pre-defined parts group or family by physically changing the structure of manufacturing processes, material flows and logistical functions including removal or adding of components.
- *Transformability* indicates the tactical ability of an entire factory structure to switch to different product groups or families. This calls for structural interventions in the production and logistics systems, in the structure and facilities of the buildings, in the organization structure and process, and in the area of personnel.
- *Agility* means the strategic ability of an entire company to respond to changing markets by opening up new markets, developing the desired products portfolio and services, and building necessary manufacturing capacity.

It should be noted that flexibility and reconfigurability may be applied on levels 2 to 4 depending on the targeted change strategic and extension. In the context of this chapter, the changeover ability needs no special attention for changeable factories since this aspect is an ongoing topic of machine tool and assembly systems design. Agility is above the factory level and is treated as a strategic approach for the design of a changeable factory. Therefore, only flexibility, reconfigurability, and transformability will be considered further.

1.7 Changeability Objectives

It is important to overview the objectives of the physical and logical components of a changeable manufacturing, with a particular focus on three objectives of flexibility as defined by Chryssolouris (2005):

- *Product flexibility*, which enables a manufacturing system to make a variety of part types using the same equipment.
- *Operation flexibility*, which refers to the ability to produce a set of products using different machines, materials, operations, and sequence of operations.
- *Capacity flexibility*, which allows a manufacturing system to vary the production volumes of different products to accommodate changes in demand, while remaining profitable.

Although these objectives are intended to describe the flexibility of manufacturing systems, they are equally applicable as changeability objectives for assembly systems and the whole factory. They will be discussed in more detail for the manufacturing, assembly and factory levels.

1.7.1 Manufacturing Level

A plethora of publications exist on flexibility, the majority of which is devoted to manufacturing flexibility (D'souza, 2000). Chryssolouris defines flexibility of a manufacturing system as its sensitivity to change and states that: "The lower the sensitivity, the higher is the flexibility" (Chryssolouris, 2005).

A literature survey performed by H. ElMaraghy (2005) identified 10 types of manufacturing flexibility. These are: 1) Machine flexibility, 2) Material handling flexibility, 3) Operation Flexibility, 4) Process Flexibility, 5) Product Flexibility, 6) Routing Flexibility, 7) Volume Flexibility, 8) Expansion Flexibility, 9) Control Program Flexibility, and 10) Production Flexibility.

"This classification promotes a better understanding of various types of flexibility, albeit some of them are interrelated. It should be noted that the expansion flexibility is consistent with the current understanding of manufacturing systems reconfigurability" (ElMaraghy, 2005).

1.7.2 Assembly Level

Eversheim defines five types of flexibility objectives for assembly systems shown in Fig. 1.7 (Eversheim, 1983).

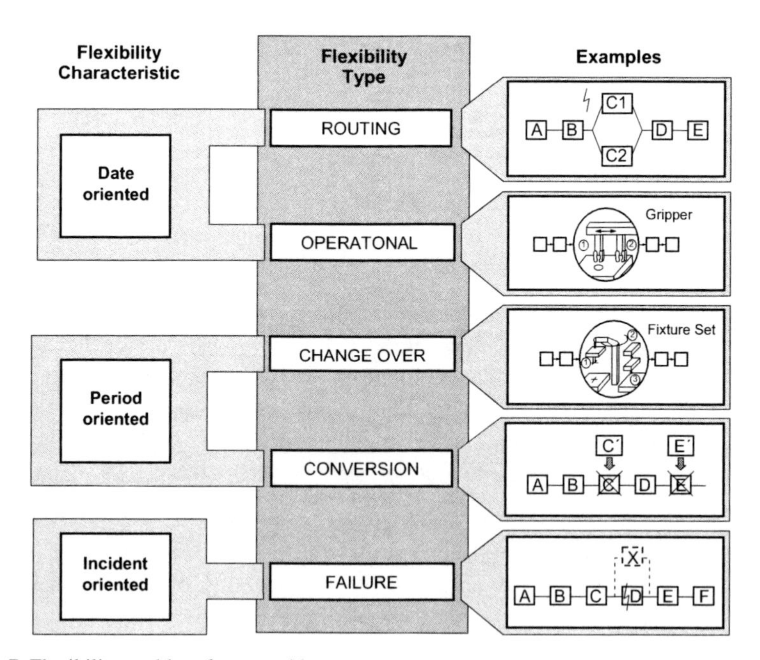

Fig. 1.7 Flexibility enablers for assembly systems

Three characteristics are used to group the flexibility types. *Date-oriented flexibility* has a short time-frame. If product variants change in a line and certain variants need to be processed at specific stations, *routing flexibility* is needed. *Operational flexibility* is necessary when several assembly operations are performed on one object in a sequence in short periods of time, e.g. by changing grippers or tools in a matter of seconds. *Period-oriented flexibility* has a longer time-frame. This is typically used in situations where batches of various products are assembled for hours or even days and a *change-over flexibility* is utilized, e.g. to exchange work piece carriers. *Conversion flexibility* is very much like reconfigurability, where complete workstations are exchanged and replaced, e.g. replacing automatic stations by manuals stations and vice versa. The third characteristic of flexibility is *incident-oriented*. The associated *failure flexibility* aims to achieve fast reaction if a station develops a serious disruption and needs quick replacement of a whole assembly unit.

1.7.3 Factory Level

On the factory level the changeability objectives include both the manufacturing and assembly levels. However, there are more possible change drivers and the necessary extent of change is higher at this level. For example, the increase in production volume in a factory is not only higher compared to the change occurring in a single system but it also affects more objects, because not only does the manufacturing system have to be adaptable, but so does the PPC, the means of transportation, and the labor organization.

The following changeability objectives are the most important at the factory level:

- *Product transformability*: the ability of a factory to produce a products portfolio consisting of variety of different products.
- *Technology transformability*: the ability to integrate and/or remove specific products and production technologies.
- *Capacity transformability*: the ability to scale up or down the production volumes of each product in response to varying demands.
- *Logistical transformability*: the ability of a factory to respond to new logistical requirements such as delivering just-in-sequence or delivering different lot-sizes.
- *Transformable degree of vertical integration* is the ability to adapt the degree of added value within the factory (e.g. by out- or in- sourcing of preceding or following production or logistical steps).

1.8 Changeability Enablers

A factory that is designed to be changeable must have certain inherent features or characteristics that will be called *changeability enablers*. They enable the physical

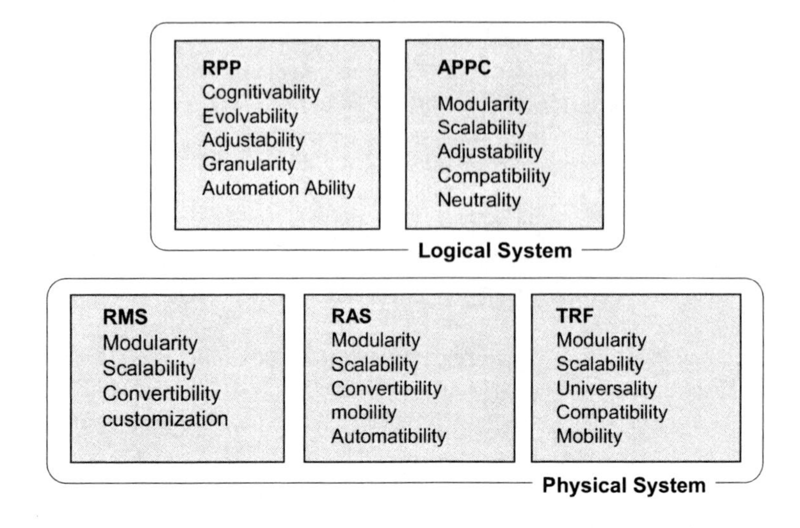

Fig. 1.8 Enablers of changeable manufacturing sub-systems

and logical objects of a factory to change their capability towards a predefined objective in a predefined time and are not to be confused with the flexibility types or objectives.

An enabler contributes to the fulfillment of a transformation process. Furthermore, the enablers characterize the potential of the ability to transform and become active only when needed. The characteristics of an enabler positively or negatively influence a factory's ability to adapt.

Figure 1.8 presents an overview of the enablers of the physical and logical subsystems of changeable manufacturing.

1.8.1 Manufacturing Level

On this level, reconfigurable manufacturing systems can satisfy some of the needs for changeability. In order to achieve exact flexibility in response to fluctuation in demands, an RMS should be designed considering certain qualitative and quantitative properties, known as key RMS characteristics: modularity, integrability, customization, scalability, convertibility and diagnosability (Koren, 2006; ElMaraghy, 2006; Hu, 2006; Abele, 2006 and Koren, 2005). Therefore, RMS is an enabler of changeability and these RMS characteristics are in turn enablers of reconfiguration. They can be divided into essential characteristics (customization, scalability and convertibility) and supporting characteristics (modularity, integrability and diagnosability) (Hu, 2006).

1.8.2 Assembly Level

The same enablers of reconfigurable manufacturing systems are applicable to assembly systems. Two specific enablers should be added: Mobility, which is important to reconfigure (add/remove) single stations or modules of an assembly system or even to move the whole system to another location, and the ability to upgrade or downgrade the degree of automation. For assembly operations, in contrast with machining operations, there is often the possibility to perform them either manually or automatically. The system should allow for adapting the ratio of manual and automated work content depending on various factors like production rate and wage levels. Mobility is also applicable to machining systems and some machine tool companies are already offering such ability.

1.8.3 Factory Level

Figure 1.9 illustrates the five main transformation enablers that the factory planner may use in the design phase for purposes of attaining changeability:

- *Universality* represents the characteristic of factory objects to be dimensioned and designed to meet the diverse tasks, demands, purposes and functions. This enabler stipulates an over-dimensioning of objects to guarantee independence of function and use.

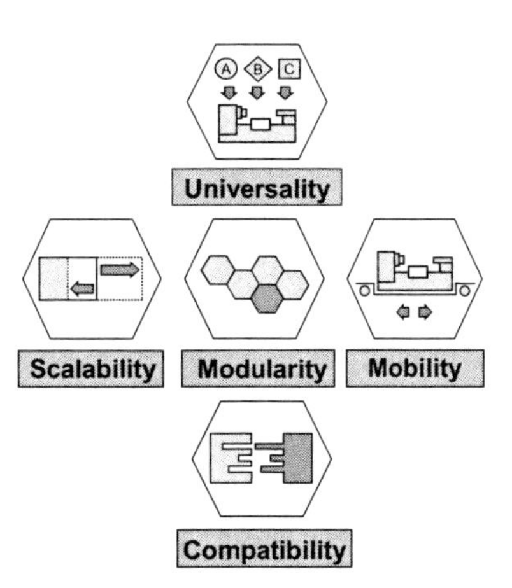

Fig. 1.9 Enablers of factory changeability

- *Scalability* provides technical, spatial and personnel incrementality. In particular, this enabler also provides for spatial degrees of freedom, regarding expansion, growth and shrinkage of the factory layout.
- *Modularity* follows the idea of standardized, pre-tested units and elements with standardized interfaces. It also concerns the technical facilities of the factory (e.g. buildings, production facilities and information systems) as well as the organizational structures (e.g. segments or function units). Modules are autonomous working units or elements designed to ensure high inter-changeability with little cost or effort, which are commonly known as 'Plug and Produce Modules'.
- *Mobility* ensures the un-impeded movement of objects in a factory. It covers all production and auxiliary facilities, including buildings and building elements, which can be placed, as required, in different locations with the least effort.
- *Compatibility* allows various interactions within and outside the factory. It especially concerns all kinds of supply systems for production facilities, materials and media. It also facilitates diverse potential relationships regarding materials, information and employees. Besides the ability to detach and to integrate facilitates, this enabler allows incorporating or eliminating products, product groups and work pieces, components, manufacturing processes or production facilities in existing production structures and processes with little effort, by using uniform interfaces.

These enablers are applicable mainly to the objects of a factory on the *segment* and the *site* levels regarding the structure, layout and system configuration.

1.8.4 Reconfigurable Process Planning Level

There are certain key enablers for achieving reconfigurable process plans and commensurate techniques for their efficient regeneration when needed. These are:

- *Cognitivability*: The ability to recognize the need for and initiate reconfiguration when pre-requisite conditions exist. This ability is imparted on process planning through the use of some artificial intelligence attributes.
- *Evolvability*: The ability to utilize the multi-directional relationships and associations between the characteristics of product features, process plan elements and all manufacturing system modules capable of producing them.
- *Adjustability*: The ability and representation characteristics that facilitate implementing optimally determined feasible and economical alterations in process plans to reflect the needed reconfiguration.
- *Granularity*: The ability to model process plans at varying levels of detail in order to, readily and appropriately, respond to changes at different levels (e.g. in products, technologies and systems).
- *Automation ability*: The availability of complete knowledge bases and rules for process planning and reconfiguration, accurate mathematical models of the

various manufacturing processes at macro- and micro-levels, as well as meta-knowledge rules for using this knowledge to automate the plan reconfiguration.

1.8.5 Production Planning and Control Level

A prerequisite for order processing is planning and control. On the one hand this needs a PPC system comprised of the PPC methods and their functional model, the data model as well as the data interfaces. On the other hand the work-flow of order processing refers to the structural and process organization.

The changeability of these elements in the PPC system is facilitated by the following five enablers.

- *Modularity*: workable functions and methods or clearly defined objects, e.g. 'plug & produce' modules exist.
- *Scalability*: applicability is independent of product, process, customers and supplier relationship complexity.
- *Adjustability*: to requirements concerning the functional logic of order processing, i.e. the weight of PPC targets and the importance of PPC functions.
- *Compatibility*: the external structure of the PPC design units supports networkability with each other.
- *Neutrality*: design of the work-flow of order processing for different requirements making the definition of the process status independent of the structural and process organization and the enterprise size.

1.9 Changeability Process

The purpose of factory design activities should not be to achieve the transformability of the manufacturing changeability objects at any cost. Quite the contrary; the objective should be to determine the degree of changeability that is *appropriate and economically justifiable* in a given situation.

It must be stated first that it is not possible to define an absolute changeability. One can imagine cases of extreme changeability, (e.g. a circus, where the equipment is completely disassembled in a few days, travels to another place and is reassembled again in days; or a theater in which a play is on stage every night and changing the scenes is done within minutes).

Changeability can be interpreted in analogy to quality. Quality in a broad sense is defined as "conformance to requirement" and is the sum of multiple separate attributes. For changeability, this means that a company has to define the changeability requirements, compare them with the actual degree of conformance with these requirements, and then aim for continuous adaptation. Figure 1.10 shows the resulting action cycles.

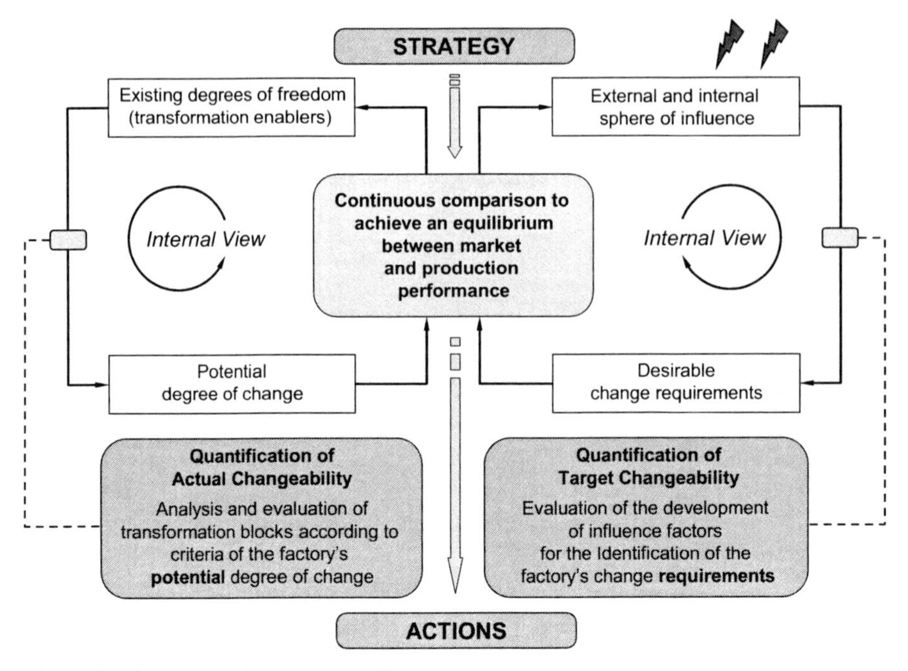

Fig. 1.10 Adaptation cycle of changeability

The *target changeability* has to be set based on external and internal factors. This refers to the scope (operational, tactical, and strategic), level (factory, segment, cell, and workplace) and objects (product, process, volume, mix) of changeability.

The result is the *desirable changeability*. On the other hand, the existing production has certain degrees of freedom to change, hence, the actual changeability offers a *potential for change*. Typically this potential is not sufficient to cope with the desired changeability level. Therefore, an economic evaluation has to be performed to determine the feasible and justifiable courses of action that are consistent with the management overall corporate strategy and vision (Kuzgunkaya and ElMaraghy 2007). This is an ongoing process typically performed as part of the corporate planning.

Changeability can be seen as a life-cycle-oriented process with two phases, as shown in Fig. 1.11. In the design and implementation phase, the necessary adoption of the transformation objects has to be determined followed by the actual transformation to the new level of changeability. The factory organization is now technically empowered to change the identified objects on the desired level. At the right moment, either a reactive or proactive change is performed. This process is similar to factory set up when facing a new production situation.

The implementation of the actual change also has typical phases. First, a work plan is developed that describes the sequence of operations and their duration. Sec-

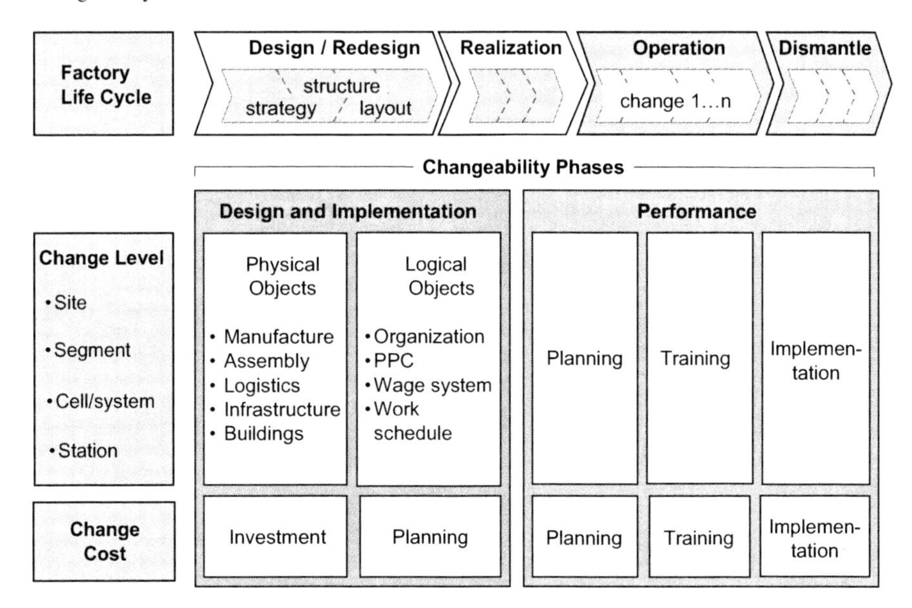

Fig. 1.11 Map of factory changeability process

ond, the people involved in the change have to be trained. Third, the changeover itself has to be implemented using the previously built-in changeability attributes. The procedure can be compared with a pit stop for a race-car in which a change of 4 wheels is required. There is a precise plan that specifies who has to do what with which tools and in which sequence. Then the team is trained by performing the procedure repeatedly while improving the process step by step. Finally the real situation arises and within 12 seconds, all wheels are changed.

But even highly transformable objects do not adapt to changes of the environment by themselves. Human input is always needed to trigger and perform the planned transformation. Therefore, other requirements for a successful change and transformation process must be identified, in addition to the technical transformability of the objects. These are related to the human operators in terms of changing competency through motivation and education, leadership by giving permissions and resources, and culture though training and incentives and using change management techniques (Nyhuis 2006). In addition, intelligence and creativity are seen as other important factors besides the ability to react (Reinhart 2000).

These factors have been extensively discussed in the field of change and innovation management but are rarely connected with the transformability of a factory and its objects, although this connection offers significant synergies. Figure 1.12 illustrates the inter-dependencies that have to be considered in order to support the necessary change processes within the factory by adequate objects, as well as to create the environment to be able to fully utilize the full potential of the transformability of the objects.

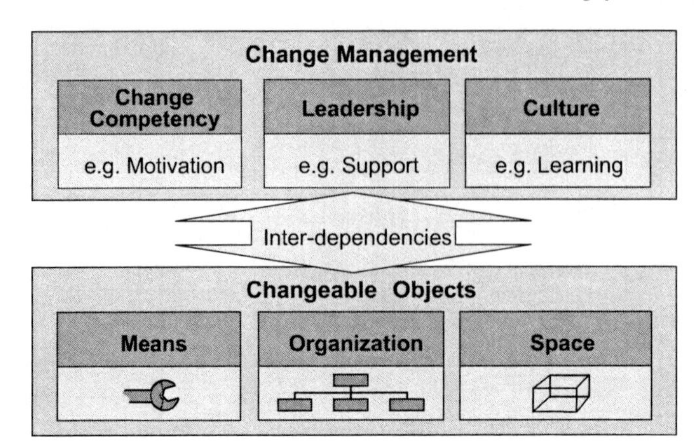

Fig. 1.12 Prerequisites for a successful change

1.10 Conclusion

"Changeability is the answer to uncertainty". This is common sense and is applicable also to manufacturing but has in the last decade gained an increasing importance. The intense study of this field has revealed that:

- There is nothing like an absolute or hundred percent changeability, instead an appropriate changeability has to be established which is on the one hand sufficient to cope with unforeseen changes either from outside or from inside a company, and on the other hand is affordable.
- The basic difference between flexibility and changeability is that the latter enables a company to react on many more levels and also beyond expected, preplanned or foreseen developments.
- Changeability is multi-faceted including the objectives, manufacturing levels, objects and enablers.
- The most important change enablers are modularity, mobility and scalability.
- Changeability impacts not only physical objects but also logical objects like planning and control as well as workforce.
- Key metrics to measure changeability are not established yet. A practical approach is to compare the desired changeability with the existing change potential and start the necessary actions to reduce the gap according to the strategy, urgency and importance to survive.

Many relevant issues and open questions regarding changeability remain solved, but they will be the subject of ongoing challenge and motivation for research and industrial practice.

References

Abele E., Liebeck T., Wörn A., 2006, Measuring Flexibility in Investment Decisions for Manufacturing Systems. Annals of the CIRP 55/1:433–440

Alexopoulos K., Mamassioulas A., Mourtzis D., et al., 2005, Volume and Product Flexibility: A Case Study for a refrigerators Producing Facility, 10th IEEE Int. Conf. on Emerging Technologies and Factory Automation (ETFA 2005), Catania, Italy, 19.–22.09.2005

Azab A., ElMaraghy H.A., 2007, Mathematical Modeling for Reconfigurable Process Planning. CIRP Annals 56/1:467–472

Chryssolouris G., 2005, Manufacturing Systems: Theory and Practice, 2nd edn. Springer Verlag, Berlin/Heidelberg

D'souza D., Williams F., 2000, Toward a Taxonomy of Manufacturing Flexibility Dimensions. Journal of Operations Management 18/5:577–593

Dashchenko O., 2006, Analysis of Modern Factory Structures and Their Transformability. In: Dashchenko A.I. (ed) Reconfigurable Manufacturing Systems and Transformable Factories. Springer Verlag, Berlin/Heidelberg, pp 395–422

ElMaraghy H.A., 2005, Flexible and Reconfigurable Manufacturing Systems Paradigms. International Journal of Flexible Manufacturing Systems 17/4:261–276, Special Issue: Reconfigurable Manufacturing Systems

ElMaraghy H.A., 2006, Reconfigurable Process Plans for Responsive Manufacturing Systems, 3rd International CIRP Conference on Digital Enterprise Technology, Setúbal, Portugal, 18.–20.9.2006

ElMaraghy H.A., 2007, Reconfigurable Process Plans for Responsive Manufacturing Systems, Digital Enterprise Technology: Perspectives & Future Challenges, Editors: P.F. Cunha and P.G. Maropoulos, Springer Science, ISBN: 978-0-387-49863-8, pp 35–44

Eversheim W., Kettner P., Merz K.-P., 1983, Ein Baukastensystem für die Montage konzipieren (Design a modular system for assembly). Industrie Anzeiger 92/105:27–30

Gerwin D., 1993, Manufacturing Flexibility: A Strategic Perspective. Management Science 39/4:395–410

Koren Y., 2006, General RMS Characteristics. Comparison with Dedicated and Flexible Systems. In: Dashchenko A.I. (ed) Reconfigurable Manufacturing Systems and Transformable Factories. Springer Verlag, Berlin/Heidelberg, pp 27–46

Koste L.L., Malhotra M.K., 1999, A Theoretical Framework for Analyzing the Dimensions of Manufacturing Flexibility. Journal of Operations Management 18/1:75–93

Kuzgunkaya O., ElMaraghy H., 2007, Economic and Strategic Perspectives on Investing in RMS and FMS, International Journal of Flexible Manufacturing Systems (IJFMS) Special Issue on Managing Change in Flexible and Reconfigurable Manufacturing Systems 19/3:217–246, doi: 10.1007/s10696-008-9038-8

Nyhuis P., Gerst D., 2006, Wandlungsfähigkeit als Grundlage der lernenden Organisation (Transformabiliy as the prerequisite of the learning organization), 19. HAB-Forschungsseminar der Hochschulgruppe Arbeits- und Betriebsorganisation (19th HAB research seminar of the university group of labor- and business organization), Karlsruhe, Germany, 13.–15.10.2006

Nyhuis P., Kolakowski M., Heger C.L., 2005, Evaluation of Factory Transformability, 3rd International CIRP Conference on Reconfigurable Manufacturing, Ann Arbor, USA, 11.–12.05.2005

Ohno T., 1988, The Toyota Production System: Beyond Large-Scale Production. Productivity Press, Portland, USA

Reinhart G., 2000, Im Denken und Handlen wandeln (Change in thinking and doing), Münchener Kolloquium 2000 (Munich colloquium 2000), Munich, Germany, 16.–17.03.2000

Shi D., Daniels R.L., 2003, A Survey of Manufacturing Flexibility: Implications for e-business Flexibility. IBM Systems Journal 42/3:414–427

Westkämper E., 2006, Digital Manufacturing in the global Era, 3rd International CIRP Conference on Digital Enterprise Technology, Setúbal, Portugal, 18.–20.9.2006

Wiendahl H.-P., 2002, Wandlungsfähigkeit: Schlüsselbegriff der zukunftsfähigen Fabrik (Transformability: key concept of a future robust factory). wt Werkstattstechnik online 92/4:122–127

Wiendahl H.-P., 2006, Global Supply Chains – A New Challenge for Product and Process Design, 3rd International CIRP Conference on Digital Enterprise Technology, Setúbal, Portugal 18.–20.9.2006

Wiendahl H.-P., Heger, C. L., 2004, Changeability of Factories – A Prerequisite for Global Competitiveness, 37th CIRP International Seminar on Manufacturing Systems, Budapest, 19.–21.05.2004

Wiendahl H.-P., Hernández, R., 2001, The Transformable Factory – Strategies, Methods and Examples, 1st International Conference on Agile, Reconfigurable Manufacturing, Ann Arbor, USA, 20.–21.5.2001

Wiendahl H.-P., ElMaraghy H.A., Nyhuis Zäh P.M.F., Wiendahl H.-H., Duffie N., Brieke M., 2007, Changeable Manufacturing – Classification, Design and Operation. Annals of the CIRP 56/2:783–809

Wirth S., Enderlein H., Petermann J., 2004, Kompetenzwerke der Produktion, IBF-Fachtagung „Vernetzt planen und produzieren", Cited from: Schenk M., Wirth S.: Fabrikplanung und Fabrikbetrieb, p 106, Berlin/Heidelberg, Springer Verlag, 2004

Womack J.P., Jones D.T., 2003, Lean Thinking. Simon & Schuster, New York

Chapter 2
Changing and Evolving Products and Systems – Models and Enablers

H.A. ElMaraghy[1]

Abstract Many manufacturing challenges emerged due to the proliferation of products variety caused by products evolution and customization. They require responsiveness in all manufacturing support functions to act as effective enablers of change. This Chapter summarizes some recent findings by the author and co-researchers that address these issues.

A variation hierarchy for product variants, from part features to products portfolios, was presented and discussed. The evolution of products and manufacturing systems is discussed and linked, for the first time, to the evolution witnessed in nature. The concept of *evolving families* for varying parts and products is presented. A biological analogy was used in modeling of *products evolution* and Cladistics was used for its classification. This novel approach was applied to the design of assembly systems layouts with the objective of rationalizing and delaying products differentiation and managing their variations.

Process planning is part of the "soft" or "logical" enablers of change in manufacturing as the link between products and their processing steps. New perspectives on process planning for changing and evolving products and production systems are presented. *Process-neutral and process-specific products variations* were identified and defined. A recently developed innovative, and fundamentally different, method for *Reconfiguring Process Plans* (RPP) and new metrics for their evaluation are presented and their significance and applicability in various domains are summarized. The merits of reconfiguring process plans on-the-fly for managing the complexity and extensive variations in products families, platforms and portfolios are highlighted and compared with the traditional re-planning and pre-planning approaches.

The conclusions shed light on the increasing challenges due to variations and changes in products and their manufacturing systems and the need for effective solutions and more research in this field.

---

[1] Intelligent Manufacturing Systems Center, University of Windsor

H.A. ElMaraghy (ed.), *Changeable and Reconfigurable Manufacturing Systems*,

Keywords Evolving Parts and Products Families, Variation Hierarchy, Products Evolution, Classification, Reconfiguring Process Plans, Assembly Systems Design

2.1 Introduction and Motivation

Frequent and unpredictable market changes are challenges facing manufacturing enterprises at present. In the short term, there are many triggers for products changes including evolving over time due to innovation. Similarly, manufacturing systems frequently undergo incremental changes due to introducing products with new features. They also experience significant evolutions in the long term due to products and technological changes as well as introduction of new paradigms. There is a need, in the meantime, to reduce the cost and improve the quality of highly customized products. Agility, adaptability and high performance of manufacturing systems are driving the recent paradigm shifts and call for new approaches to achieve cost-effective responsiveness and increase the ability to change at all levels of the enterprise. It is important that the manufacturing system and all its support functions, both at the physical and logical levels, can accommodate these changes and be usable for several generations of products and product families.

Modern manufacturing paradigms aim to achieve these multi-objectives through: 1) pre-planned generalized flexibility as in Flexible Manufacturing Systems (FMS) designed and built-in a priory for pre-defined anticipated product variants over a period of time, 2) limited/focused flexibility to suit a narrower scope of products variation, or 3) customized flexibility on demand by physically reconfiguring a manufacturing system (RMS) to adjust its functionality and capacity. Many enablers are required for the successful implementation of these paradigms and achieving the desired adaptability. The flexibility, reconfigurability and changeability at the system hardware level are available to varying degrees today. However, the most challenging tasks encountered during their implementation include changes required, in light of the encountered variations, in the soft/logical support functions such as product/process modeling, process planning, production and capacity planning, control of processes and production, and logistics. These support functions must not only be in place but should also be adaptable, changeable and well integrated for any successful and economical responsiveness to changes in manufacturing to be realized (ElMaraghy, 2005).

A number of novel strategies and solutions to manage the inevitable products variations and related manufacturing changes are presented including new methods for modeling products evolution in manufacturing and designing their manufacturing systems accordingly as well as for reconfiguring process plans. An important contribution and a common theme utilized in the presented strategies is the use of natural evolution principles to develop new methods and solution to cope with variations and changes.

This chapter overviews the evolution of products and proliferation of their variants and highlights the need to effectively respond to these variations and the importance of modeling their evolution. It is essential to manage these changes in order to mitigate the resulting complexities as well as to prolong the life of their manufacturing systems and use their capabilities more effectively to produce the desired products variations. The concept of evolving parts and products families in changeable manufacturing is introduced as well as a modeling technique, inspired by laws of nature, to capture the evolutionary products changes and help design manufacturing systems accordingly; with an application for design of assembly systems. Focus is also placed on process planning and its functions as an important link between the features of generations of products/product families and the features, capabilities and configurations of manufacturing systems and their modules throughout their respective life cycles. A recently introduced innovative approach to re-configure process plans is highlighted as an enabler of the necessary changes in response to products and parts variations. It represents a fundamentally new concept of process planning as an effective means of managing the pervasive products variations while minimizing the resulting changes on the shop floor. Its rationale, characteristics, features and merits are discussed.

2.2 The Hierarchy of Parts and Products Variants

Customers' demands, innovation, new knowledge, technology and materials, cost reduction, environmental concerns and legislation's and legal regulations, drive the evolution of products. Product versions are developed over time in response to these requirements. Derivatives and variations in function, form and configuration lead to new product classes including Series of Products with different Functions, Series of Components with different Configurations and Series of Features with different Dimensions. This gives rise to product families that contain variants of the products and their parts, components and configurations.

It is informative to capture and classify the resulting products hierarchy, outline concisely the types and degrees of variation that occur at different hierarchy levels and consider ways of modeling them and their consequential effects on soft change enablers such as products and systems modeling and design and process planning among others.

An industrial example of automotive products is used, where information about the various products are readily available in the manufacturer's products information and open literature. Typical products, components and parts are selected and arranged/classified according to the suggested *variation hierarchy* for illustration as shown in Fig. 2.1.

There are eight distinguishable levels in the hierarchy: 1) Part Features, 2) Parts/Components, 3) Parts Family, 4) Product Modules or Sub-Assemblies, 5) Products, 6) Products Families, 7) Products Platform, and 8) Products Portfolios.

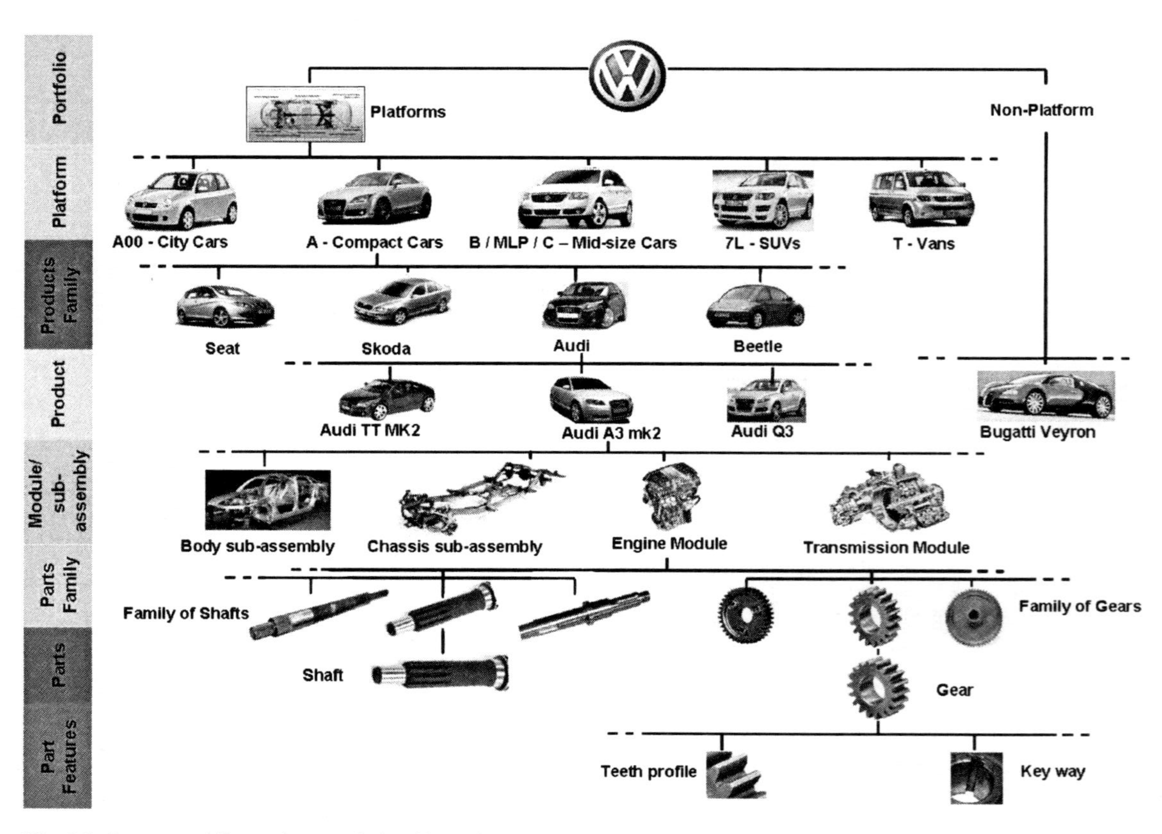

Fig. 2.1 An automobile products variation hierarchy

1) *Part Features* are either geometric features (such as flat, cylindrical and conical surfaces) or functional features (such as holes, slots/grooves, gear teeth, keyways, pockets, chamfers and threads). Features variations are easily illustrated; for example, holes may vary in dimensions, geometry and shape; they may be round or prismatic, smooth or threaded, stepped or having a constant diameter. The characteristics of the geometric and functional features are best-captured at the design level using variation geometry and parametric modeling techniques that reflect the changes within the features while respecting the geometric and dimensional constraints that express the functional requirements and designers intent. Subsequent analysis and manufacturing applications make good use of the similarities and parametric representation. Logical/soft support functions, at the process and machine levels, are directly affected by these variations. For example, in metal removal, micro/detailed process planning and tools/machines selection would utilize these models to account for the change in features.

2) *Parts/Components* are objects that are non-decomposable/non-divisible without loss of function. They contain both functional and non-functional features. Change at this level leads to parts/components variants within a class. The addition, removal and/or modification of part features require adaptation to these changes in upstream design and analysis applications as well as in downstream manufacturing logical support functions: a) at the process level such as macro- process planning (sequencing), planning of set-ups, and CNC programming, and b) at the system level such as the make/buy decisions.

3) *Parts Family* is a concept that was first introduced along with Group Technology (GT), where parts are grouped according to similarities in geometry and/or processing requirements. The objective is to capitalize on these similarities to increase the efficiency of many applications such as modeling and design, planning of fixtures and work holders, tools, production processes, parts/machine assignments, parts grouping into batches for production, and production flow management (e.g. manufacturing cells).

4) *Product Modules and Sub-Assemblies*. Modules represent functionally independent units that consist of more than one part or component and are meant to fulfill one or more technical function. A Sub-Assembly represents a number of strongly connected components and/or parts that may be considered as a single entity and is stable, in at least one direction, once assembled. A sub-assembly does not necessarily have an independent function, but is rather a convenient way of grouping parts and components into an intermediate assembly unit. Figure 2.1 illustrates instances such as the engine and transmission modules and the body and chassis sub-assemblies. A drive system for example contains many modules such as the engine, gearbox, stick shift, cooling and exhaust systems, electrical system, engine mounts, etc. (Shimokawa et al., 1997). These in turn contain common components; a gearbox for example consists of many parts such as the housing, gears, shafts, bearings, etc. that can be standard and modular. Determining the collection of parts/components that will form modules and sub-assemblies and defining their boundaries are important decisions, as they affect the extent of modularity and commonality and the

subsequent ability to interchange and combine modules into different products to offer the desired customization. In addition, they will affect the design and efficiency of the corresponding manufacturing systems. The choice of modules can also help manufacturers protect their intellectual properties by carefully planning the modules and making decisions to produce in-house, purchase or sub-contract their production.

5) *Products* are a collection of sub-assemblies and modules, the variation of which leads to different instances of that product. The dominant manufacturing activity at this level is the joining and assembly of modules and sub-assemblies into a final product. The same notion of grouping, based on similarities in features or processing steps, does apply to the modules and sub-assemblies that make up the product. At the process level, applications such as work holding, palletizing, and fixturing, parts feeding and orienting and assembly planning should benefit from the modularity and similarity between modules. Automation solutions at the system level can also be streamlined and rationalized as a result recognizing the nature and extent of similarities and variations.

6) *Products Families* is a concept similar to that of the parts families where variations in parts, sub-assemblies and modules produce different instances/members of a product family. The product family consists of related products that share some characteristic, components and/or sub-assemblies. These product families are meant to satisfy a variety of customers' demands and markets. This concept has been used more often in the context of products design and related analysis, and later for planning manufacturing processes, products platforms and market strategies. Examples of product families, some instances of which are shown in Fig. 2.1, include: the Audi Family [Audi A3 (3 and 5 doors), the Audi TT Coupe and Audi TT Roadster], the Seat Family [Toledo, Coupe, Station Wagon & Convertible], the VW Beatle Family, and the VW Golf Family.

Products variants within a family, as with parts variants, result from the modification, addition or removal of one or more modules. Macro-process planning, which determines the best sequence of assembly operations while respecting the logical and technological constraints, is dominant at the level of product modules/sub-assemblies, products and products families. The need for effectively changing macro-process plans at these levels did not receive much attention in literature to date. It can benefit greatly from novel methods for dealing with the variations while minimizing the consequential changes or disruptions in the manufacturing system as discussed in Sect. 2.6. At the system level, managing the products variation such as to provide as much variety to the consumer with as little variety as possible between the products manufacturing methods is very important to remain competitive. Delayed products differentiation is a key strategy that has been adopted to achieve this objective. A novel assembly system design method that exploits the similarities and commonalities among the product variants in a family is discussed in Sect. 2.5.

7) *Products Platform* is set of sub-systems/modules and their related interfaces and infrastructures, which forms a foundation used to produce a number of products that share common features. The platform features, parts and components remain

unchanged within a product family. Modules added to the platform serve to differentiate various products. This concept was originally introduced on the products level then extended to the products modules and component levels to achieve economies of scale through higher volume production of common product constituents. Modularity, standardization and commonality figure strongly at this level as manufacturers adopt the philosophy of products platforms to satisfy the desire for both products differentiation and customization. VW planned the "A" platform to produce 19 vehicles including all product variants of the VW Golf, VW Bora, VW Beetle, Skoda Octavia, Seat Toledo and Audi passenger cars. The chassis for the front wheel drive mid-size compact VW cars represents a fundamental module in a platform used to produce many product variants by adding and interfacing both common and different sub-systems. The body sub-system with its many components acts as one of the strong products differentiators. It is estimated that VW would save more than $1Billion/yr in capital investments, product development and engineering cost by using platforms.

Modularity promotes the exchange and re-use of components, helps the rapid introduction of new technologies, facilitates outsourcing and encourages more flexible allocation of production facilities locally and globally. This "Plug and Produce" approach to products design and manufacture supports more extensive variations in chasing customers' satisfaction through maintaining maximum flexibility to achieve truly differentiated products while enabling controlled evolution of products identities.

The design and planning of these products platforms present many challenges throughout the life cycle of both products and manufacturing systems and have significant financial impact on the manufacturer. Products platforms must be planned, managed and updated over time to ensure the success of its derivative products and the efficiency of their production. Product-specific platforms limit the potential synergy and leveraging in products development, manufacturing technologies and processes, re-tooling, procurement of parts and equipment and marketing. Understanding products evolution and its impact on manufacturing systems design is discussed in Sect. 2.3 and 2.4.

8) *Products Portfolio* represents the range of different products offered by a company. It contains several, and sometimes quite different products and may include more than one product platform as well as non-platform products, such as the Bugatti in Fig. 2.1. The special niche products satisfy demands in certain smaller segments of the market, they do not benefit from the economies of commonality, they cost more to produce and sell for higher price and profit margins. A company decides on its products portfolio depending on its strategic goals, growth plans, market opportunities, market demands and emerging segments, competing products, risk tolerance, leverage possibilities, and economic considerations.

As the utilization and benefits of products platforms are directly influenced by the scope of its members and families, it is important to carefully plan the products, technologies or sub-systems selected for the platform and their degree of products similarity and differentiation they can support. However, many companies design new products individually without formal consideration of the whole range of products and families in their portfolios. This does not promote commonality, modularity and compatibility among products or ensure the best business justification. Ultimately it is a trade-off between commonality and distinctiveness, whereas extensively diversified platforms lead to product derivatives that lack distinctive character and do not serve well either the high or low end products, while sparsely populated platforms become inefficient, costly and thus unjustified.

In light of the above discussion and the presented products variation hierarchy, it is imperative to find ways of understanding and managing the formation of parts/products families and their variation and evolution, as well as capitalizing on their commonalities to achieve economic advantages in related activities such as product design, process planning, production planning, design of manufacturing systems and supply chain management.

2.3 Evolving and Dynamic Parts and Products Families

The classical notion of a *Static Parts/Products Families* was established in conjunction with the concept of Group Technology (GT) where members of the family have similarities in the design and/or manufacturing features. Flexible manufacturing systems relied on this definition of pre-defined and pre-planned parts and products families with non-changing boundaries for pre-planning the manufacturing system flexibility, processes and production plans according to the defined scope of variations within the family. Classification and group technology codes were introduced, such as OPITZ (1970), to make information retrieval and modification easier. In this case, a "Composite Part" that contains all features of the family members is considered and a "Master Process Plan" is devised and optimized, in anticipation of the pre-defined variations, for use in "Variant Process Planning" and other manufacturing related activities (Groover, 2008 and ElMaraghy, 1993 and 2006). The parts' family concept is a pre-requisite for the success of flexible manufacturing where the similarity among members of a well-designed family helps achieve the economy of scale while realizing a wider scope of products.

In the current dynamic and changeable manufacturing environment, the products are frequently changed and customized, and it is also possible to reconfigure the manufacturing systems by changing their modules and hence their capabilities. Therefore, the notion of constant parts/products families is changing. This presents new challenges for related activities such as systems design and process planning to cope with both the variations on the product side and the changes in resources and their capabilities on the manufacturing side.

ElMaraghy (2007 and 2006) proposed a new class of "*Evolving Parts/Products Families*" where the boundaries of those families are no longer rigid or constant. The features of new members in the evolving families of parts/products overlap to varying degrees with some existing features in the original families; they mutate and form new and sometimes different members or families similar to the evolution of species witnessed in nature. Species is the theoretical construct that biologists use to explain why one population of organisms should be considered different from another. Species are the highest-ranked category of individuals, above which, all classes are abstract groupings of different species. Therefore, species are considered the unit of diversity in nature. This is illustrated for manufactured parts and products in Fig. 2.2. Since adding, removing, or changing manufacturing systems' modules changes their capabilities and functionality, a reconfigured system would be capable of producing new product features that did not exist in the originally planned product family. This allows the manufacturing system to respond to the rapid changes in products, their widening scope and faster pace of customization and support the evolving parts/products families.

Products evolution may be time or function based. Chronological Evolution develops gradually over time and represents a unidirectional natural progression as more knowledge and better technologies become available. It is unidirectional because as new and better solutions are obtained, there is no need to revert to older inefficient or flawed product designs. Functional Evolution is caused by significant and major changes in requirements, which are normally forced by many factors. It is often selective and discrete although a major overhaul is also possible. This type of change may be bi/multi-directional as the new product would fulfill different functional requirements but would not necessarily render previous designs obsolete.

In summary, the introduced natural evolution metaphor (ElMaraghy, 2006 and 2007) is useful in explaining the concept of evolving parts/products families and finding solutions to the associated challenges. Static parts family members are seen as closely knit, having a strong core of common features where all parts/products variations are within the pre-defined boundaries (as would be applied in FMS). The concepts of Composite Parts, Master Plans and Retrieval/Variant Process Planning are both valid and useful in this case. After some parts/products generations, new parts (species) emerge and parts families gradually lose their roots as some features disappear and new (additional features) and different (modified features) branches are developed. The extent of difference between parts generations depends on the number and nature of features' changes until a clear differentiation of characters develops. The same evolution notion applies to products where parts, modules or

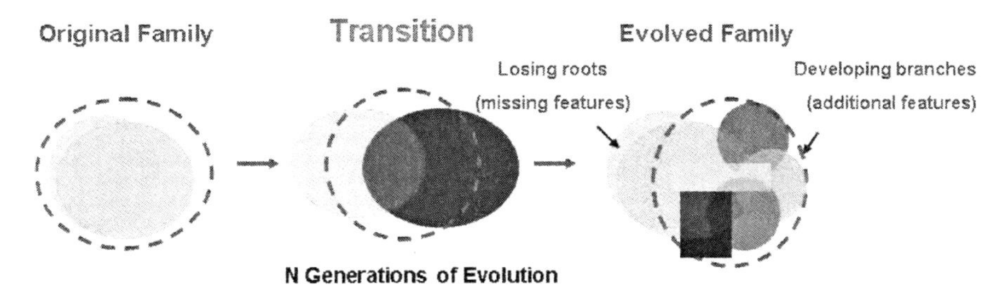

Fig. 2.2 Evolving parts/products families

sub-assemblies may be added, removed or changed causing the product and its family members to evolve. After many and different products generations, new product features and different products (species) and product families emerge with much less resemblance to the original parent family. In this case, many of the previously used and familiar rules and methods (e.g. for process planning) do not apply any longer. The magnitude of change and distance between new and old members of the parts/products families significantly influences the characteristics of the process plans in this new setting.

In light of the above discussion, the concept of "*Evolvable and Reconfigurable Process Plans*", which are capable of responding efficiently to both subtle and major changes in "*Evolving Parts/Products Families*" and changeable and reconfigurable manufacturing systems, was introduced (Azab and ElMaraghy, 2007a and 2007b) and is discussed in Sect. 2.6.

2.4 Modeling Products Evolution – A Biological Analogy

The importance of understanding and managing the variation and evolution of parts/products families has been emphasized in previous sections. A novel biological analogy was introduced (ElMaraghy et al., 2008) and used for modeling evolution in manufactured products with the aim of extending it to their manufacturing systems and understanding the relationship between them.

Evolution does not only mean change, it marks modifications occurring over time, which can be inherited by descendants, in the process of developing new species. "Adaptation" is the main driver of evolutionary changes, which we contend can be observed in both nature and manufacturing. This approach was first proposed (ElMaraghy et al., 2008), to study evolution in the context of manufactured products, was demonstrated using the Cladistics analysis originally introduced by Hennig (1966) but only used to date in biological analysis. A family of engine cylinder blocks, two instances of which are shown in Fig. 2.3, was used as an exam-

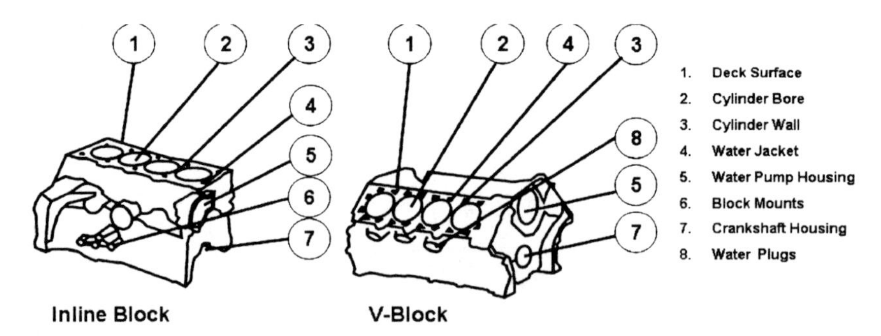

Fig. 2.3 Two members of the automobile engine cylinder blocks family – Inline and V-types (ElMaraghy et al., 2008)

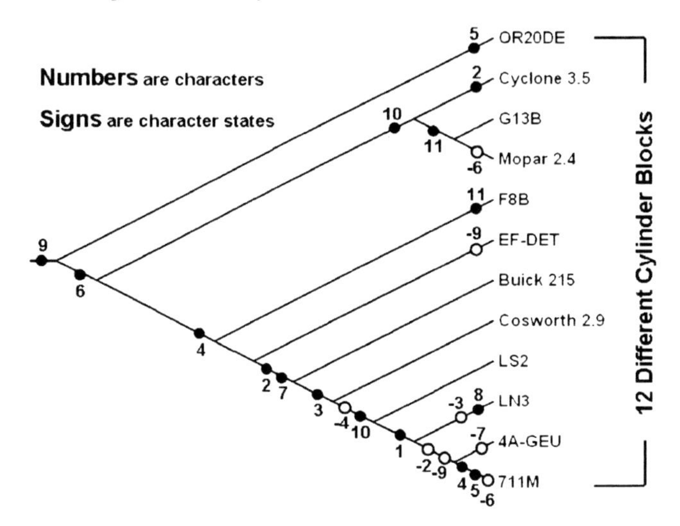

Fig. 2.4 Cylinder blocks Cladogram and product groups (ElMaraghy et al., 2008)

ple to demonstrate the developed application of Cladistics analysis and its merits. The cylinder block variants belong to automotive engines of different makes, material and types from Japan and North America, ranging in capacity from half a liter to six liters. The cylinder blocks are made of either Aluminum or Cast Iron. They belong to In-line or V-type, High-deck or Low-deck, Front or Rear Wheel Drive, Over Head Cam (OHC) or Over Head Valve (OHV) engines.

The Cladistics classification technique was shown to be capable of determining a logical representation of a group of automobile engine cylinder block variants and showing their path of evolution, in the most efficient way, using the parsimony analysis (ElMaraghy et al., 2008). The resulting Cladogram (Fig. 2.4) can yield additional useful information. These include potential possibilities of re-arranging existing product families to form more logical groupings, tracking their evolution trends, easily generating composite parts corresponding to a given set of product variants, identifying potential design and manufacturing latitudes, enhancing product design decisions and encouraging simplification, determining relevance of new variants to existing families, and anticipating future evolution directions of products design and development.

2.5 Design of Assembly Systems for Delayed Differentiation of Changing and Evolving Products

Customized and modular products allow manufactures to offer rich varieties to customers; however, this increases the complexity of both the products and manufacturing systems design. Delayed Product Differentiation (DPD) is a strategy introduced and adopted by companies to achieve the desired products variability while ensuring more manufacturing efficiency (e.g. He et al., 1998 and Xuehong et al., 2003). The objective is to postpone the stage in manufacturing where each of the products becomes differentiated and begins to have its own separate manufacturing path. Little work in literature has contributed methodologies for designing a physical manufacturing system that follows and implements the DPD strategy.

Clustering techniques are basic tools for establishing the different families of products. They are used to define the boundaries between the different families of products resulting in a number of differentiated sets, each containing a number of parts, components or products that are manufactured similarly, or have geometric likeness. Such techniques are used extensively in Group Technology (GT) and Cellular Manufacturing (CM). However, it has recently been proposed to use the commonality analysis as a fundamental method for analyzing each individual family of products for suitability to being produced in a DPD environment (AlGeddawy and ElMaraghy, 2008). Commonality analysis is mostly used with complete products composed of different parts, modules and sub-assemblies, rather than individual parts with different features. Its objective is to recognize commonality; it results in a metric of likeness among the products rather than identifying different sets of products. It should be noted that Cladistics were never used in the DPD literature, moreover, there is a lack of research in applying commonality analysis to products families in general, and in areas related to the DPD environment in particular.

The new framework offers a novel application of Cladistics applied to assembly lines design for Delayed product Differentiation. It: 1) uses products commonality schemes, and 2) complies with the precedence constraints that must be respected in sequencing assembly steps. It effectively links products design with the assembly line design. This model produced a set of unique Cladorams, as shown in Fig. 2.5,

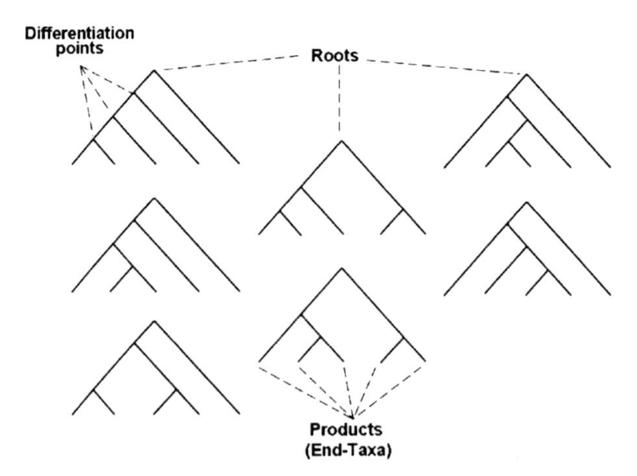

Fig. 2.5 Unique Cladograms representing assembly system layout schemes for the studied family of products (AlGeddawy and ElMaraghy, 2008)

which in fact schematically represent the possible physical assembly flow lines and their assembly steps/stations that can produce the analyzed family of five household products used for boiling water. It indicates where branching (i.e. product differentiation) takes place in each system alternative. The resulting flow line schemes would be further analyzed and compared using other performance and cost criteria and the best assembly system design would be selected.

Analysis of the resulting assembly line patterns reveals further information that should be considered to enable future improvements in the assembly line design and better management of the products variation while maximizing the delay in differentiation, including: 1) Re-sequencing of assembly operations by relaxing some precedence constraints for those similar and repeated assembly steps of closely related products, to allow these common steps to be moved up in the assembly line, and avoid their repetition, and 2) Re-designing products by adding commonly occurring characters/features to products that lack those characters, hence delaying branching out into different products on the assembly line.

The presented approach is a novel manufacturing system layout design method and a decision support tool for its further improvement. This design technique is not limited to assembly lines in one physical location but may be extended to the whole supply chain where products differentiation may be delayed till the point of delivery at distributed geographical locations. It helps manage the witnessed wide scope of products variation at the assembly level.

2.6 Process Planning – The Link Between Varying Products and their Manufacturing Systems

The process plans and planning functions are important links between the features of various generations and variations of products /product families and the features, capabilities and configurations of manufacturing systems and their components throughout their respective life cycles. One of the challenges for process planning in an environment characterized by change is to define methodologies and constructs that can be used consistently to respond to the variations observed in parts, products and families as well as changes in manufacturing resources and their availability on the shop floor. The efficient generation and adaptation of process plans is an important enabler for changeable and responsive manufacturing systems.

2.6.1 Existing Process Planning Concepts

The process planning activities have seen significant growth and development since the nineties. Manufacturing process planning seeks to define all necessary steps required to execute a manufacturing process, which imparts a definite change in shape, properties, surface finish or appearance on a part or a product, within given constraints while optimizing some stated criteria (ElMaraghy, 1993).

Process planning techniques are now being applied in many domains such as metal removal, assembly/disassembly, inspection, robotic tasks, rapid prototyping, welding, forming, and sheet metal working. The process planning concepts and approaches are classified based on their level of granularity into: 1) Multi-Domain Process Planning to select the most suitable manufacturing technology to produce the part/product, 2) Macro-Process Planning, which selects the best sequence of multiple different processing steps and set-ups as well as the machines to perform those operations, and 3) Micro-Process Planning, which details each individual operation and optimise's its parameters. Process planning may be done manually or using Computer-Aided Process Planning (CAPP) systems. Automated process planning varies according to the type and degree of automation and includes: 1) Retrieval/ Variant Process Planning, which capitalizes on the similarity in design or manufacturing features among parts grouped into families, and revises existing master plans, 2) Semi-Generative Process Planning that benefits from retrieved "Master Process Plans" to make some "part-specific" decisions, but also optimize the operations to be performed and their parameters using algorithmic procedures assisted by CAD models, databases, decision tables or trees, heuristics and knowledge rules, and 3) Generative Process Planning that aims to generate an optimized process plan from scratch. Its success is predicated on the availability of complete and accurate models of the parts and processes, and their behavior, constraints and interactions. Automated reasoning, knowledge-based systems and Artificial Intelligence techniques are essential in this approach. A truly generative process planning system in any domain is yet to be realized. The major challenge is the availability of complete and reliable mathematical models of the various manufacturing processes and their characteristics as well as complete process planning knowledge and rules.

2.6.2 Process Plans Changeability

A change in products and/or manufacturing systems would not necessarily result in changes in process plans; the nature of change matters. The nature and extent of change in process plans depend on the type and degree of parts/products variations. Hence different process planning schemes would be needed for different scenarios. The products variation hierarchy shown in Fig. 2.1 can be used to illustrate the need for changeable process plans at the various parts/products families and levels. The following types of products variations can be identified along with the corresponding required changes in process plans.

Process-Neutral Products Variations that help create product identities and differentiation visible to the customer without changing the manufacturing process steps such as changing of automobile body colors, the color and material of the interior finish or type of special modules such as audio equipment. These and similar variations are observed at the products platforms, product families, products and sub-assemblies/modules levels where the macro-assembly process sequence and steps would not normally be affected by such changes in these cases.

Process-Specific Products Variations tend to be seen in the features of products modules, sub-assemblies and parts. These variations may affect the manufacturing process, as in the case of abandoning brushes and adopting brush-less technology in electric motors that would require major changes in the manufacturing processes, or affect the process sequence (macro-plan) as parts and features are changed, added or removed. The variations can be of a parametric nature where the detailed micro-plan would need to be adjusted accordingly. These parametric changes may be *small* and hence the same technology (e.g. metal removal) would still be used but with adapted parameters, or they may be *extreme* so as to call for a completely different manufacturing method. Some dimensional variations can lead to significant changes in the method of fabrication and would therefore require major process re-planning. For example, small variation, within limits, in the features and dimensions of a gear in the family of gears shown in Fig. 2.1 (e.g. gear teeth profile, key-way, and inner and outer diameters) would not lead to significant changes in the method of manufacture. Existing metal removal process plans can still be changed/adapted effectively using parametric variations, where group technology, composite parts and retrieval/variant process planning would be used. The addition or deletion of features affects the sequence of operations; and significant changes in macro-process plans would be required where all types of precedence constraints must be checked and satisfied. In addition, extreme reduction in dimensions may require micro-machining of the gear, and very large gears may have to be cast or forged first then machine finished. Both types of extreme variations call for different fabrication method/technology and complete process re-planning rather than adaptation.

Since not only products variations are increasing in scope and frequency and the families of manufactured parts are evolving, but also manufacturing resources on the shop floor and their functionalities are becoming changeable and reconfigurable (ElMaraghy, 2005), then "Reconfigurable Process Plans" are becoming an essential enabler of change.

There are some key criteria for reconfiguring process plans and commensurate techniques for their efficient re-generation when needed: 1) The utilization of the multi-directional and multi-faceted relationships and associations between the characteristics of product features, the process plan elements and all manufacturing system modules capable of producing them, 2) The process plan representation characteristics that facilitate adjusting and implementing optimally determined feasible and economical alterations in process plans to reflect the needed reconfiguration, 3) The ability to model process plans at varying levels of detail and granularity in order to, readily and appropriately, respond to changes at different levels (e.g. in products, technologies and systems), and 4) The availability of complete knowledge bases and rules for process planning and reconfiguration, accurate mathematical models of the various manufacturing processes and resources as well as meta-knowledge rules for using this knowledge to automate the plan reconfiguration. The optimality (time, quality, cost, etc.) of the evolved and reconfigured process plans should always be verified and maintained.

Some examples of newly developed approaches and methods for process planning for variation, based on the author's and her group's research are presented next for illustration.

2.6.3 Reconfiguring Process Plans (RPP) and Its Significance

The Reconfigurable Process Planning (RPP) approach represents an important enabler of changeability for evolving products and manufacturing systems. It addresses the new problem that arises due to the increased frequency and extent of changes in products and systems and the need to manage these changes cost effectively and with the least disruption of the production activities and their associated high cost.

A hybrid retrieval/generative reconfiguration model and algorithms for process planning (RPP) have been developed (see Azab and ElMaraghy, 2007a and 2007b for more details). The parts/products family, closest to the new part, would be identified and its composite part and corresponding master process plan are retrieved and missing features/operations are removed. Novel 0–1 Integer Mathematical Models and Mathematical Programming for reconfiguring these macro-level process plans were formulated and applied, for the first time, to the process planning/sequencing problem. It fundamentally changed it from an optimal sequencing to an optimal insertion problem. Reconfiguration of precedence graphs to optimize the scope and cost of process plans reconfiguration is achieved by inserting/removing features/operations iteratively in their string representation to determine the best location for added features and related operations in the operations sequence. This is akin to inserting new genes in a chromosome using the genetic evolution metaphor and lends itself to modeling and capturing the evolution of the parts/products features and corresponding processing operations and plans. The proposed RPP macro-process plan reconfiguration methodology readily supports evolving part families and manufacturing systems. Mathematical programming and formulations were presented, for the first time, to generate process plans that would account for changes in parts' features as they evolve beyond the scope of their original product families.

Two criteria were used in Reconfiguring Process Plans. First, the parts handling and re-fixturing time, when no value is added to the product, is minimized to arrive at a reconfigured and optimal process plan that minimizes the extent of reconfiguration and hence its implications. Second, a process plan Reconfiguration Index (RI), which is a Changeability Metric that captures the extent and cost of changes in the plan, was introduced as a new criterion to evaluate the reconfigured process plans. It can be used for choosing among alternate process sequences with substantially similar total cost by opting for the one that causes the least changes and disturbances on the shop floor (i.e. smallest RI). This saves other direct and indirect costs such as those related to changes in set-ups, tools, re-programming and associated errors and related quality issues. This tends to favor *limiting/localizing* the extent and effect of plan reconfiguration compared with the initial process plan, and it is *done by design*.

The weight given to the above two planning criteria, in practice, depends on the cost component that matters most in a given situation and different emphasis on initial vs. running cost in large volume and small series production as they are affected by frequent changes. The developed RPP model can use either criterion or a combination of both. Thus, the process planner would have the opportunity to consider which criterion matters most, based on experience and available data.

The RPP model has been applied in the metal removal domain, at the parts family level, for a family of single-cylinder aluminum engine front covers (Azab and ElMaraghy, 2007b) where the parts features changed. It was also applied in the assembly domain, at the products family level, for a family of small kitchen appliances (kettles) where the product features changed (Azab, 2008). These test cases clearly demonstrated the effectiveness of the proposed RPP methods.

The RPP approach is more advantageous than existing methods for dynamic, adaptive and non-linear process planning that utilize either pre-planning or re-planning methodologies. For the pre-planning methods, alternate process plans are developed and documented ahead of time in anticipation of future changes. In addition to the obvious cost and computational burden involved in this approach, future changes in products and technology cannot be fully predicted a priory. Moreover pre-planned process plans would likely become obsolete as the products, resources and technologies are changed. In re-planning, a whole new plan is created from scratch, without benefiting from currently available plans, set-ups, tooling, etc., every time some changes are made with the obvious added cost of not only re-planning, but also more importantly the potential major changes and disturbances on the shop floor as a result.

This new methodology for Reconfiguring Process Plans (RPP) is applicable to macro-process planning where determining the optimal sequence of operations and satisfying precedence constraints are important on the parts, modules and products levels. Effective macro-process planning, involving all manufacturing fabrication and assembly steps and their logical sequence, is important if the potential to offer greater product variety rapidly while reducing cost and risks is to be achieved.

2.6.4 Process Planning for Reconfigurable Machines

The RPP approach deals with variations in the process plans as a result of changing parts and products. Changes in process plans might require different machines assignment, depending on the available machines and their capabilities. Changes in machines, through purchase, replacement or reconfiguration would also trigger changes in process plans to utilize and benefit from the new capabilities. Therefore, a two-way mapping between the features of products and machine tools was devel-

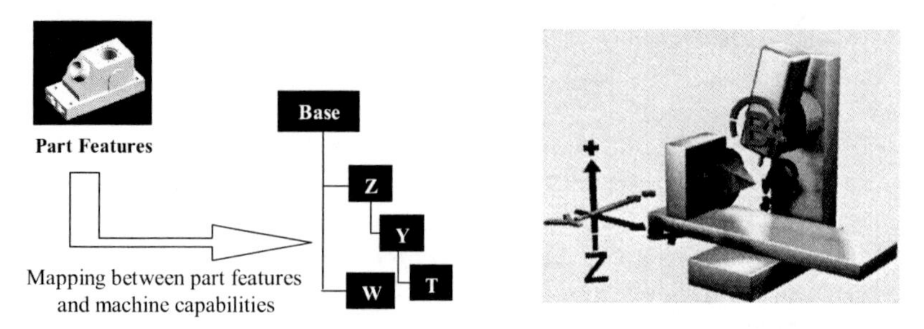

Fig. 2.6 Mapping between part features and machine capabilities (Shabaka and ElMaraghy, 2007)

oped and used for re-planning. The selection of the different types of machine(s) and their appropriate configurations to produce different types of parts and features, according to the required machine capabilities, is a fundamental building block in generative planning of manufacturing processes (Shabaka and ElMaraghy, 2007). The machine structure is represented as kinematic chains that capture the number, type and order of different machine tool axes of motion, which are indicative of its degrees of freedom and ability to produce certain geometric features as well as the size of workspace (Fig. 2.6). Operations are represented by a precedence graph and clustered according to the logical, functional and technical constraints.

Optimal process plans are generated using Genetic Algorithms (GAs) based on a constraint satisfaction procedure that ensures the feasibility of all produced plans. A rule-based semi-generative Computer-Aided Process Planning approach was introduced to adapt existing process plans through re-planning and account for changes in product requirements and/or availability of system resources. This approach minimizes the required hard-type reconfiguration on both the manufacturing system and machine levels if less costly soft-type adaptation of existing process plans can be performed instead. This research work advances the existing knowledge about process planning in the RMS domain with regards to macro-process planning (sequencing), operation selection and selection of machines and their configurations. It supports the process planer's decision making regarding the machine assignment/selection and sequencing activities at the initial stages of manufacturing systems design and subsequent changes in products features and scope. The developed approach is not limited to RMS and can be applied to other manufacturing systems such as FMS.

2.7 Discussion and Conclusions

The proliferation of products variants is wide spread due to the natural products evolution, which has been on the rise to satisfy customers' needs and specifications and benefit from advances in new materials and technologies as well as comply with imposed environmental legislation's and legal regulations. Products innovation and mass customization introduce many changes aimed at achieving products differentiation, which is an important key to surviving globalization and ensuring a competitive advantage. In addition, many engineering changes in products take place frequently throughout the product life and affect all types and sizes of manufacturers, from job shops, tool and die makers, to large automotive or aerospace companies. All product changes and revisions result in costly and significant changes in the design and manufacturing steps, setups, process plans, tools, fixtures and the used machines.

The increased products customization has also lead to a wider scope of products variants and increased their complexity as well as that of their manufacturing methods and systems. New manufacturing systems paradigms, such as flexible and reconfigurable manufacturing, evolved to achieve maximum products variety while remaining competitive, profitable and responsive to the frequent changes in markets and products.

This chapter presented a number of novel strategies and solutions to manage the inevitable products variations and related manufacturing changes.

A *Variations Hierarchy* was presented to classify variations at different levels from products families and platforms to individual parts and their features, and the implications of variation and commonality for both design and manufacturing were discussed.

A new class of "*Evolving Parts/Products Families*" was presented and contrasted with the traditional notions of static parts families. The implications of such evolution on planning products families and platforms were explored and its effects on downstream manufacturing support functions, such as process planning and assembly systems design were highlighted.

A novel approach for *modeling evolved products*, utilizing mechanisms analogous to those observed in nature, was presented. This innovative concept has the potential for modeling not only the evolution of products or their manufacturing systems but also their symbiotic co-evolution relationship. Its application, using Cladistics, to recognize and classify the commonalities among products and to design assembly systems layouts for delayed products differentiation was illustrated. The obtained results provide a promising foundation for future research in this domain.

An innovative method for *Reconfiguring Process Plans (RPP)* was presented. The new RPP method is also akin to the genetic evolution metaphor in manipulating the strings of ordered operations. It helps manage the complexity and variation of changing and evolving parts and products families and introduces innovative planning techniques that were demonstrated for both parts fabrication and products assembly. One of the main contributions of this method is the development of a new mathematical model for solving the classical problem of process planning through reconfiguration rather than sequencing. It limits the changes on the shop floor resulting from changing process plans by seeking a localized optimally reconfigured plan. Hence, process plans reconfiguration can be performed only when needed, where needed, and to the extent needed. This is done by design. It introduces an efficient way of coping with the frequent changes and allows this reconfiguration to take place on the fly. Hence, it reduces the need to keep, maintain and manage a huge number of process plans variants by manufactures. It demonstrates that process plans reconfiguration on demand is an effective management strategy to cope with variation.

In conclusion, the designers of products, processes and manufacturing systems as well as production planners should be cognizant of the coupling between generations and variations of products and manufacturing systems, its special nature and characteristics, and capitalize on its potential benefits for improving the productivity of the whole enterprise. This chapter presented a number of research contributions, which utilize this notion, towards providing enablers of change and achieving these goals.

Acknowledgements The author acknowledges the research support from the Canada Research Chairs (CRC) program and the Natural Sciences and Engineering Research Council (NSERC) of Canada.

References

AlGeddawy T., ElMaraghy H.A., 2008, Assembly System Design for Delayed Product Differentiation, 2nd CIRP International Conference on Assembly Technologies and Systems (CATS), 21.–23. September 2008, Toronto, Canada

Azab A., 2008, Reconfiguring Process Plans: A Mathematical Programming Approach, Ph.D. Dissertation, University of Windsor

Azab A., ElMaraghy H.A., 2007a, Mathematical Modelling for Reconfigurable Process Planning. CIRP Annals 56/1:467–472

Azab A., ElMaraghy H., 2007b, Sequential Process Planning: A Hybrid Optimal Macro-Level Approach, Journal of Manufacturing Systems (JMS) Special Issue on Design, Planning, and Control for Reconfigurable Manufacturing Systems 26/3:147–160

Azab A., Perusi G., ElMaraghy H. and Urbanic J., 2007c, Semi-Generative Macro-Process Planning for Reconfigurable Manufacturing, Digital Enterprise Technology: Perspectives & Future Challenges, Editors: P.F. Cunha and P.G. Maropoulos, Springer Science, ISBN: 978-0-387-49863-8, pp 251–258

ElMaraghy H., AlGeddawy T., Azab A., 2008, Modelling evolution in manufacturing: A biological analogy. CIRP Annals -Manufacturing Technology 57/1:467–472, doi: 10.1016/j.cirp.2008.03.136

ElMaraghy H.A., 2007, Reconfigurable Process Plans for Responsive Manufacturing Systems, Digital Enterprise Technology: Perspectives & Future Challenges, Editors: P.F. Cunha and P.G. Maropoulos, Springer Science, ISBN: 978-0-387-49863-8, pp 35–44

ElMaraghy H.A., 2006, Reconfigurable Process Plans for Responsive Manufacturing Systems, Keynote Paper, Proceedings of the CIRP International Design Enterprise Technology (DET) Conference, Setubal, Portugal, 18.–20. September 2006

ElMaraghy H.A., 2005, Flexible and Reconfigurable Manufacturing Systems Paradigms. International J of Manufacturing Systems (IJMS) 17/4:261–276, Special Issue

ElMaraghy H.A., 1993, Evolution and Future Perspectives of CAPP. CIRP Annals 42/2:739–751

Groover M.K., 2008, Automation, Production Systems, and Computer-Integrated Manufacturing, 3rd Ed., Chapter 18, Pearson/Prentice Hall (Publisher)

He D., Kusiak A., Tseng T., 1998, Delayed Product Differentiation: a Design and Manufacturing Perspective. Computer-Aided Design 30:105–113

Hennig W., 1966, republished in 1999, Phylogenitic Systematics, Urbana, University of Illinois Press.

Opitz H., 1970, A Classification System to Describe Work Pieces. Pergamon Press, Oxford, England

Shabaka A.I., ElMaraghy H.A., 2008, A Model for Generating Optimal Process Plans in RMS. International Journal of Computer Integrated Manufacturing (IJCIM) 21/2:180–194

Shabaka A.I., ElMaraghy H.A., 2007, Generation of Machine Configurations based on Product Features. International Journal of Computer Integrated Manufacturing (IJCIM) 20/4:355–369, Special Issue

Shimokawa K., Jurgens U., Fujimoto T., (eds) 1997, Transforming Automobile Assembly. Springer, New York

Wiendahl H.-P., ElMaraghy H.A., Nyhuis P., Zaeh M., Wiendahl H.-H., Duffie N., Kolakowski M., 2007, Changeable Manufacturing: Classification, Design, Operation. CIRP Annals 56/2:783–809, Keynote Paper

Xuehong D., Jiao J., Tseng M., 2003, Identifying Customer Need Patterns for Customisation and Personalization. Integrated Manufacturing Systems 14/5:387–396

Chapter 3
Focused Flexibility in Production Systems

W. Terkaj, T. Tolio and A. Valente[1]

Abstract Manufacturing flexibility is seen as the main mechanism for surviving in the present market environment. Companies acquire systems with a high degree of flexibility to cope with frequent production volume changes and evolutions of the technological requirements of products. However, literature and industrial experience show that flexibility is not always a well-defined concept. Therefore it is really complex to understand and use flexibility during system design process. Indeed, the development of structured approaches to support the system design by considering basic flexibility forms is still an open issue. This work presents an Ontology on Flexibility aiming at providing a standard method to analyze flexibility. Firstly, it contributes in systemizing the large number of existing flexibility definitions and classifications. Secondly, it can be used to analyze real systems and to better understand their characteristics in terms of flexibility. Finally this ontology represents a key point of a general approach to design production system with the right level of flexibility.

Keywords Focused Flexibility Manufacturing Systems (FFMSs), Manufacturing system design, Ontology on Flexibility

3.1 The Importance of Manufacturing Flexibility in Uncertain Production Contexts

Companies producing mechanical components to be assembled into final products produced in high volumes, in order to remain competitive, must deal with critical factors such as: tight tolerances on the parts, short lead times, frequent market changes and pressure on costs (Matta et al., 2000; Tolio, 2008). Obtaining optimal-

---

[1] Dipartimento di Meccanica, Politecnico di Milano, Milan, Italy

ity in each of these areas can be difficult and companies often define production objectives as trade-offs among these critical factors (Chryssolouris, 1996).

The flexibility degree of a manufacturing system represents a critical issue within the system design phase. On the one hand, it is considered a fundamental requirement for firms competing in a reactive or a proactive way. On the other hand, flexibility is not always a desirable characteristic of a system. This point needs to be clarified since in many cases flexibility can jeopardize the profitability of the firm. It is rather frequent to find in the literature descriptions of industrial situations where flexible manufacturing systems have unsatisfactory performance (Koren et al., 1999; Landers, 2000), cases where the available flexibility remains unused (Sethi and Sethi, 1990; Matta et al., 2000), or cases where the management perceives flexibility more as an undesirable complication than a potential advantage for the firm (Stecke, 1985).

From the scientific perspective, focusing the flexibility of a production system on the specific needs represents a particularly challenging problem. In fact, the customization of system flexibility provides economical advantages in terms of system investment costs, but, on the other hand, tuning the flexibility on the production problem reduces some of the safety margins, which allows decoupling the phases of manufacturing system design (Tolio and Valente, 2008). One of the key issues is that focused flexibility asks for a very careful risk appraisal. To reach this goal all activities ranging from the detailed definition of the manufacturing strategy to the configuration and reconfiguration of production systems must be redesigned and strictly integrated, thus highlighting the need of combining and harmonizing different types of knowledge which are all essential to obtain a competitive solution.

3.1.1 Focused Flexibility Manufacturing Systems – FFMSs

The simultaneous need of flexibility and productivity is not well addressed by available production systems, which tend to propose pre-selected types of flexibility to introduce in the system. Traditionally, rigid transfer lines (RTL) have been adopted for the production of small families of part types (one or few part types) to be produced in high volumes (Koren et al. 1998). Since in RTLs scalability is low, RTLs are usually designed according to the maximum market demand that the firm forecasts to satisfy in the future (volume flexibility); as a consequence, in many situations RTLs do not operate at their full capacity since their designed volume flexibility is frequently over-sized. On the other hand, flexible manufacturing systems (FMSs) and parallel machine-FMSs (PM-FMSs) have been adopted for the production of large mixes of parts to be produced in small quantities (Grieco et al., 2002; Hutchinson and Pflughoeft, 1994). FMSs are conceived to react to most of the possible product changes. The investment to acquire a FMS is very high and it considerably affects the cost to produce a part. Indeed their flexibility is frequently too large and expensive. This is extremely evident for instance in the case producers of components for the automotive industry (Sethi and Sethi 1990) where even if the types of products are rather stable still fully flexible FMSs are frequently adopted.

Therefore, in many situations, there is a need to address the trade-off between productivity and flexibility by means of manufacturing systems having the minimum level of flexibility required by the production problem on hand (Tolio and Valente, 2007). This new class of production systems can be named Focused Flexibility Manufacturing Systems (FFMSs) (Tolio and Valente, 2006).

The flexibility degree in FFMSs is related to the required ability to cope with volume, mix and technological changes, and it must take into account both present and future changes. The required level of system flexibility impacts on the architecture of the system and the explicit design of flexibility often leads to hybrid systems (Matta et al. 2001), i.e. automated integrated systems in which parts can be processed by both general purpose and dedicated machines. This is a key issue of FFMSs and results from the matching of two different features that characterize respectively FMSs and Dedicated Manufacturing Systems (DMSs) (Tolio, 2008). Another way FFMSs reach their goal is by combining in the same system old and new machines. In other words, in FFMSs the customization of the flexibility for a certain production problem explicitly addresses the trade-off between flexibility and productivity and tries to maximize system profitability.

As it can be noticed, FFMSs differ from Reconfigurable Manufacturing Systems (RMSs) (Koren et al., 1999; Ling et al., 1999; Landers, 2000) in the timing of flexibility acquisition and in the explicit analysis of the cost of flexibility. Indeed the key idea of RMSs is to provide in each moment the production system exactly with the capabilities required by the production problem on hand and to modify the system if the needs change with time. Frequently, the reconfigurability option needs to be considered at system level since in many cases available hardware and software devices are not mature enough to support reconfigurability at machine level (Wiendhal et al., 2007). The FFMSs consider reconfigurability and flexibility as two options and mix them on the basis of their costs. For instance, it could be cheaper to acquire more flexibility than the amount strictly required by the present production in order to avoid possible future system reconfigurations and ramp-ups. Another example to pursue the extra-flexibility option, involving lower economical investments, is to design the system introducing, among the others, old machines that have been totally depreciated but are still very flexible. In this case, FFMSs have some extra-flexibility designed to cope with future production changes, i.e. a degree of flexibility tuned both on present and future part families. The strategic decision of designing the reconfigurability option or the extra-flexibility option depends on the result of costs analysis (Tolio et al., 2007; Tolio, 2008).

Although the concept of FFMS would fit particularly well in the current production context, frequently the tradition and know-how of both machine tool builders and production system users play a crucial role in hindering the exploitation of this idea. In fact even if firms often agree with the focused flexibility vision nevertheless the lack of a clear definition of the flexibility design problem prevents the exploitation of this approach. In order to overcome this limitation new frameworks for defining the manufacturing system flexibility have to be developed together with methodologies and tools to design the degree of flexibility on the basis of the specific production problem.

3.2 Literature Review

Flexibility is the ability to change or react with little penalty in time, effort, cost or performance (Upton, 1994) in order to cope with a set of production requirements (De Toni and Tonchia, 1998). On the one hand internal requirements strictly related to production call for *internal flexibility*; on the other hand, when flexibility represents a competitive advantage for the company in relation to external turbulence, *external flexibility* is required.

A dominant feature of the literature, and an important step in providing a better understanding, is the use of taxonomies of flexibility, which classify different types of manufacturing flexibility. These categories are useful since they provide general types that can be used to distinguish one form of flexibility from another, as stated by Upton (1994). In order to characterize each important type of flexibility, Upton suggests that, if flexibility is an issue, questions regarding which changes and how often they happen should be asked. These drivers force the manufacturers to evaluate their ability to change their manufacturing systems and the penalty to face the change. This is a complex task in dynamic manufacturing contexts (Beach et al., 2000), as for instance the automotive, semiconductor, electronics and high tech markets, because products are affected by frequent changes in volume and technology.

The key issue highlighted in the literature is the multidimensional nature of flexibility. Many efforts, over time, have been dedicated to the development of taxonomies in which all the possible forms of flexibility are classified and characterized. Sethi and Sethi (1990) gave order to the exiting literature by proposing a classification where 11 different dimensions of flexibility are identified. Later on, Gerwin (1993) reduced to 9 forms of flexibility the framework provided by Sethi and Sethi (1990). Gupta and Somers (1996) developed an instrument to measure manufacturing flexibility and they also analyzed the relation among business strategy, manufacturing flexibility and performance: moreover, they carried out an empirical study to validate the dimensions of flexibility defined by Sethi and Sethi (1990). De Toni and Tonchia (1998) definitely contributed to the activity of conceptual systemization of the earlier works on flexibility. Their work proposes a classification framework consisting of six main aspects of manufacturing flexibility, such as the definition of flexibility, factors which determine the need for flexibility and classification (dimensions) of flexibility (hierarchical, by phases, temporal, by object of variation, or based on a mixture of the previous dimensions). This framework has been used to classify more than twenty years of research contributions on the topic.

A further contribution is proposed, by Zhang et al. (2003) where manufacturing flexibility and its sub-dimensions are described as integral components of value chain flexibility. ElMaraghy (2005) links the concept of manufacturing system life-cycle to manufacturing systems flexibility and reconfigurability; this paper presents the most recent views of a panel of experts from Academia and Industry on the comparisons between flexible and reconfigurable manufacturing.

Although much effort has been devoted in the literature to the analysis of flexibility, as a solution to cope with uncertainty in the market and to support the manufacturing strategy, the link between the need for flexibility and the design of manufacturing systems is still very weak (Tolio and Valente 2006). In this area, examples are provided of the relation between the level of flexibility embedded into the system and corresponding system performance (Koren et al., 1999; Landers, 2000) as well as critical analyzes of production systems characterized by extra-flexibility (Sethi and Sethi, 1990; Matta et al., 2000). However, methodologies to design systems with predefined levels of flexibility are still almost missing.

3.3 Proposal of an Ontology on Flexibility

In the previous sections, the importance of designing manufacturing systems endowed with the right degree of flexibility has been underlined. This task is complex because it requires addressing internal and external issues; in particular, product and processes are easily and frequently changed by market and manufacturing strategies, while production systems must cope with relevant inertia to changes. The goal of this work consists of providing a contribution to fill the modeling gap between a production problem and the manufacturing system best suited to face it. This gap, for instance, consists of a lack of proper knowledge concerning the logical framework required to deal with the problem, the type of information to be collected and the methodologies and tools to be applied to jointly consider information of different nature. Considering the state of art for system design and system flexibility analysis, three main issues can be identified to reach the final goal:

1. Identification of Basic Flexibility Forms which can lead the solution of the System Design problem;
2. Integration of new concepts (e.g. Reconfigurability and Changeability) in the Flexibility theory;
3. Design of Production Systems characterized by the right degree of flexibility, translating Flexibility Forms into System Specifications.

In this section, the first two issues will be addressed, while the third one is dealt with in Sect. 3.5. Considerable effort has been devoted to the definition of different forms of flexibility to describe the characteristics of a manufacturing system (see Sect. 3.2). Some authors have pointed out that a given form of flexibility is the capability of reacting to a well-defined type of "stimulus", which can be experienced by the manufacturing system (Upton, 1994; Correa and Slack, 1996; De Toni and Tonchia, 1998; Grubbstrom and Olhanger, 1997). Other authors have shown that a given form of flexibility may support various proactive strategies of the firm (Gupta and Goyal, 1989; Sethi and Sethi, 1990; Hyun and Ahn, 1992; Gerwin, 1993; Gupta and Somers, 1996). Since both the stimuli acting on the firm and the proactive strategies of the firm may differ, there is a need for various forms of flexibility. The result is that the number of types of flexibility proposed in the literature is growing and, even if some rationalization has taken place, still the number is very high. Also, different authors tend to assign to a given type of flexibility slightly different meaning given the different fields of application they take as reference. Moreover, there is ambiguity among flexibility forms and other concepts (e.g. Expansion Flexibility vs. Reconfigurability). This situation cannot be overcome since it depends on the fact that a given form of flexibility is actually an answer to a very specific problem. Since the number of problems is uncountable, the number of forms of flexibility, which may be devised, is also in principle uncountable.

In this chapter, an ontology is proposed where each form of flexibility is considered as a recipe to tackle a specific situation. According to the first issue (e.g. the identification of Basic Flexibility Forms which can lead the System Design problem), the key question is whether there are some basic *dimensions* of flexibility from which all the various *forms* of flexibility may be obtained by means of a specific combination tuned for specific problems. *Dimensions* are general theoretical concepts and should not find a direct implementation. Instead, dimensions are embedded in the various *forms* of flexibility, which can be found in specific applications. For this reason, *dimensions* should not be measured but should be treated as logical categories.

In this view, to solve specific problems there is the need of system specific *forms* of flexibility, which may be implemented and measured but in turn they incorporate a combination of the basic *dimensions* of flexibility.

If such a set of *dimensions* can be defined, one desirable property is that each *dimension* in the set should be orthogonal to the other *dimensions* in the sense that the *dimensions* in the set are independent and that one *dimension* cannot be obtained as a combination of the other ones. Another desirable property is completeness, i.e. each *form* of flexibility should be derived as a specific combination of the given *dimensions*. A set of flexibility *dimensions* is proposed, as reported in Table 3.1, to answer to these requirements.

Table 3.1 Basic Flexibility Dimensions

Basic Flexibility Dimension	Definition
Capacity	The system can execute the same operations at a different scale
Functionality	The system can execute different operations (different features, different level of precision, etc.)
Process	The system can obtain the same result in different ways
Production Planning	The system can change the order of execution or the resource assignment to obtain the same result

The proposed set is orthogonal, indeed it is impossible to obtain one dimension as a combination of the others A good combination of these *dimensions* makes the *form* of flexibility a valuable answer to the specific problems a company may encounter.

Each *dimension* needs to be further specified by attributes, as proposed by Upton (1994): attributes can be used for each of the four *dimensions*. Table 3.2 provides the *attribute* definitions. The goal here is not to derive a metrics to measure the *dimensions*; the idea is that the concepts contained in the *dimensions* cannot be completely defined if the described attributes are not introduced. Therefore, attributes are treated here at a conceptual level.

To completely define the various forms of flexibility another concept must be introduced. A given flexibility *form* specified by its *dimension* and *attributes* may be present in a given system. However, another system may exist where the considered *form* is not present but it can be acquired so that, after this acquisition, the two systems have similar capabilities. This second situation differs from the first one in the fact that the system is one step behind because to obtain the same capability some actions must be taken. However, the fact that these actions can be taken means that the system has some pre-disposition, which makes it different from a system, which cannot be modified. This pre-disposition is normally called in the literature "Reconfigurability". The fact that a system is one step behind under a given *form* suggests the concept of a ladder with different *levels*. At the top level of the ladder the given *dimension* considered is fully operational. At the lower levels of the ladder

Table 3.2 Flexibility Attributes

Attribute	Definition
Range	Range expresses the extension of the differences among the various ways of behaving of the system under a given dimension. Range increases with the diversity of the set of options or alternatives, which may be accomplished. For example in the Functionality dimensions it represents how diverse is the set of different operations, which can be executed by the system.
Resolution	Resolution expresses how close are the alternatives within the range of a given dimension. Resolution increases with the number of viable alternatives if they are uniformly distributed within the range. For example in the Functionality dimensions it expresses how small is the distance between similar but different operations, which can be done by the system.
Mobility	Mobility within the range. Mobility expresses the ease with which it is possible to modify the behavior under a given dimension. In fact, in order to start operating at a different point on a given dimension of change, there will be some transition penalty. Low values of transition penalties imply mobility. For instance in the Functionality dimension it may represent how easily it is possible to move from doing one operation to performing another one.
Uniformity	Uniformity within the range. Uniformity expresses how the performance of the system varies while moving within the range. If the performance is similar then the uniformity is high. For example in the Functionality dimension it may represent the difference in capability or costs while executing different operations.

more steps must be taken in order to reach the top level. The *levels* of the ladder are defined in Table 3.3.

Through the definition of Basic Flexibility *Levels*, the proposed ontology allows to unify the concepts of Flexibility, Reconfigurability and Changeability, coping with the second issue previously identified ('Integration of new concepts in the Flexibility theory'). All these concepts deal with modifications in Production Systems and the difference among them consists of the timing, cost and number of steps necessary to implement a modification.

A basic flexibility *form* is the combination of a specific *dimension* (specified by its attributes) and of a specific *level* of the ladder. Therefore various basic *forms* of flexibility can be derived. By combining basic *forms* of flexibility, compound *forms* of flexibility can be obtained. A graphical representation of the proposed ontology through an UML Class Diagram is reported in Fig. 3.1.

To test the viability of the proposed framework two analyzes are carried out and presented. Firstly, the possibility to map all the *forms* of flexibility described in the existing literature through the four basic *dimensions* is investigated. Secondly, the attention is focused on the *forms* of flexibility applied in the industrial production context. Herein, some industrial cases are analyzed using the framework to understand how the basic *dimensions* of flexibility are combined.

Table 3.3 Basic Flexibility Levels

Basic Flexibility Level	Definition
Level 1 (Flexibility)	The system has the ability
Level 2 (Reconfigurability)	The system can acquire the ability already having the enablers
Level 3 (Changeability)	The system can acquire the enablers

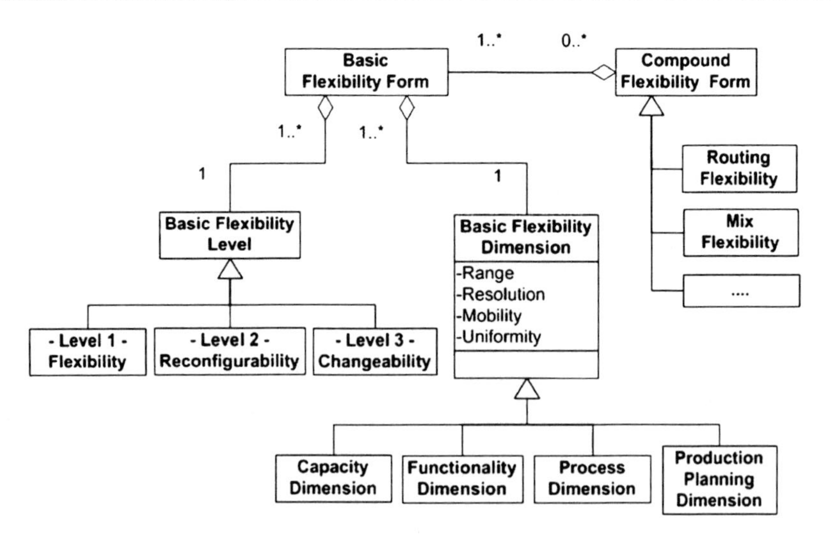

Fig. 3.1 Structure of the Ontology on Flexibility

Table 3.4 Extract of the *flexibility forms* found in the literature mapped according to the proposed ontology

Compound Flexibility Form	Paper	Capacity Flexibility	Functionality Flexibility	Process Flexibility	Production Planning Flexibility
Expansion	(Sethi and Sethi 1990)	Level 2, Level 3	Level 2, Level 3	–	–
Expansion	(Parker and Wirth 1999)	Level 2, Level 3	–	–	–
Volume	(Sethi and Sethi 1990)	Level 1	–	–	–

The flexibility *forms* defined in the literature have been mapped according to the basic *dimensions* of flexibility defined above. Globally, 109 *forms* (both basic *forms* and compound *forms*) of flexibility have been mapped by Tolio et al. (2008). Three examples of flexibility *form* analysis are reported in Table 3.4. The flexibility *forms* have been mapped defining which *Basic Flexibility Dimensions* are involved and at which *Basic Flexibility Levels*. For instance, Volume Flexibility as reported by Sethi and Sethi (1990) corresponds to Capacity Flexibility at Level 1.

3.4 Analysis of Real Systems

The proposed Ontology on Flexibility has been validated by analyzing some real production systems. The goal was to verify whether the requirements of flexibility addressed by these systems could be described using the proposed flexibility *dimensions* and *levels*; moreover, the cases have been studied paying attention to the topic of Focused Flexibility, finding out how different manufacturing system solutions cope with the need of flexibility. The following case studies have been considered:

1. Lajous Industries SA
2. Riello Sistemi S.p.A.

These industrial cases have been selected since they exemplify the need of rationalizing system flexibility. Starting from the description of the production context in which firms operate as well as the related designed production system, the current system solution will be evaluated using the provided ontology on flexibility.

3.4.1 Lajous Industries SA Case Study

Lajous Industries SA, which belongs to the industrial group Peugeot-Japy, is the French leader in the market of production of metal components for automotive industry. The components produced by Lajous can be divided into the following categories:

- Engine related components (e.g. manifolds, engine supports and accessories, pump bodies, fly wheels, oil cups, *etc.*);
- Chassis/suspensions/brakes (e.g. brake drums, pivot supports, *etc.*);
- Transmission components (e.g. synchronization rings, differential boxes, *etc.*).

Many important firms in the automotive sector are customers of Lajous, as for instance Audi, Pegeout-Citroen, Ford France SA, Magneti Minarelli, New Holland and Renault. Lajous pays a strong attention to the problem of total quality and important efforts are directed towards the introduction of Lean Production and Total Production Maintenance concepts.

The plant of Lajous at Compiegne (France) was studied within the Mod-Flex-Prod European project (EU Project BE96-3883); the plant consisted of approximately 600 machine tools. A part of the global production problem was characterized by a family of metal components, which can be clustered in few families of product types described by technological and volume evolutions. In particular, in the portion of the system studied, at a first stage the firm had to produce a family of three products: codes A, B, C (Fig. 3.2). The firm forecasted that a new product type, code D, could be produced at a later time. In this case, technological and mix changes would characterize the part family evolution.

Therefore, Lajous decided to install a new type of FMS (Fig. 3.3a) proposed by MCM S.p.A. to address the production problem characterized by frequent technological and volume changes of products.

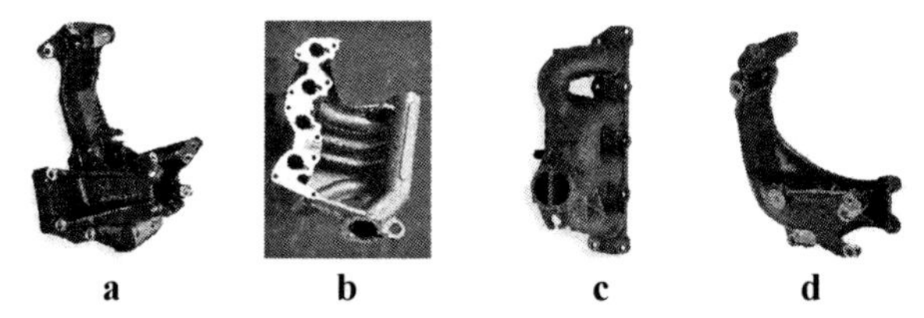

Fig. 3.2 **a** Code A, an engine support; **b** Code B, an alloy manifold; **c** Code C, a cast-iron manifold; **d** Code D, an engine support

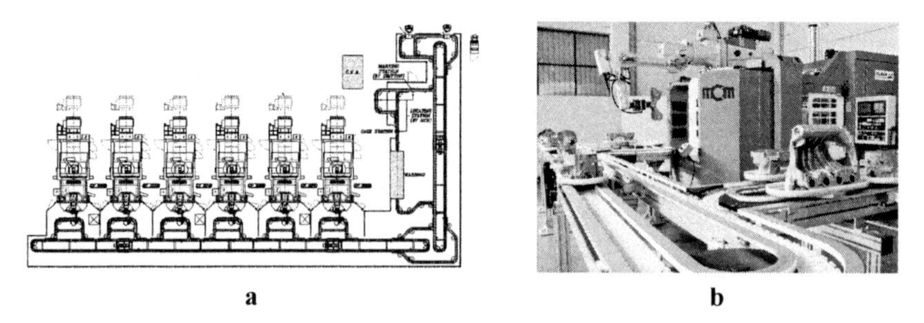

Fig. 3.3 **a** Lajous production system layout; **b** Clamping robot and conveyor belt by MCM S.p.A.

In the system, parts flow on a conveyor belt and are loaded by a clamping robot on the fixtures, which equip the various machining centers (Fig. 3.3b). This solution allows a significant reduction in the number of pallets and fixtures in the system and eliminates manual load and unload of the parts on/from the fixtures. Therefore, the implemented system can work during unmanned shifts, and the production is not limited by the number of fixtures available in the system as it happens in traditional FMSs. This solution, however, does not allow a frequent change of product mix because fixtures cannot be loaded/unloaded automatically on/from the machines.

Therefore, the rationale adopted by the firm consisted of purchasing a high level of system flexibility in order to face changes in product demands and in product versions coming from different customers. At the same time an effort was made to focus the flexibility taking into account that at a given time the number of products of the mix is rather small while production volumes are high. The production system is composed by identical machining centers; therefore, Functionality flexibility at Level 1 is guaranteed because general purpose machining centers allow executing a wide range of operations with a good capability and require very short setup times.

Considering the size and the weight of the fixtures the manual change of fixtures is a rather complex and time consuming operation lasting more than one shift. Therefore the functionality mobility of the system is not extremely high which is coherent with the production problem where the mix is stable in the short time. The savings in terms of pallets and fixtures result in a reduction of system flexibility, which has been focused on the specific production problem on hand.

The conveyor belt allows moving the parts between any couple of stations, therefore, the Functionality flexibility of the transport system is guaranteed at Level 1. Also, the availability of a conveyor belt to connect the various machines allows to easily add/remove machines form the system since it is rather easy to modify the layout of the transport system. Therefore, the Capacity dimension is addressed at Level 2, which again is coherent with a situation where volume changes can be foreseen in advance and may be rather significant.

Both the characteristics of the machines and of the conveyor belt give the system also Production Planning flexibility at Level 1. However, the fact that fixtures are stable on the machines limits the way parts can be assigned to machines. The analysis of the manufacturing system according to the ontology on flexibility (see Sect. 3.3) is reported in Table 3.5, where the basic flexibility *dimensions* embedded into the Lajous manufacturing system are represented.

Table 3.5 Lajous case flexibility analysis

Capacity Flexibility	Functionality Flexibility	Process Flexibility	Production Planning Flexibility
Level 2	Level 1	–	Level 1

3.4.2 Riello Sistemi Case Study

Gruppo Riello Sistemi S.p.A. is a machine tool manufacturer whose plants are distributed in Europe, North America and China. The set of products designed and manufactured by Riello Sistemi consists of production lines, in particular rotary table transfer machines (TTRs), flexible transfer machines (VFX) and machining centers (MC). Riello Sistemi proposes highly customized solutions, which always present a mix of standardized and specialized components. Therefore, each line solution can be considered as unique. In this environment, the phase of design of the product-line plays a key role. Some examples of transfer lines produced by Riello Sistemi are reported in Fig. 3.4. Typical features of transfer lines are:

- Presence of up to 14 stations, with up to 3 main spindles in each station;
- Many tools contemporary working (up to 36 if single tools are used or even more with multi-spindle unit);
- Every spindle has usually 1 or 2 controlled axes, but may have up to 4;
- The line is dedicated to a single component or to a family of similar components and often it integrates special devices;
- Very high production rate over investment costs ratio;
- Low space occupation in the workshop.

Given the high turbulence of the market in which its customers operate (automotive, aeronautics, electronic devices sectors), Riello Sistemi S.p.A. decided in the last decade to endow its production line with a certain degree of flexibility, starting from its conception and design. The idea was to change the characteristics of transfer lines, which were generally rigid solutions for high production volume, in order to include the possibility of modifying the line structure when some changes in the market happen. In particular, common customer requirements are easy and quick machine set-up and machine adaptation to geometrical shape modification of the parts to be machined.

In order to achieve this goal, a set of technical solutions has been introduced, involving both the software and the hardware of the machine. Regarding the software, the adoption of flexible control, through the use of programmable CNC controls, allows to rapidly change sequences and priorities. Regarding the hardware, devices (e.g. linear or angular slides manually operated), which can modify the access direc-

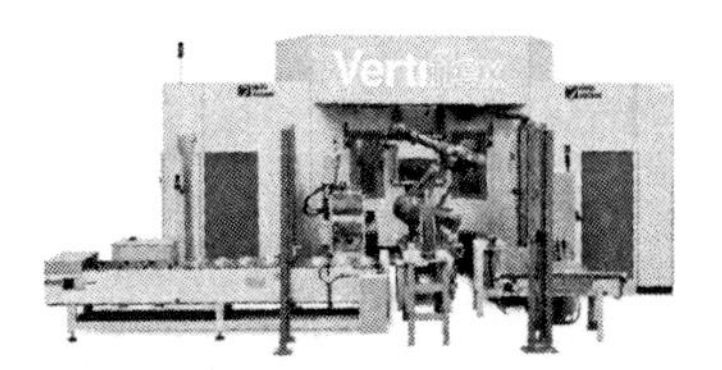
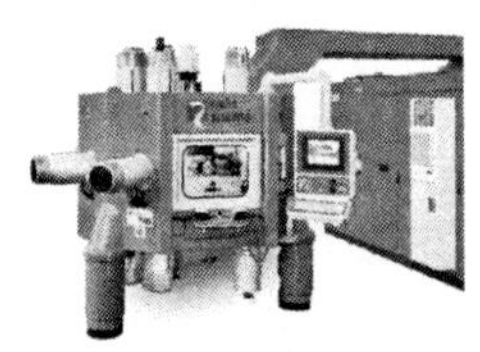

Fig. 3.4 Examples of transfer machines produced by Riello Sistemi

tion of working spindles, have been introduced together with technical solutions to allow the implementation of additional rotary axes when needed. Some of the modular components enabling the transfer line reconfiguration are reported in Fig. 3.5.

To show the importance of the design phase while proposing a customized reconfigurable transfer line solution, a real transfer reconfiguration case is reported. In this case, the customer needed special equipment to produce three product types named part A, B and C. While the customer assumed that part A would remain constant in the future, both in terms of technological features and in terms of volumes (1 000 000 parts/year), products B and C were supposed to change after 18 months, in terms of technological features, while remaining constant in terms of volumes (500 000 parts/year for B and C). All the products were steering gear holders and the modification of products B and C into D and E consisted of the elimination of the part named "top hat" and the reorientation of some features. The sketches of product B and C and the modification of the codes into product D and E are reported in Fig. 3.6.

The knowledge about the new product variants expected after 18 months allowed the system designer to propose a reconfigurable solution enabling the system modification with low cost. Two transfer lines were designed, one dedicated to product A, which was stable over the system life cycle, and one dedicated to products B and C, which were expected to evolve. Each transfer line consisted of two machines in parallel, in order to guarantee the satisfaction of the throughput constraint. Indeed, in this type of systems, raw parts enter the system at the first station and must visit all the stations in the line before exiting as a finished product.

After the analysis of the impact of product modifications on the manufacturing process, the technical solutions, which allow modifying the structure of the line

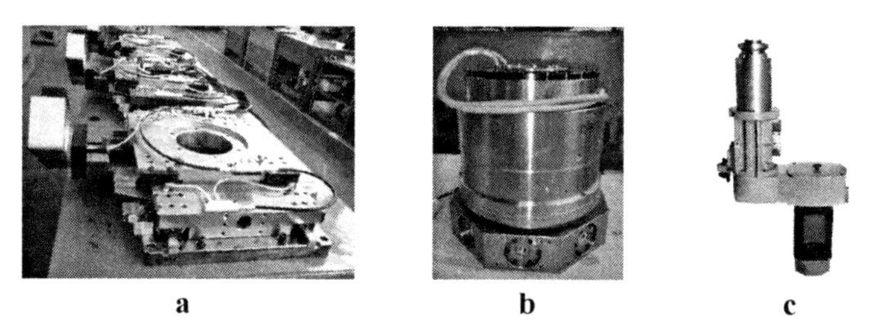

Fig. 3.5 **a** Cross slide; **b** Rotary table; **c** Unit head

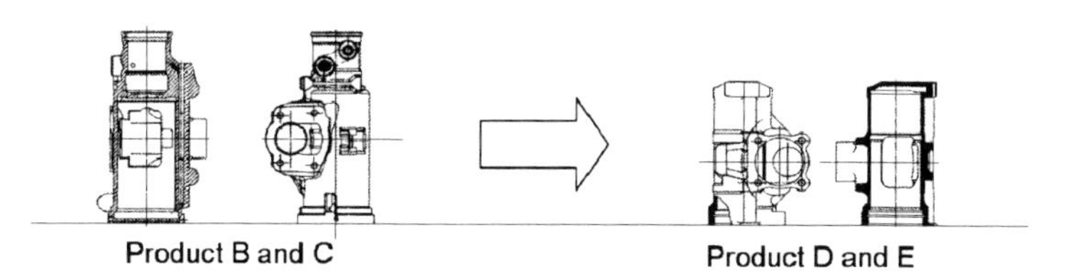

Fig. 3.6 Modification of product *B* and *C* into products *D* and *E* expected after 18 months

Table 3.6 Riello case flexibility analysis

Capacity Flexibility	Functionality Flexibility	Process Flexibility	Production Planning Flexibility
Level 1	Level 1, Level 2	–	–

to tackle the process modifications were analyzed. In particular, the possibility of changing the loading slide and the clamping fixtures were introduced together with the possibility of modifying the position of some working units using hydraulic slides. Finally, the possibility of changing multi-spindle heads was included.

The second line of the manufacturing system designed by Riello for product B and C has been endowed with the ability of doing different products, thus working different product features, with the same set of resources; therefore, the case study is characterized by Functionality flexibility at Level 1 since the second line can process both product B and C requiring short setup times.

The system has been designed to be easily reconfigured thanks to a set of technical solutions, which allow rapidly changing its configuration, without high costs and time. This means that the proposed case study is an example of Functionality flexibility at Level 2 as well.

Regarding the other basic dimensions, few information are available to evaluate if Capacity flexibility has been designed; despite the production volumes of the different codes seem to remain unchanged over time it is reasonable that the customer asked to design a system with some overcapacity (Capacity flexibility at Level 1). Finally, both Process and Production Planning flexibility can be hardly considered while dealing with transfer lines. In fact, since the flow of parts is rigid it is a challenge to modify the production sequence or to provide alternative processes to realize the same product. The analysis of Riello system solution is summarized in Table 3.6.

While in Sect. 3.4.1 it was shown how it is possible to reduce the investment of flexibility in a FMS, in this section it has been shown how a rigid transfer line can be endowed with a certain level of flexibility to cope with production changes. Therefore, also in this case it is possible to say that the proposed solution is an example of Focused Flexibility. The level of vocalization is very high because the system and its possible reconfigurations have been tailored to a defined set of parts.

3.5 Using the Ontology on Flexibility to Support System Design

Traditionally, models to support the design of production system embedding flexibility are based on the definition and implementation of flexibility forms, as is represented in Fig. 3.7 (Sethi and Sethi 1990; Gerwin 1993; Upton 1994; Chryssolouris

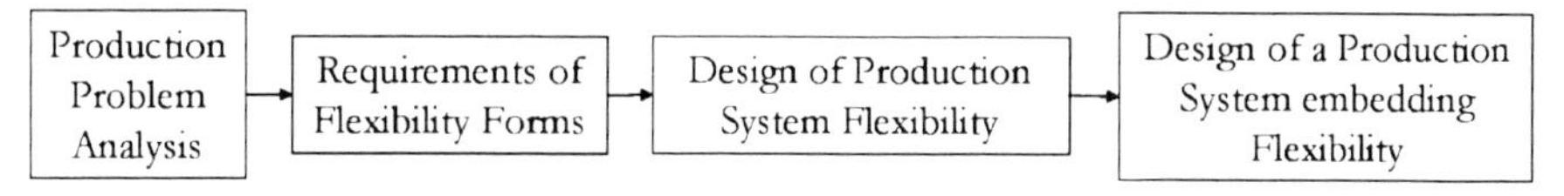

Fig. 3.7 Design of Flexibility

2006). It can be difficult to adopt this kind of approach because it requires the definition, measurement and implementation of abstract concepts such as flexibility forms. Section 3.3 and other works (e.g. Tolio et al. 2008) have shown that many definitions of flexibility forms are available and all of them can be the right ones if applied to a particular context; therefore, how should the flexibility forms be chosen?

The measurement of flexibility forms is critical as well, because there is no standardized measurement unit and the measure itself tends to be subjective. Finally, even if precise measures of the required flexibility were obtained, it would be necessary to translate these abstract values into a real production system. But how to carry out this task is not clear.

Indeed, in practice, it is very difficult to use synthetic flexibility values to design complex systems because, due to the interaction among system components, there is no simple mapping between the required flexibility and the physical components that are able to provide it. Therefore, in this section a different approach for system design is presented (Fig. 3.8). This general approach does not try to design

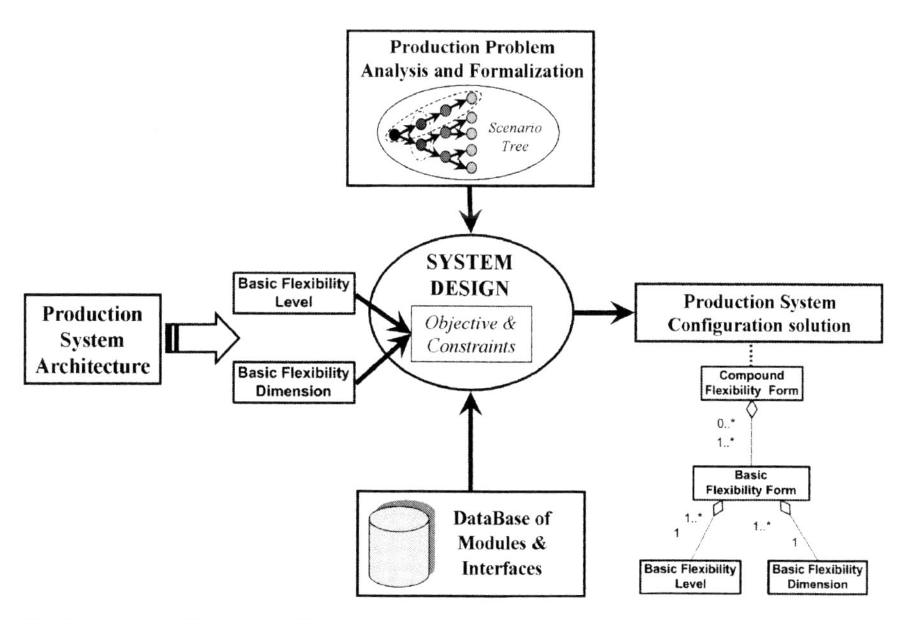

Fig. 3.8 Design of Production System with Focused Flexibility

a system with a predefined level of flexibility, but it aims at designing production systems, which are able to face the production problem on hand with the minimum economical effort over the system life-cycle.

The proposed approach is based on a careful analysis of the manufacturing environment. In fact, a Production Problem Analysis and Formalization activity (upper box in Fig. 3.8) is necessary to gather data about products and processes (for instance production volumes and product technological requirements), which define the present and future production problems to be addressed (Tolio and Valente 2007; Cantamessa et al. 2007). The different nature of the information to be handled during the production problem analysis highlights the need to support this activity by a proper formalism; the book by Bernard and Tichkiewitch (2008) presents some contributions about this topic. It must be noted that if no future information is available then the whole System Design approach collapses and the task of designing the right degree of flexibility becomes meaningless. Probably in this case the tendency would be to buy a system incorporating the maximum amount of flexibility as an answer to complete uncertainty about the future.

Therefore, the formalization of production problems should take into account also dynamic aspects. In fact, production problems are not static and can evolve during time (SPECIES 2008); system flexibility can be seen as a means to answer to this variability. A possible way to formalize the evolution of the production problem characteristics consists in dividing the planning horizon into periods and adopting a scenario tree approach (Ahmed 2003; Tolio and Valente 2008).

Once the production problem has been formalized, the structured information is used by a solution engine, here named *System Design* (central circle in Fig. 3.8). Together with formalized information, the engine also receives as input the Database of Modules and Interfaces (lower box in Fig. 3.8). This database contains information about devices that can be used inside a system (e.g. machining centers, transporters, pallets, buffers, etc.); moreover, since these modules must be integrated within a system, the database includes the interfaces among these devices. The *System Design* works finding a solution by matching system characteristic with the production problem at hand. This matching is achieved thanks to the objective and constraints, which are generated following the *Basic Flexibility Levels* and *Basic Flexibility Dimensions*. In fact, the main goal of the proposed Ontology on Flexibility (see Sect. 3.3) consists of supporting the system design phase. *Basic Flexibility Levels* and *Dimensions* lead the definition of the structural constraints while Production Problem Formalization and the Database of Modules and Interfaces provide the numerical data to be inserted in these constraints. Therefore, the ontology helps to match the system requirements, expressed by the Production Problem Formalization, with the available resource options, expressed by the Database of Modules and Interfaces. In this way, the flexibility degree required by the system is not explicitly dimensioned but it becomes an implicit output of the System Design phase.

Constraints can be divided into four groups according to the *Basic Flexibility Dimensions*. Acting on the constraints it is possible to implicitly define the Flexibility embedded in the System. For example:

- *Process Constraints* limit the choice of the process plan to adopt to process a type of product;
- *Functionality Constraints* deal with the assignment of operations to resource types;
- *Production Planning Constraints* deal with the assignment of operations to specific resources;
- *Capacity Constraints* limit the selection of the number of resources to be introduced in a system.

Beyond the Basic Flexibility Dimensions, a system design approach should consider also the *Basic Flexibility Levels*. This means that it is possible to cope with the evolution of the production problems by means of Flexibility, Reconfigurability or Changeability. Mathematical methods such as Stochastic Programming (Birge and Louveaux 1997) and Real Options Analysis (Copeland and Antikarov 2001) allow exploiting the concept of levels by clearly separating the configuration decisions which must be taken immediately from those which can be taken at a later time. In this way, constraints and decision variables incorporate the concept of Flexibility Levels.

The generation of objectives and constraints described so far is influenced by the decisions about the Production System Architecture. An architecture defines the general structure of a system and how the modules of the system can interact. Examples of production system architectures are Rigid Transfer Lines, Flexible Transfer Lines, Flexible Manufacturing Systems, Focused Flexibility Manufacturing Systems, Reconfigurable Manufacturing Systems, etc. Within the proposed *System Design* approach, architectural information (left box in Fig. 3.8) is filtered by *Basic Flexibility Levels* and *Basic Flexibility Dimensions*; in this way, constraints are built following the structure imposed by the selected architecture.

The output of *System Design* is the Production System Configuration solution (right box in Fig. 3.8). This solution can be analyzed according to the proposed ontology (see Sect. 3.4), which in this case can be seen as a classification and evaluation tool. In fact, each production system can be linked with a Compound Flexibility Form, which is the aggregation of different Basic Flexibility Forms. Indeed a designed system is a specific solution to a specific production problem.

3.6 Conclusions and Future Developments

The introduction of Focused flexibility may represent an important means to rationalize the way flexibility is embedded in manufacturing systems. Especially for mid- to high-volume production of well-identified families of products in continuous evolution focused flexibility may represent the missing species in manufacturing system evolution. However, in order to reap the benefits deriving from the acquisition of flexibility at the best time and in the right quantity many obstacles should be overcome. At first, a deeper understanding of the nature of flexibility asks for a clear definition of the dimensions of flexibility and for a formalization of the information required to describe future scenarios together with the risk connected with alternative choices. This could also simplify the system flexibility assessment, supporting the decision maker in evaluating the benefits coming from the use of such flexibility options, for instance by considering the make-to-stock option or capacity renting strategies. Secondly, a stronger ability to design the required flexibility should be developed. To this aim multistage decisions methodologies that explicitly take into account uncertain information about future scenarios are extremely valuable. Thirdly, the realization of new system architectures, new machines, devices and modules, new system supervisors are required in order to take advantage of the possibility of designing exactly the required flexibility and make focused flexibility a reality.

Some interesting solutions in this direction have been already provided by the most advanced companies. For instance, with more or less clear intents, machine tool builders are trying to create new system architectures, which to some extent allow to focalize manufacturing flexibility. The aspects which in the long run can convince machine tool builders to provide innovative solutions to the customers depend on the profitability of FFMSs compared to traditional FMSs or to RMSs. Finally, another interesting aspect concerns the attention that both the system designer and user are showing concerning the analysis and formalization of present and future information. In fact, it often happens that the system user does not provide accurate forecasts to the system designer, jeopardizing the system design process. Therefore, on the one hand the support of a formalism allows the system user to collect and analyze data in a more structured way. On the other hand, the developed tool could guarantee that the system designer starts the system design process from the basis of a more comprehensive production problem description.

Acknowledgements The research has been partially funded by VRL-KCiP (Virtual Research Lab for a Knowledge Community in Production), a Network of Excellence in the 6th Framework Program of the European Commission (FP6-507487-2).

References

Ahmed S., King A.J., Parija G., 2003, A Multi-Stage Stochastic Integer Programming Approach for Capacity Expansion under Uncertainty. J Glob Optim 26:3–24

Beach R., Muhlemann A.P., Price D.H.R., Paterson A., Sharp J.A., 2000, A review of manufacturing flexibility. Eur J Oper Res 122:41–57

Bernard A., Tichkiewitch S., 2008, Methods and Tools for Effective Knowledge Life-Cycle-Management. Springer

Birge J.R., Louveaux F., 2000, Introduction to Stochastic Programming. Springer-Verlag, N Y

Cantamessa M., Fichera S., Grieco A., Perrone G., Tolio T., 2007 Methodologies and tools to design production systems with focused flexibility. Proc of 4th Int Conf on Digit Enterp Technol, Bath, UK, 19.–21. Sept. 2007, pp 627–636

Chryssolouris G., 1996, Flexibility and its Measurement. Ann CIRP 45/2:581–587

Chryssolouris G., 2006, Manufacturing Systems: Theory and Practice, 2nd edn. Springer-Verlag, N Y

Copeland T., Antikarov V., 2001, Real Options: a practitioners-guide. Texere, N Y

Correa H.L., Slack N., 1996, Framework to analyse flexibility and unplanned change in manufacturing systems. Comput Integr Manuf Syst 9/1:57–64

De Toni A., Tonchia S., 1998, Manufacturing flexibility: a literature review. Int J Prod Res 36/6:1587–1617

ElMaraghy HA., 2005, Flexible and reconfigurable manufacturing systems paradigms. Int J Flex Manuf Syst 17:261–276

Gerwin D., 1993, Manufacturing Flexibility: a strategic perspective. Manag Sci 39/4:395–410

Grieco A., Pacella M., Anglani A., Tolio T., 2002, Object-Oriented modeling and simulation of FMSs: a rule-based procedure. Simul Model Pract Theory 10/3:209–234

Grubbstrom R.W., Olhanger J., 1997, Productivity and Flexibility: Fundamental relations between two major properties and performance measure of the production system. Int J Prod Econ 52:73–82

Gupta Y.P., Goyal S., 1989, Flexibility of manufacturing systems: Concepts and measurement. Eur J Oper Res 43:119–135

Gupta Y.P., Somers T.M., 1996, Business Strategy, Manufacturing Flexibility, and organizational performance relationships: a path analysis approach. Prod Oper Manag 5/3:204–233

Hutchinson G.K., Pflughoeft K.A., 1994, Flexible process plan: their value in flexible automation systems. Int J Prod Res 32/3:707–719

Hyun J.H., Ahn B.H., 1992, A Unifying Framework for Manufacturing Flexibility. Manuf Rev 5/4:251–260

Koren Y., Hu S.J., Weber T.W., 1998, Impact of Manufacturing System Configuration on Performance. Ann CIRP 47/1:369–372

Koren Y., Heisel U., Jovane F., Moriwaki T., Pritschow G., Ulsoy G., Van Brussel H., 1999, Reconfigurable Manufacturing Systems. Ann CIRP 48/2:527–540

Landers R.G., 2000, A new paradigm in machine tools: Reconfigurable Machine Tools. Japan-USA Symp on Flex Autom, Ann Arbor, Michigan

Ling C., Spicer P., Son S.Y., Hart J., Yip-Hoi D., 1999, An Example Demonstrating System Level Reconfigurability for RMS. ERC/RMS Technical Report 29

Matta A., Tolio T., Karaesmen F., Dallery Y., 2000, A new system architecture compared with conventional production system architectures. Int J Prod Res 38/17:4159–4169

Matta A., Tolio T., Karaesmen F., Dallery Y., 2001, An integrated approach for the configuration of automated manufacturing systems. Robotics Comput Integr Manuf 17:19–26

Parker R.P., Wirth A., 1999, Manufacturing flexibility: Measures and relationship. Eur J Oper Res 118:429–449

Sethi A.K., Sethi S.P., 1990, Flexibility in Manufacturing: A Survey. Int J Flex Manuf Syst 2:289–328

SPECIES, 2008, Robust Production System Evolution Considering Integrated Evolution Scenarios. http://www.species.polimi.it. Accessed 27 March 2008

Stecke K.E., 1985, Design, planning, scheduling and control problem of flexible manufacturing systems. Ann Oper Res 3:1–12

Tolio T. (ed), 2008, Design of Flexible Production Systems - Methodologies and Tools. Springer-Verlag

Tolio T., Valente A., 2006, An Approach to Design the Flexibility Degree in Flexible Manufacturing Systems. Proc of Flex Autom and Intell Manuf Conf, Limerick, Ireland, 25.–27. June 2006, pp 1229–1236

Tolio T., Valente A., 2007, A Stochastic Approach to Design the Flexibility Degree in Manufacturing Systems with Focused Flexibility. Proc of 4th Int Conf on Digit Enterp Technol, Bath, UK, 19.–21. Sept. 2007, pp 380–390

Tolio T., Valente A., 2008 A Stochastic Programming Approach to Design the Production System Flexibility Considering the Evolution of the Part Families. To appear in Int J Manuf Technol Manag – Special Issue on Reconfigurable Manuf Syst. In press.

Tolio T., Terkaj W., Valente A., 2007, Focused Flexibility and Production System Evolution. Proc of 2nd Int Conf on Chang, Agile, Reconfigurable and Virtual Prod, Toronto, Canada, 23.–24. July 2007, pp 17–41

Tolio T., Terkaj W., Valente A., 2008, A Review on Manufacturing Flexibility. In: Tolio T. (ed), Design of Flexible Production Systems – Methodologies and Tools. Springer-Verlag

Upton D.M., 1994, The Management of Manufacturing Flexibility. Calif Manag Rev 36/2:72–89

Wiendahl H.-P., ElMaraghy H.A., Nyhuis P., Zäh M.F., Wiendahl H.-H., Duffie N., Brieke M., 2007, Changeable Manufacturing - Classification, Design and Operation. Ann CIRP 56/2:783–809

Zhang Q., Vonderembse M.A., Lim J.-S., 2003, Manufacturing flexibility: defining and analyzing relationships among competence, capability, and customer satisfaction. J Oper Manag 21:173–191

Part II
Physical Enablers

Manufacturing Machinery and Hardware

Chapter 4
Control of Reconfigurable Machine Tools

G. Pritschow, K-H. Wurst, C. Kircher and M. Seyfarth[1]

Abstract Changes in manufacturing requirements and market demands call for increased flexibility, adaptability and sometimes reconfiguration. This is true for manufacturing systems and their components such as machines and robots. This chapter discusses both the physical hardware reconfiguration of these components as well as their logical reconfiguration manifested in the control system. The challenges involved in the physical reconfiguration are detailed and solutions are presented. The requirements for reconfigurable control systems are discussed and the state-of-the-art implementations and systems are described. The remaining obstacles and challenges for future research and industrial adoption are highlighted.

Keywords Reconfigurable robots and machine tools, hard- and software interfaces, modular design, mechatronic components, field-bus systems, configurable control systems, configuration procedure, self adapting control systems

4.1 Introduction

With the introduction of PC's for numerically controlled machines control engineering made considerable progress in regard to velocity for path interpolation, multi-channels and functionalities. Today, the control device product lines of large manufacturers allow the configuration of machines of varying designs with a multitude of axes and special functions according to customer specifications without having to make special adaptations.

Machine tools as well as robots are nowadays designed as modular systems, so that a multitude of variants can be realized to match the respective application.

[1] Institute for Control Engineering of Machine Tools and Manufacturing Units, Universität Stuttgart, Stuttgart, Germany

The adequate control technology is therefore built like a function toolbox the many parametrizable modules of which can be adapted to the control task using configuration tools.

So what is so special about a reconfigurable machine tool compared to the existing modular and configurable machine tools? To answer this question, first the difference between these two types of machine tools has to be analyzed then the requirements of a control for a reconfigurable machine tool can be deduced. These requirements lead to developing adequate control systems and it will be described how this system can be implemented.

4.1.1 Basic Idea for Reconfigurable Machine Tools and Systems

The research work regarding the design and requirements of reconfigurable manufacturing systems in the past ten years resulted in a state-of-the-art definition that Y. Koren (Koren, 2005) summarized as follows:

'A RMS is a system designed at the outset for rapid changes in structure, as well as in its machines and controls, in order to rapidly adjust production capacity and functionality (within a part family).'

The basic messages here – in contrast to configurable systems – are 'rapid changes' or 'rapid adjustment' which must happen in relatively short time ranging between minutes and hours and not days or weeks.

For the design of machine tools this means a highly customized flexibility required for a part family, i.e. the machine structure is not fixed but adjustable to different demands rapidly (Koren et al., 1999).

4.1.2 Initial Situation in Machining Systems and Machine Tools

In production engineering, results of the latest research in the field of modular robots (Wurst, 1991, Wurst et al., 2006)) and developments for reconfigurable CO_2 laser machines revealed the potentials and problems of reconfigurability. The robot basis is represented by active robot modules (one- or multi-axes integrated joint drives) and passive robot modules (robot arms), which are the complementary structural modules (Fig. 4.1). Why did this concept of reconfigurability for industrial robots not become widely accepted?

The required design elements were specified as follows:

- A mechanical interface for the exchange of passive and active modules within seconds, and
- An integrated bus system for the communication with optical interfaces in the module adapter.

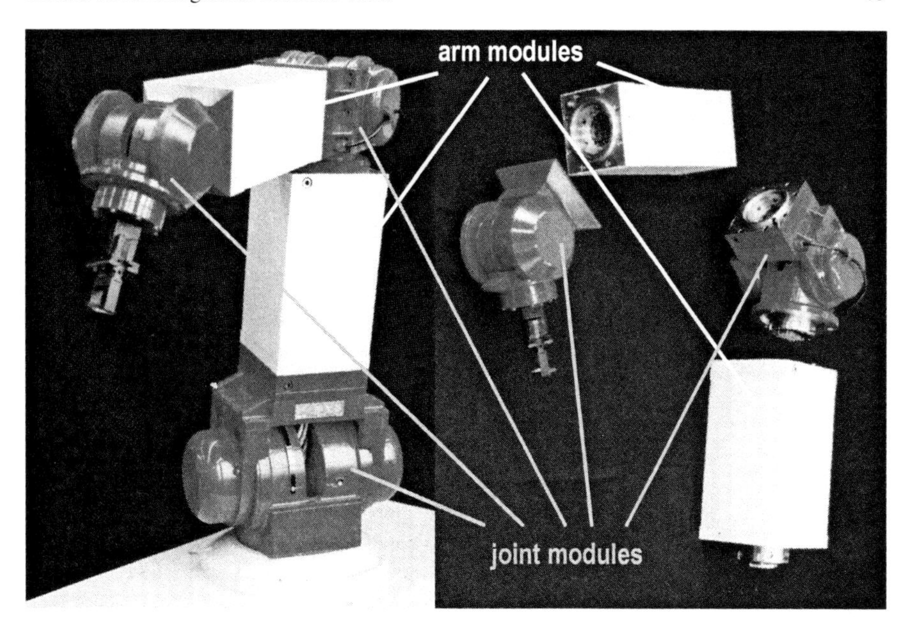

Fig. 4.1 Reconfigurable modular robot system

However, structural defects were discovered as for example:

- The device-internal guidance of many highly movable wires (three-phase AC wires and in addition two for the position measurement system for every drive) gives a potential for wire breakages, and
- The limited adaptability of the robot control in response to the changing kinematic conditions.

Also in addition, deficiencies in the system had been recognized during the reconfiguration process:

- The system user was not supported by simulation and implementation tools, which should have also provided the option for reconfiguration.
- Adequate diagnostic methods for ascertaining the machine capability were lacking.

In the area of laser machining, for example, a problem-free (re)configuration of the guiding machine was attempted by the generation of modular machines and laser components (Fig. 4.2). In the research phase the feasibility of this approach was realizable, but in the industry only a minor part of the modular beam-guiding components was accepted.

The reasons why "reconfigurable" laser machines have not established themselves in the industry are comparable to those that prevented a widespread use of the reconfigurable robot. Though, in that case, a configurable control system

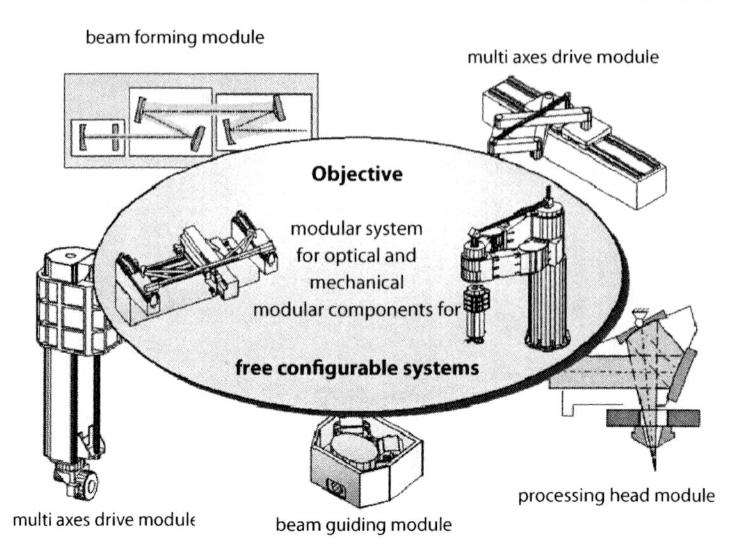

Fig. 4.2 Reconfigurable modular CO_2 laser machining system

with a high-speed communication system was already available. But, the process-determining modules like beam guiding and beam forming components worked parallel to the machine tool modules. This resulted in considerable technical efforts required when the machine had to be calibrated, adjusted or when the machine capability had to be identified.

The best known functioning and (re)configurable devices are represented by the PC systems. The PC as an active basic module can be configured by means of active process modules (camera, printer, scanner etc.) into a higher system without any major problems. The software-technical adaptation of the control and the communication is guaranteed. The essential difference compared to the earlier mentioned systems is that the functional features of the self-sufficient, autonomous modules are not coupled. Also a more complex mechanical coupling of components such as that found within a machine tool does not occur. The purely communicative coupling also does not affect the "machine capability".

The previous examples indicate how manufacturing systems and machinery in production engineering have to be designed, so that they can be easily configured using a "plug and play" method similar to that in the PC world:

- All module interfaces have to be reduced to a minimum and they must have a basic design, in order to allow (re)configuration in minimum time.
- The specifications of functional and structural system boundaries have to be coordinated, so that useful and manageable interfaces result.
- The functional range of the modules should allow an easy proof of machine capability of the system or assume warranty of it.
- Meshed net mechanically coupled systems should be avoided.

- Wherever possible, autonomous, self-sustaining modules should be created and applied.
- The mechanical calibration of interchangeable reconfigurable modules should be easy and feasible to accomplish without special knowledge and in a short time.
- Configuration tools in the form of simulation tools and implementation programs, which reflect reconfigurable machines and plants in their structure, should be available.
- Fast methods for ascertaining the machine capability have to be provided.

Hence, it can also be conceived, that a machine tool or a robot is composed of a certain number of autonomous modules (motion module, process modules, transport modules) without the need to possess the conventional supportive structures of machines. A reconfigurable machine system would then consist of the coupled modules, or as yet of inter-linked single machines (largest conceivable module in a conventional plant structure).

4.2 State of the Art

In general, the changeability of hardware as well as software always requires system configurations that consist of exchangeable components. For a long time, modular systems were known in the case of machine tool components (for example "purchased parts") as well as tools and device systems. Also, modular transfer lines or interlinking devices based on modular systems were well known. All these modular systems are based on defined mechanical interfaces, which allow an exchange either within a certain product range or a general exchange in case of standardized interfaces. Just as in the case of mechanical interfaces, also in the field of information and control technology as well as in the area of energy transfer with machine tools a multitude of standardized interfaces are common practice (Wurst, Heisel and Kircher, 2006).

But modularity alone does not meet the demands of (re)configurability. (Re)configurability requires a rapidly realizable adaptation of the production process, which has an effect on the functional and capacity requirements of the machining systems. Surveys have shown that current machining systems meet these demands only to a certain extent. Today's machining systems that allow a certain degree of configurability or reconfigurability are characterized by a modularity, which permits changes in function or extensions of a machine system only within limits. Further, they are characterized by a very limited adaptability of the control technology to possible configuration changes of machines.

The amount of time needed for reconfiguration could be a problem, because high demands on accuracies and rigidities of the machines have to be observed. Up to now the definition of suitable system boundaries for reconfigurable components is still missing. Furthermore, in the case of reconfiguration, the machine tool's behavior is to a large extent unknown, because methods for the process description and its effects are yet to be developed. Last but not least, there is also a lack of adequate configuration and planning tools (aids) for continuous simulations of reconfigurable machine tools.

In the present discussion about possible solutions, different views exist. From a system point of view; reconfiguration into new interconnections can be achieved by re-organizing or exchanging individual machining units or machines. This results, for instance, in new capabilities in regard to work sequence and capacity by parallel or serial arrangement of the individual machining units. Examples exist in both assembly systems and metal cutting machining systems (Fig. 4.3).

Self-sustaining machine units such as these shown in Fig. 4.3 can be easily combined during reconfiguration, because the mechanical interface is standardized and focused on the transport system so that it can be easily exchanged and connected. Furthermore, the communication interconnection with other units and the central control system is realized via a high-capacity plug-gable bus system (e.g. Ethernet).

Fig. 4.3 Reconfigurable system “Teamos” for assembly and finishing operations [team-technik, Germany]

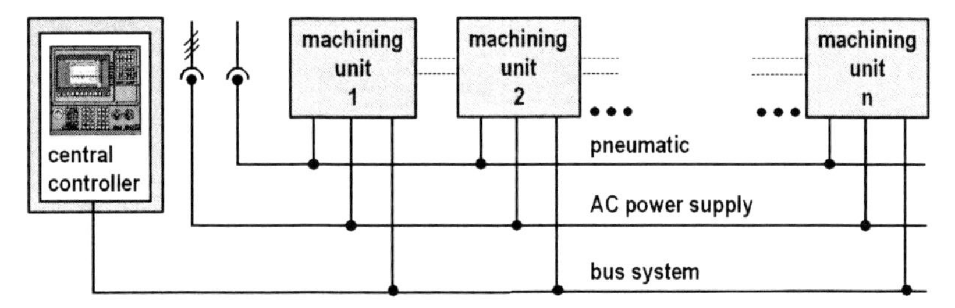

Fig. 4.4 "Wiring" interfaces of a reconfigurable manufacturing system

Further flexible and plug-gable interfaces are used for the electric and pneumatic power supply. A hydraulic "wiring" is in general very time-consuming, which is not consistent with the "rapid" requirement. Besides, a central hydraulic supply system cannot be designed for customers who have not been defined yet. If a hydraulic supply unit is needed for a machine unit, it will be built-in. Thus, the basic structure is relatively simple regarding wiring. This can be seen in Fig. 4.4 where there is a universal electric and, where required, pneumatic energy supply as well as a connecting field bus system. This way, the exchange of units is feasible in the shortest time possible.

4.3 Configurable and Reconfigurable Machine Tools

4.3.1 Development of (Re)configurable Machine Tools

Modular designed machine tools and robots of leading and innovative machine manufacturers are company-specifically standardized, so that modular systems for planning and design tools can be created (Fig. 4.5). The machine can be configured according to customer specifications and its individual modules selected using computer-aided software planning tools. The chosen modules are then manufactured and assembled, if they are not already pre-assembled. The assembly is mostly performed from the "inside" to the "outside", which means that the motion-controlling modules on the machine bed are supplemented by process modules (e.g. spindle), additional drive systems (e.g. hydraulic power unit), tool and pallet change systems, machine casing, covering, etc. The energy supply for the electrics, hydraulics and pneumatics are attached securely and firmly installed in a separate cabinet. Consequently many machine components are rigidly coupled with each other. The control is then especially adapted to the machine configuration. For this, experienced specialists who can adapt the control using configuration tools are needed for the initial operation. The control has to be reconfigured each time for different machine configurations.

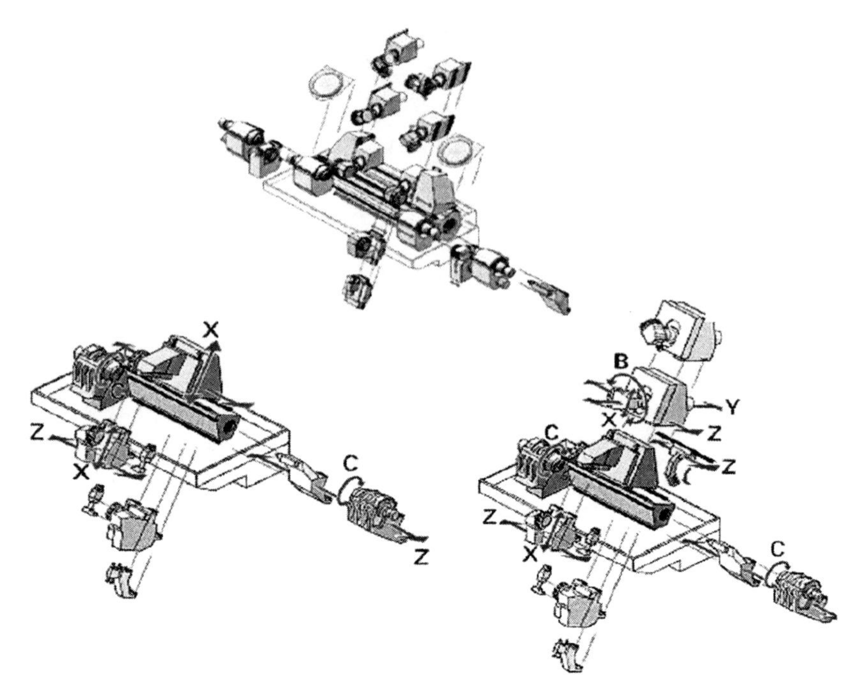

Fig. 4.5 Selected section of a modular system for machine tools [INDEX]

Due to the applied interfaces and the cables and tubes going through the system, even machine tools constructed as modular systems can only be changed with extraordinary effort. Auxiliary functions like hydraulic or pneumatic power supply units must be adapted to the specific consumer requirements. Changes in the type or number of the consumers in a later expansion are thereby made difficult. An order-related reconfiguration of a machine tool is usually possible only in a limited and time-consuming manner.

The challenges in the field of reconfigurable machine tools, therefore, lie in the generation of adequate interfaces. A first step in the direction of reconfigurable machine tools and robots in this connection has to be the determination of new system boundaries and the reduction of interface elements.

Figure 4.6 shows the typical schematic design of a machine tool and electric cabinet according to Fig. 4.5. Since the components for energy supply and control are concentrated in the cabinet, a multitude of connections to the various machine elements as, for example, axes, tool changer or transport systems are needed. In spite of the use of field bus systems like SERCOS, Profibus or Real-Time (RT) Ethernet the number of connections could not be reduced until this day.

Minimizing these interfaces may be realized if all connecting elements, except the electric power supply and a bus system connection, were integrated in a mechatronic module. This idea is illustrated by the dotted line in Fig. 4.7. With this, PLC,

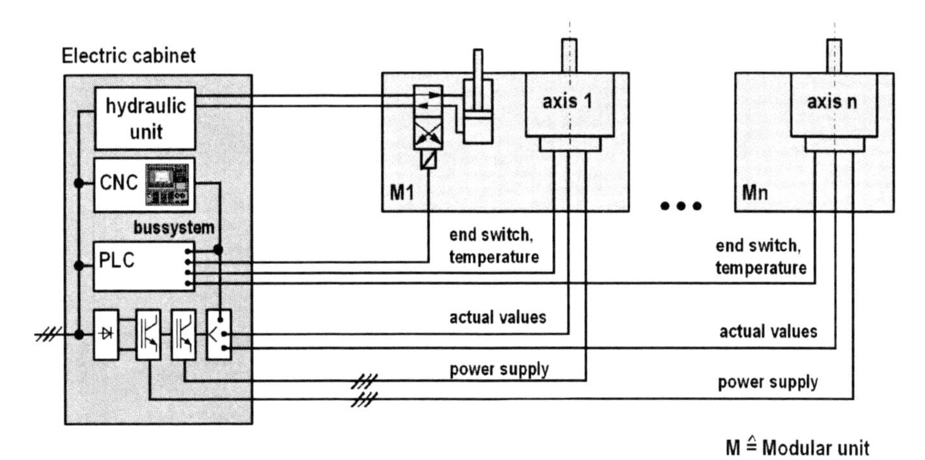

Fig. 4.6 "Wiring" interface of a conventional machine tool or robot system

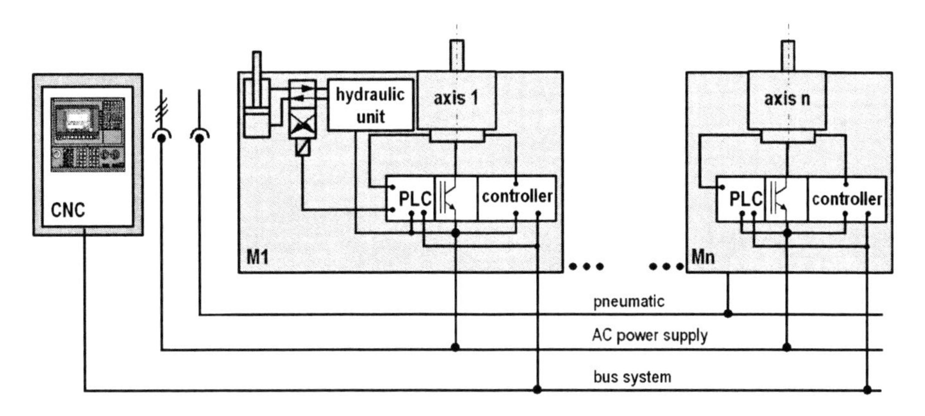

Fig. 4.7 "Wiring" interface of a reconfigurable machine tool or robot system

drive controller, converter, rectifier and hydraulic power unit would become parts of the mechatronic system: A venturous step for the designer, but feasible today, because miniature electric elements are already used in robots and machine tools. They are in fact part of the 'state of the art' and in the automotive industry or in medical engineering a variety of such components is being offered in the area of hydraulics or pneumatics.

The wiring of such a reconfigurable machine tool leads to the structure shown in Fig. 4.7, which is basically not much different from that in Fig. 4.5.

The design of a reconfigurable machine tool or robot with only mechatronic modules that have one interface for the power supply and one for the communication would solve the problem of 'wiring in minute intervals'.

4.3.2 Conception of a Reconfigurable Machine Tool

A basic concept for designing a reconfigurable machine tool is shown in Fig. 4.8. Mechanical passive and active modules form the machine structure. Active modules are those that create motion by drives or other adaptive elements, whereas passive modules have static and supporting function.

Each module aggregates a main- or auxiliary function like motion generation or work piece exchange. The connection of the modules can either be movable or fixed. The mechanical interfaces (i.e. attachment and load transmission), power interfaces (i.e. electricity, pneumatics) and information interfaces (i.e. bus systems like PROFIBUS, SERCOS Interface, RT-Ethernet) of the modules are well defined in order to guarantee their free exchange.

Furthermore, the reconfigurable machine tool does not only consist of the mechanical modules but also of decentralized linked controllers integrated into the modules. Thereby a module in a Reconfigurable machine is no longer only mechanical but it has become a mechatronic component.

In order to meet the requirements set out in Sect. 4.1.2, the presented concept is not sufficient. Particularly the manageability and the operability of the reconfiguration process would still be very complex because of the lack of methods for testing whether the modules are connected correctly and checking the machine's functionality and accuracy after reconfiguration. Therefore, measurement methods and particularly measurement devices have to be developed. In addition, tools for the (re)configuration planning and execution are needed. But these tools and systems are beyond the scope of this chapter.

The following sections describe the requirements for field bus systems and controllers for reconfigurable machine tools in detail.

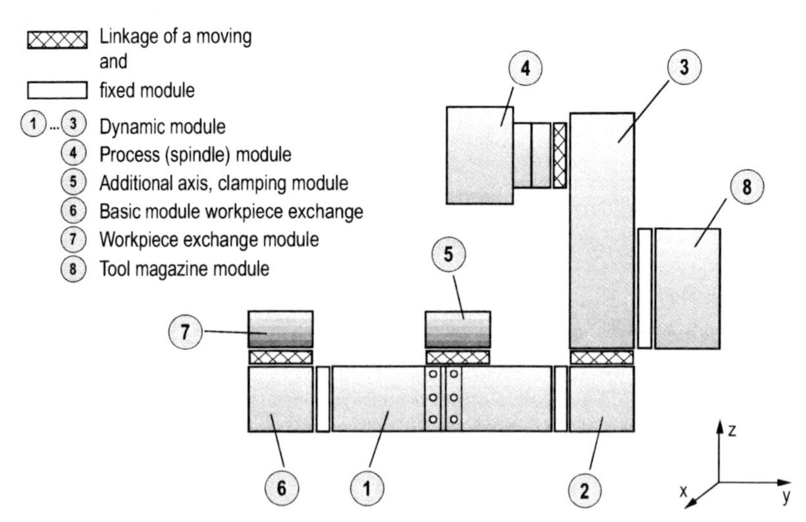

Fig. 4.8 Arrangement of a machine tool with a modular design

4.4 Field Bus Systems Requirements

Based on the state-of-the-art, feed drives for machine tools and robot systems are supplied with a set of given values by the controller, produced by a control module for geometric data and interpolation. The interpolation cycle today is approximately 1 ms and the synchronization between the axes should lie in the range of $\pm 1\,\mu s$ (jitter). These are the real-time requirements for a field bus system. The transmission bandwidth is characterized by the interpolation cycle, the data format (usually 64 Bit, 8 Byte) as well as the number of axes that have to be supplied periodically and synchronously.

In addition, there is a multitude of asynchronous messages and feedback signals that can be interchanged by the axes as well as by a decentralized PLC, which present requirements that have to be met. In recent years, for this purpose the widely spread low-cost Ethernet Bus System has been developed further as a Real-time (RT) Ethernet, that is able to fulfill all requirements because of its high transmission rate of $>$100 Mbit/s. In addition, the high bandwidth available allows that in addition to cyclic and acyclic process data as well as the exchange of secure data between machines. For this purpose, real-time communication systems are extended by safety protocols so that requirements up to SIL3 (Safety Integrity Level) according to IEC 61508 are met. Consequently, separate safety busses are not needed that would require additional interface costs and higher maintenance and engineering efforts.

The risk of collision of communicating participants sending simultaneously as it is permitted in the standard Ethernet procedure has to be ruled out, of course. For this purpose, solutions in the full duplex mode are available, in which the network is operated in a time slot mode or where "switches" enable the resolution of the access and hence preventing collisions. But the switches – except for the added cost and wiring disadvantages – lead to additional run-times. Furthermore, considerable delays result sometimes in forwarding the data packets ('Queuing') if several data packets aim at the same exit. In order to secure a synchronous processing and acquisition of the process data in spite of these variable time lags, adequate synchronization methods (e.g. IEEE 1588 or IEC 61588) are needed. With hardware support for generating exact timestamps, synchronization accuracies around 100 nanoseconds are achieved (Fig. 4.9). The shortest cycle-time is around 300 microseconds. SERCOS III or Ethercat offer an alternate solution with a master that provides a connection to the different participants using the time slot method.

The SERCOS III solution can be operated in a ring topology offering hardware redundancy for increased availability of a machining system. Even in case of a cable break or a node failure the communication is maintained. The shortest cycle time is around 30 microseconds. To reach these very short cycle times, the delay and transmission times in the network have to be minimized. This is achieved by using specific hardware controllers that allow an 'on-the-fly' processing of real-time data – while the real-time frames pass through the nodes – and the merging of real-time

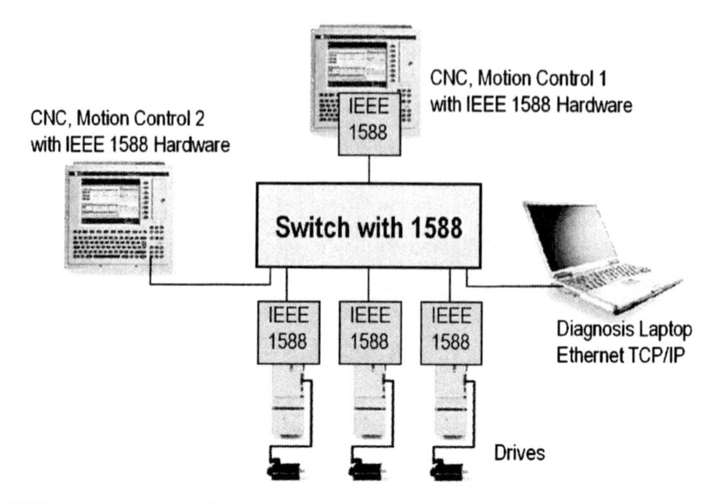

Fig. 4.9 RT Ethernet on the basis of switches

data in one Ethernet frame shared by several nodes. This also has the advantage of a higher efficiency of the bandwidth as the overhead caused by the Ethernet frames is reduced (Examples: SERCOS III and Ethercat).

A plug and play version only works if the used protocols and data formats are standardized. Besides a standardized protocol based on Ethernet IEEE 802.3 and standardized communication protocols such as Profinet or SERCOS III (IEC 61784 and IEC 61158), standardized machine profiles are also used, like Profidrive or the SERCOS drive profile (IEC 61800-7) for the data exchange between controls and drives. A standardized Ethernet TCP/IP message that can nowadays be sent by any PC, must therefore be coupled into the Real-time Ethernet by a corresponding Gateway. For some of the real-time Ethernet systems the cycle time is normally divided up in a time domain for the real-time communication, which has to be cyclic and synchronous for the drive systems and a domain for the asynchronous communication of the Ethernet system according to the TCP/IP standard (Fig. 4.10).

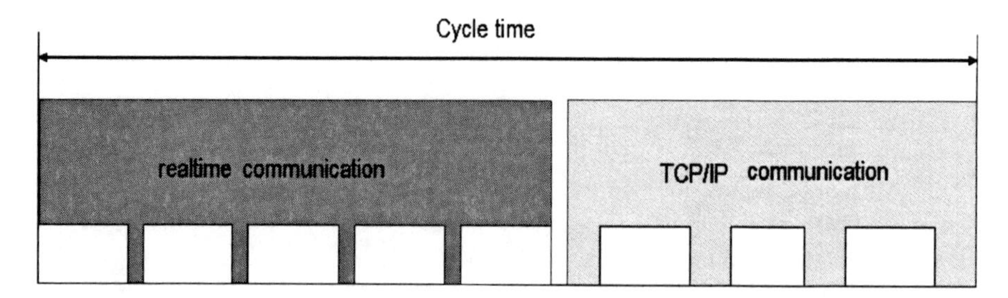

Fig. 4.10 RT Ethernet cycle-time with real-time and asynchronous part

This last part allows the complete combination with the standard Ethernet system. The real-time part has different solutions and in summary today there are more than 10 different proposals in the IEC 61158/61784 standard, for real-time Ethernet solutions, featured in most cases by different companies like Ethercat (Beckhoff), PowerLink (B and R), Profinet (Siemens). That means: beside the 10 existing field bus standards we have more than 10 real-time Ethernet specifications, which in fact will not support the idea ‘One bus for all requirements’.

Luckily, today the chip technology is able to offer a solution to this problem. For example, the chip netX from the company Hilscher (Germany) offers a free choice of the most wanted field bus systems of previous design and, in addition to two channels with freely selectable real-time protocols. The free market can decide in the future, which protocol will be of most benefit for the user. In any case, the real-time Ethernet solutions offer the possibility to transmit via a cable highly dynamic drive signals as well as the signals between the PLC components of the individual modules, which means that the communication interface between the modules of a reconfigurable machine may be reduced to a minimum.

4.5 Configurable Control Systems

Besides the design of the physical modules interfaces, the self-adapting control system lies in the center of attention. The basics for this were established by the well-known research activities concerning open multi-vendor platform-based configurable control systems as, for example, OMAC (Open Modular Architecture Controls) in the North America, OSEC/FAOP (Open System Environment for Controllers/FA Open Systems Promotion Forum)(earlier JOP) in Japan or OSACA (Open System Architecture for Controls within Automation) in Europe. With OSACA, an open (multi-vendor) object-oriented control system was developed that can be configured during the run-up of the control system via a configuration run-time system by the interpretation of a text-based configuration file. The basic idea of these approaches is the introduction of a platform with a defined user API (application programming interface), which conceals the hardware- and operating system-specific features of the controller from the user software. The platform provides the user with communication mechanisms for the data exchange between differing user application modules (AM) whereas the middle-ware for the communication was based on a specific OSACA protocol (Fig. 4.11).

These concepts also specified a coupling of these platforms to a distributed control system. The communication of these decentralized systems took place, for example, via Ethernet with TCP/IP. One disadvantage of the communication mechanisms was the lack of real-time capability, i.e. the non-deterministic data exchange. This theme and the integration of the open middle-ware CORBA lead to the project OCEAN (Open Controller Enabled by an Advanced Real-Time Network), which was supported by the European Union (Meo, 2008 and Pritschow et al., 2004).

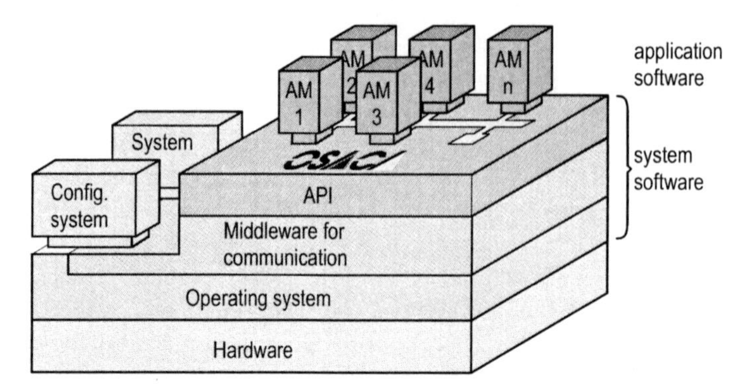

Fig. 4.11 Configuration of an OSACA platform for open architecture, adaptable and reconfigurable control system

4.5.1 *Middle-Ware*

4.5.1.1 Basics

Centrally controlled automation systems become distributed systems by the integration of functions in self-sustaining mechatronic modules. Here the control software components are distributed on different hardware platforms (nodes). Thus interfaces between these components are produced beyond the hardware boundaries, which have to be bridged by a communication system. For a universal application, the communication system has to be transparent and standardized, because otherwise a direct access to resources (data) or services (functions) would have to be programmed for each distributed software component. The application would become more complex and harder to maintain. The most important requirements to achieve such a transparency are:

- **Transparency of location and access:** The place where a service or a resource can be found is unknown to the application. The access is carried out via a certain name that does not include information about the location. There is no difference between a local or a remote access to another node.
- **Transparency of concurrency:** In case of several simultaneous accesses to services and resources, the system provides exclusive and synchronized accesses.
- **Transparency of programming language:** The communication between the software components is standardized and independent of the programming language used.

Middle-ware is an application-independent layer on level 7 of the OSI reference model (ISO/IEC 7498-1, 1994) that moderates between applications so that their complexity and infrastructure is concealed. Middle-ware offers mechanisms for the communication between distributed applications: It organizes the transport of data and communicates function calls between the components (so-called Remote Procedure Calls).

If a software component "A" communicates with a software component "B", the corresponding calls are passed on by the middle-ware using a network. The Common Object Request Broker Architecture (CORBA) is a specification for an object-oriented Middle-ware whose kernel is an Object Request Broker (ORB) (Object Management Group, Inc, 2004). This ORB has the task to accept method calls, identify the receiver and to pass it on to it. Thus the desired transparency of location is achieved. For this purpose, each hardware platform has to have an ORB. While the ORB must be implemented hardware-oriented according to the platform, because it communicates with the hardware below it, the mechanisms are provided by the ORB independent of platforms and programming languages.

For real-time applications the specification RT-CORBA (Real-time CORBA) represents a substantial progress where mechanisms are defined that ensure deterministic communication behavior of the Middle-ware. This includes mechanisms for the resource management (choice of connection, assurance of bandwidth, processor and memory capacity), the scheduling of processes, the awarding of priorities and the synchronizing of parallel processes (Schmidt, Kuhns, 2000 and Emmerich, Aoyama, Sventek, 2007). Requirements for these mechanisms are real-time operating systems as, for example, RT-Linux or RTAI (Meo, 2008) on which the Middle-ware can be based.

4.5.2 Configuration

A configuration run-time system is conducive to the configuration of a control system during the run-up phase. This run-time system provides mechanisms to instantiate the configured software components during the start-up of the control system, based on the configuration files. They contain information about the type of configured software components, the communication connections between the components and about the parametrization of the individual software components.

4.5.3 Adjustment Mechanisms for Control Systems

There are basically three different mechanisms for the adjustment of control software to a specific machine tool configuration (Daniel, 1996) (Fig. 4.12).

- **Parametrization:** With this method a defined program sequence is influenced by parameters. In the area of control engineering internal program variables that can

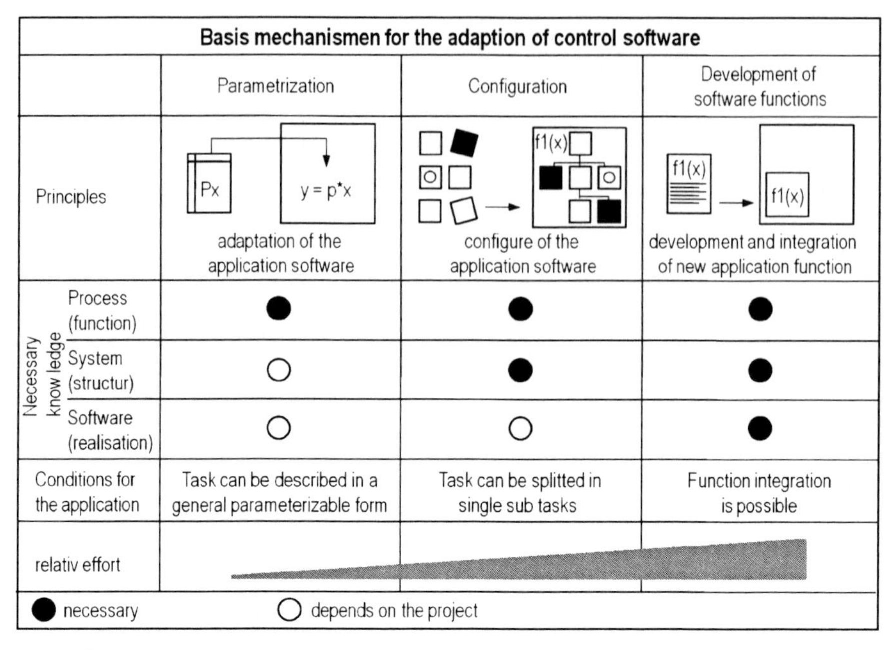

Fig. 4.12 Adjustment of control software

be used, for example, as coefficients of a position control loop are deducted from machine parameters. The description of the control task in a general parametrizable form is necessary for the application of this method.

- **Configuration:** This is understood as the combination of individual components into an integrated whole. In reference to the control software this describes the generation of the control software structure of various functional units – the outcome of this process is an application-specific configuration. Logically, this mechanism is then applied when the range of tasks can be described by varying combinations of individual functions. The process of configuration is facilitated by the use of adequate tools and it requires from the user knowledge of the system but no software-specific knowledge. This means that the functionality and the principles of structure have to be known, but not the software-specific implementation.
- **Development of functions:** If the required functionality is not yet available in software, it has to be developed. The possibility to integrate the newly developed functions into the complete system without any problems is a pre-requisite. Open control systems based on a platform provide this option.

Ideally, a configuration tool adequately supports all three adjustment mechanisms – parametrization, configuration and development of function.

4.5.4 Configuration Procedure

It is the job of the user to adjust the control system to the required machine tool configuration. In a modular configuration there are two stages. First is the provision of a modular system, i.e. a project-independent description for various machine types, like a turning, milling or laser machine and its components (e.g. axes). Next, a specific configuration based on the modular system is generated according to the requirements (Fig. 4.13).

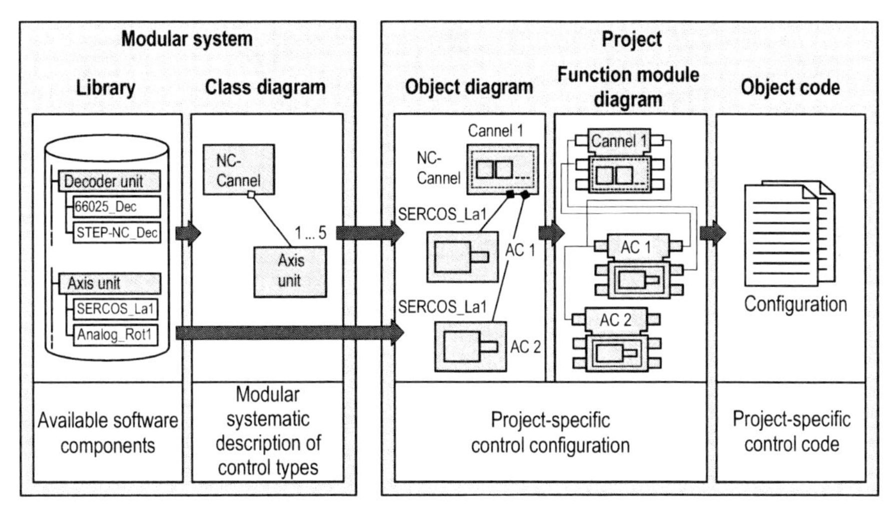

Fig. 4.13 Principle of a configuration process

4.5.4.1 Definition of a Modular System

A modular system consists of a library of available components as well as the modular systematics, which describe how the individual components can be combined. The modular systematics is a sort of 'building plan' for certain machine types.

In the library, the available components are stored in a structured manner as 'classes'. The principle 'abstraction' and 'inheritance' are used for structuring. Abstract classes (e.g. axis unit, NC decoder) provide a basis for the classification. By establishing sub-classes instantiatable classes (e.g. DIN66025 Decoder, STEP-NC Decoder) can be generated, which results in a hierarchical library structure. The following information is needed for specifying a class:

- General management data (e.g. manufacturer, version, name),
- Specifications of the resources required for operation (e.g. operating system, processor),

- Links to the existing implementations, i.e. the encoded, executable functions,
- Information regarding existing communication interfaces and
- Parametrization data.

In addition, descriptions of the available control devices (e.g. SPS S7-400, industrial PC with Windows NT, etc.) are filed in the library. These descriptions include amongst others the type and the hardware of control device (processor, memory resources, ...), the operating system, the available communication channels, etc.

In order to describe the control software for a manufacturing unit in a cross-project way using such a library, design knowledge is required. A class of manufacturing units always has the same basic structure and only varies in optional and/or alternative components. This means that modular systematics describes the basic structure of a manufacturing unit and the project-independent interrelation of the single components. The description is made with class diagrams as they are defined in the Unified Modeling Language (UML) (UML Notation Guide, 2001). The used relationship types include the "consists of" (aggregation) relation, as well as two new types of relationship, "alternative" (either... or) and restriction (if... then), which are often required in engineering applications.

4.5.4.2 Generating a Control Configuration

For the project-specific generation of a control configuration a "building plan" of a control configuration of a manufacturing unit is used as a basis. An object diagram model is created, based on the class diagram, which contains placeholders for the components to be integrated. The replacement of the placeholders is executed by instantiating a concrete class from the library, while checking the correctness of the object class selection. The inheritance structures, as well as the cardinalities and restrictions of the class diagram are considered. After this process all software components for fulfilling the control function are defined. The selected components are instantiated on the platform and displayed to the user as shown as in Fig. 4.14.

For the completion of the control configuration the user has to select from the library the required control devices, i.e. the hardware platforms where the control functionalities are to be processed. Next, the software components are allocated to the control units. Each instantiated control function has to run on exactly one control device. For error prevention this allocation is supported by testing algorithms: only if a software component has an implementation for a certain control unit, can the component be allocated to this control unit, etc.

As described in Sect. 4.5.3, parametrization is another option for the adaption of control software. The configuration tools also support this process by offering the possibility to select every software component and to parametrize it after-wards with a component-specific parameter editor.

The final step of the configuration process is the establishment of communication connections between the software components. Here the components are pre-

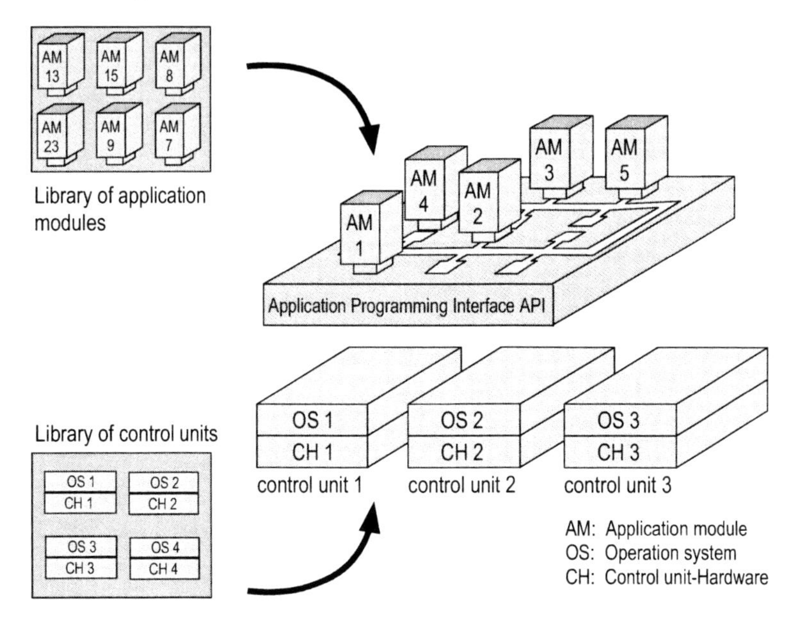

Fig. 4.14 The instantiated components on the platform

sented as function modules, with input interfaces on the left and output interfaces on the right. The user connects the necessary communication relationships graphically, with testing algorithms checking the operation sequence. Tested criteria include data type, parameters, priorities or time requirements for data exchange.

Assistance systems with optimization algorithms support the manual configuration tasks and help with the optimal allocation of software components to control units, or help find an adequate communication interface during the configuring of the communication relationships.

4.5.4.3 Generation of the Object Code

After the configuring is completed, the object code for the control is generated based on the configuration. For open control systems that comply with the OSACA specification, this takes the form of configuration files that contain the following information:

- A list of all instantiated software components and their target platform,
- Parametrization lists for all software components,
- A list of the communication relationships.

These lists can be edited into manufacturer-specific formats via post-processors. In the control units, the configuration files are read in, interpreted and the control software designed accordingly by a configuration run-time system.

4.5.5 Development of a Control Configuration Tool

A prototypical implementation of the presented configuration principle was undertaken within the joint projects MOWIMA (Modeling and Reuse of Object-oriented Machine Software) (MoWiMa, 1998) and HÜMNOS (Development of Cross-Manufacturer Modules for the User-oriented Application of Open Control Architecture) (Lutz, 1999). The prototypically implemented tool components allow the standardized, continuous and user-oriented application of the integrative and adaptive potentials provided by the openness of the control systems. The complexity and decentralized feature of current control systems can thus be efficiently handled.

4.5.6 Configuration of a Control System by an Expert

Actual control systems can be configured once for a designated mechanical structure. This is traditionally done manually by experts who know the rules of generating a valid control system configuration. They have a mental model of the "needs"- and "excludes"-relations between modules (machine and software modules). Using a configuration tool the expert generates a formal configuration file, which is interpreted by the configuration run-time system. It instantiates the necessary software components and parametrizes them according to the formal configuration description (Fig. 4.15). After a physical modification of the machining system, the control system must be configured again. The reuse of the control system, configured once, is not automatically possible because of the lack of a reconfiguration method. In

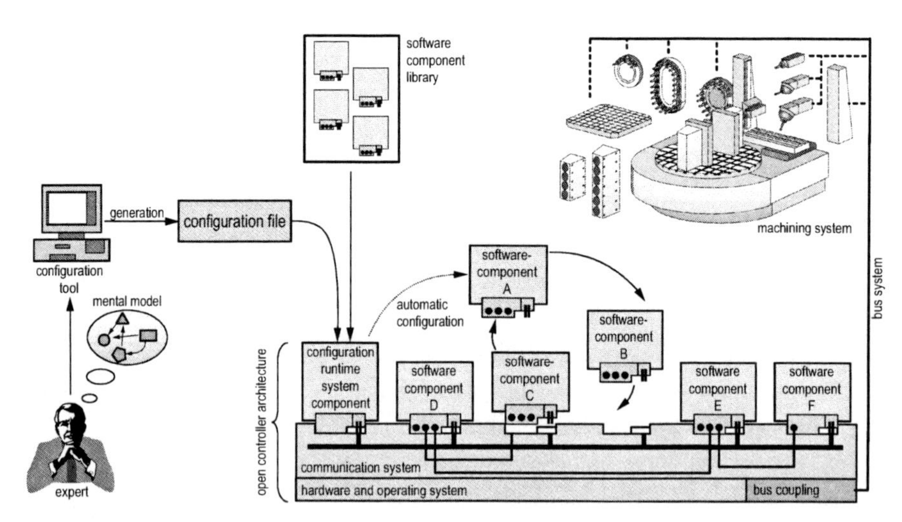

Fig. 4.15 Traditional configuring of a control system by an expert

order to automatically generate the formal configuration file by a self-adapting control system, the rules (mental model of the expert) must be described in a formal manner so that algorithms can check the constraints.

In summary, it can be ascertained that easy-to-reconfigure, modularized control systems are needed as well as a reconfiguration method and a formal description of the reconfiguration knowledge in order to obtain self-adaptable control systems.

4.6 Self-Adapting Control System for RMS

4.6.1 Elements of a Self-Adapting Control System

The idea of a self-adapting control system is the ability to detect changes in the mechanical structure of the manufacturing system after a reconfiguration. The control system then adapts itself automatically according to the new mechanical structure based on mechatronic modules. The basis of such a self-adapting control system is an open and modular control system architecture (see Fig. 4.12 and Sect. 4.4), with well-defined interfaces allowing the parametrization and configuration of software modules according to the mechanical changes in the structure of the manufacturing system. This leads to encapsulated, well-defined and adjustable software components each representing special functionalities of the control system.

A self-adaptation process begins with the detection of a change in the mechanical assembly of the machine (Fig. 4.16, step 1). Via the bus system the bus manager software component detects this change (Fig. 4.16, step 2) by a monitoring mechanism, which is similar to known mechanisms in the field of multi-media or computer technology (plug & play, FireWire/IEEE1394).

The identification software component starts the identification algorithm, which identifies the modules by their identification-ROM (a read-only memory with identification information, e.g. electronic data sheet) (Fig. 4.16, step 3). After the registration and identification of the mechanical modules, a configuration file (formal description of the control system configuration) is automatically generated based on the mechanical structure of the manufacturing system (Fig. 4.16, step 4). This description contains information about the necessary software components, their initialization sequence, their parametrization information and their interconnections. This information is extracted from an information model called Mechatronic Integrating Model. It represents the expert's knowledge about a valid configuration. It describes the control software components, the mechanical modules, and most important, the relations between them. The formal configuration description is interpreted by the configuration run-time system which instantiates and parametrizes the necessary software components (Fig. 4.16, steps 5, 6).

In order to support the described self-adapting process, the modules of the manufacturing system must be mechatronic modules (Wurst, Kircher, Seyfarth, 2004 and

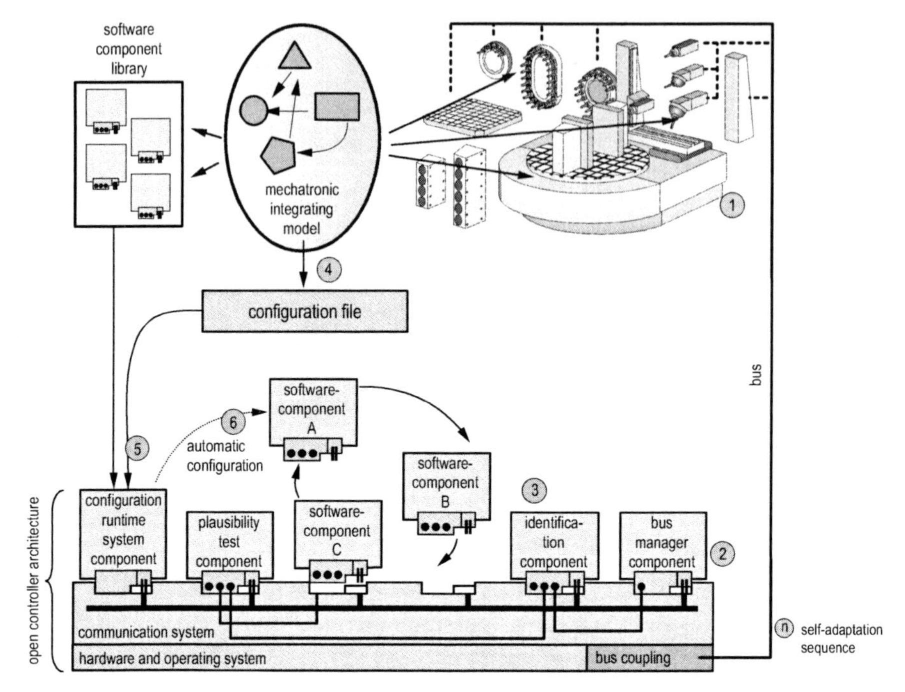

Fig. 4.16 Self-adapting control system and its elements

Pritschow et al., 2003) interconnected with a bus system, which is available anyway because of the electrical drives.

Recapitulating, a self-adapting control system ideally supporting reconfiguration consists of the following elements:

- A modular and open system architecture allowing the exchange, integration and interconnection of software components,
- Control software components implementing the mechanisms for monitoring and identification of mechanical components,
- A Mechatronic Integrating Model for the description of the machining modules as well as for the control software components and the relations between them,
- An integrated methodology for the reconfiguration process based on the actions undertaken by a human control system specialist.

4.6.2 Extensions of Self-Adapting Control Systems

In addition to the above described control software components, which provide the fundamental functionality of a control system (motion generation, logic control, etc.), self-adapting control systems need additional extension components. These software components are responsible for monitoring the manufacturing system, the identification of modules, the generation of reconfiguration orders (configuration file) and the pre-testing of the reconfigured system for plausibility and integrity. These extension components enable the system to be self-adaptable. The knowledge of reconfiguring, i.e. the consideration of the necessary software components, their dependencies and their parametrization enables algorithms to do it automatically, but only if this knowledge is represented in a formal and usable manner.

In order to reproduce the expert's knowledge about the reconfiguration process, the machine modules and the control software components as well as their interdependencies, are described in an information model called Mechatronic Integrating Model (Pritschow et al., 2006).

4.6.2.1 Software Components

The extension components of the control system are (Fig. 4.17):

- The bus manager software component, which detects newly attached (mechanical) modules, replaced (mechanical) modules or removed (mechanical) modules during the reconfiguration process.
- The identification software component, which identifies the above detected modules and the whole mechanical configuration. It uses the module's identification tag (stored in the identification-ROM) and checks it up in the Mechatronic Integrating Model, which contains the knowledge about available modules, possible combinations of modules and their dependencies to software components and parameter sets. Based on this information, it generates a reconfiguration order for the configuration run-time system of the control system.
- The plausibility test software component tests the configured control system in regard to completeness, plausibility and contradictions. This component also uses the Mechatronic Integrating Model, as there are rules in this model characterizing

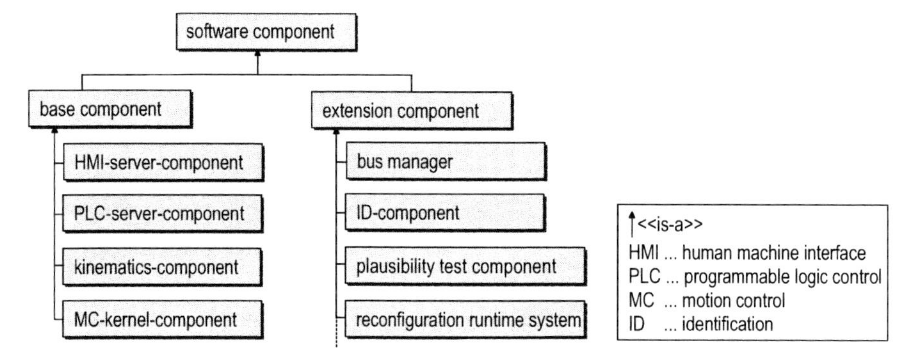

Fig. 4.17 Software components of a self-adapting control system

a correct configuration. They represent the possibilities of the cross linking of the modules and describe the dependencies between mechanical modules and control software components.

- An advanced configuration run-time system, which only stops the necessary parts of the control system according to the reconfiguration order in which the new software application components are loaded from a library. It removes unused software application components and starts up the control system automatically.

4.6.2.2 Mechatronic Integrating Model

The Mechatronic Integrating Model (Fig. 4.18) consists of two libraries: The component library and the configuration pattern library. In the Mechatronic Integrating Model the machine modules and the relations between them, the control software components and their configurations are represented in a formal manner so that algorithms can interpret them. The Unified Modeling Language (UML) (Pritschow et al., 2003) is used for the description of the modules and their relations and object-oriented principles like abstraction and inheritance are used for structuring the set of modules (see also Sect. 4.5.4.1).

The component library is a pool of modules, which can be used for the (re-)configuration of a manufacturing system, machine or robot. It represents the mechanical, electrical and control software components of a reconfigurable machine tool. This classification reflects the traditional view of the departments that build a machine tool. But most components integrate mechanical, electrical and software components and form a new, so called mechatronic component with integrated functionality. Therefore, mechatronic components are also represented in the Mechatronic Integrating Model. As they cannot be assigned to a traditional production department, there may exist three different points of view for a single module. Therefore the mechatronic modules consist of other components (mechanical, electrical, software), which are already represented in the information model. They are described using "consists-of"-relations. That means, mechatronic components consist

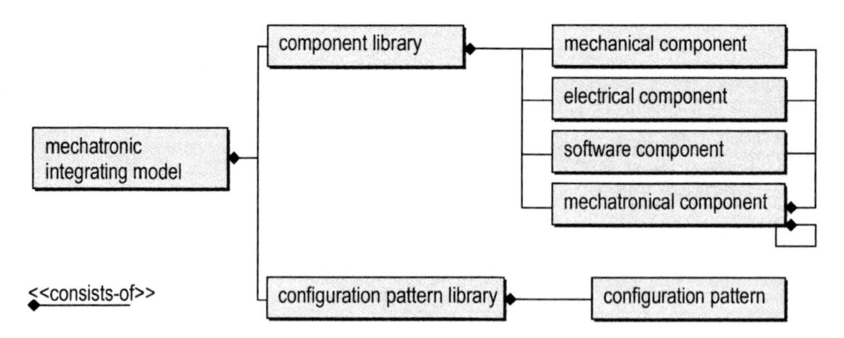

Fig. 4.18 Mechatronic integrating model

of mechanical, electrical and software components, which are already modeled in the Mechatronic Integrating Model. This same approach was applied to the modeling of manufacturing systems (Müller, 1997). An Axis is an example of a mechatronical component as shown in Fig. 4.19. It is a linear axis, which consists of a linear drive, a limit switch, a measurement system (electrical components) and guide way (mechanical component). Besides it may include an appropriate axis controller (software component).

The component library also contains the relations between modules. There may be incompatibilities between modules (e.g. spindle 'A' does not work together with axis from vendor 'Y') or dependencies between modules (e.g. an axis always needs an axis-controller). These relations are represented in the component library using "includes"- and "excludes"-relations. In order to describe global relations that only occur if modules are used together in a special configuration (e.g. two coordinated axes need a software component that generates command values for both axes) configuration patterns are used.

The configuration pattern library describes how the modules must be combined for an executable configuration. Configuration patterns are comparable to construction plans, which provide a framework of module classes belonging to a manufacturing system's configuration (modular systematic). They describe, in an abstract manner, which classes of modules are needed for a special configuration. During configuring, the classes of the configuration pattern must be filled with components from the component library taking into consideration the restrictions and relations between components (for details see also Sect. 4.5.4).

Figure 4.19 (on the right) shows an example of configuration patterns. The 2-axes-configuration pattern, which is sub-divided into two other patterns, demon-

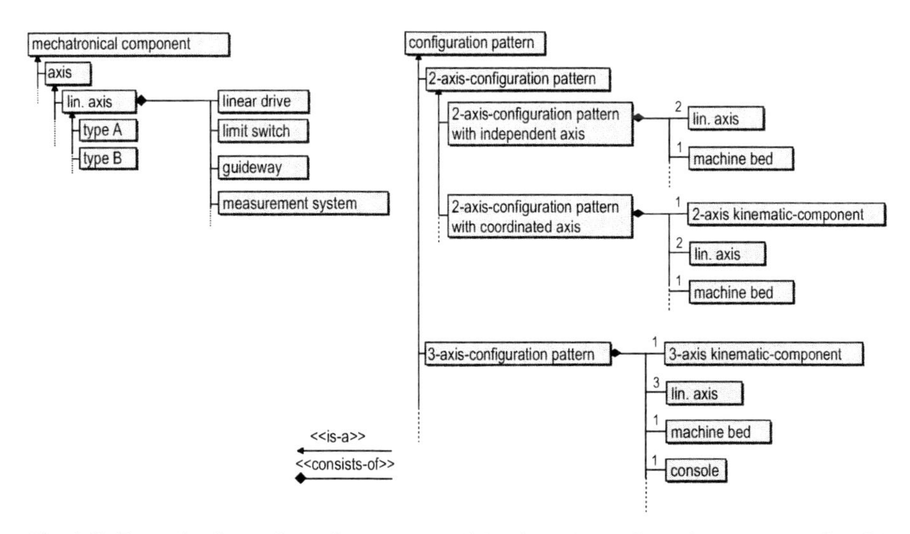

Fig. 4.19 Example of a mechatronic component (*left side*) and a configuration pattern (*right side*)

strates the necessity of different descriptions. If a machine has two axes, they can be coordinated or independent axes. For coordinated axes, the control system needs a corresponding software component (2-axes-kinematic component), which is responsible for the coordination of the axes and the generation of command values. Otherwise, no overall software component is needed. The two different configurations can be distinguished by their configuration patterns: 1) the 2-axes-configuration pattern with independent axes, and 2) the 2-axes-configuration pattern with coordinated axes and the corresponding 2-axes-kinematic component.

4.6.3 Method for Reconfiguration of the Self-Adaptable Control System

Traditionally, the human specialist compares the old and the new manufacturing system's configuration and deduces the adaptation of the control system. In order to save time, he tries to reuse the old configuration of the control system as much as possible. He adds a new software component or removes one if it is no longer needed and keeps the remaining software components. Additionally, he has to adapt parameters, which have changed because of the exchange of mechanical modules. He compares the configurations of the old and the reconfigured machine tools, and knows what to adapt in the configuration of the control system. During the whole reconfiguration process he has to consider the dependencies between the modules of the mechanical configuration and the software components of the control system. Additionally, the dependencies between software control components and initialization sequences must be considered. There are implicit rules between software components. For example, a special software component 'A' only works together with the software component 'B', since a data exchange between them is necessary. It is therefore clear that reconfiguring a control system requires specialized knowledge and methods, which to date existed only in the mind of the experts.

4.6.3.1 The Model Based Configuration Process

The first adaptation of a control system is the configuration process. In terms of object orientation a machine tool's configuration corresponds to an instance of a configuration pattern. All classes belonging to a configuration pattern are placeholders for classes of the component library (electrical, mechanical, software and mechatronic), which can be instantiated through the configuration process where the placeholders are filled with concrete objects from classes of the component library.

4.6.3.2 The Model Based Reconfiguration Process

One example of a model-based reconfiguration is shown in Fig. 4.20. The existing machine configuration (i.e. the instance "2-axes-machine") has to be reconfigured into another machine configuration (i.e. modification into a "3-axes-machine") by adding a third axis. First, the machine modules of the reconfigured machine must be identified via the field bus system and the identification software component (see Sect. 4.6.2.1). Then a suitable configuration pattern for these identified machine modules must be instantiated from the configuration pattern library. This process is just like configuring. This means that the abstract classes of the configuration pattern are created marking the placeholders for components of the component library. They must be filled with components of the existing machine tool and the add-on components of the new configuration. The automated reconfiguration thus differs from the earlier configuration only by the fact that structural changes are considered.

A placeholder may remain empty because the existing reconfigurable machine tool does not have a suitable module or modules could not be taken from the existing reconfigurable machine tool because the chosen configuration pattern does not provide an adequate placeholder. This indicates that the machine modules does not exist as a pattern in the system architecture and must be defined. Hence, add-on software for automatic reconfiguration of the controller is necessary.

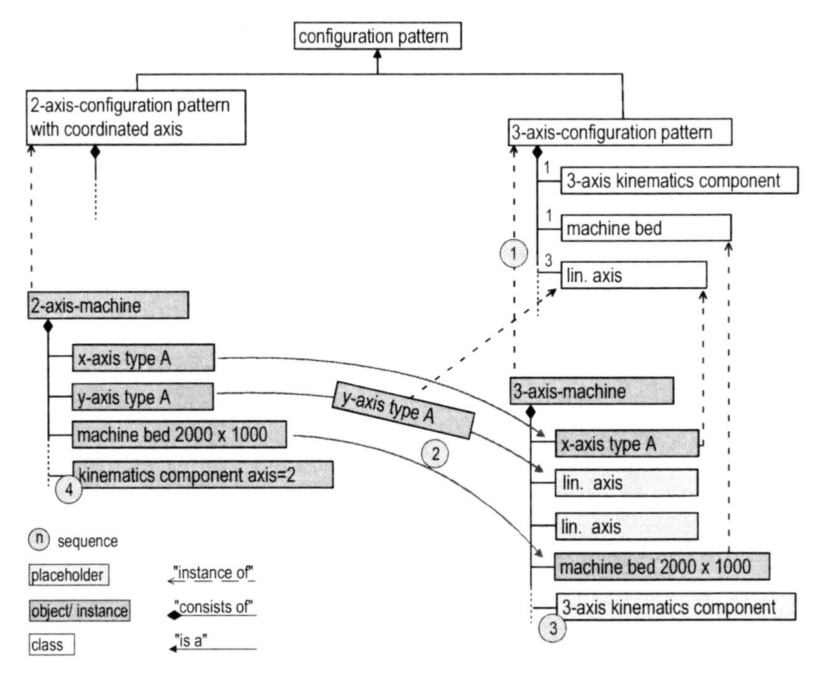

Fig. 4.20 Model based reconfiguration

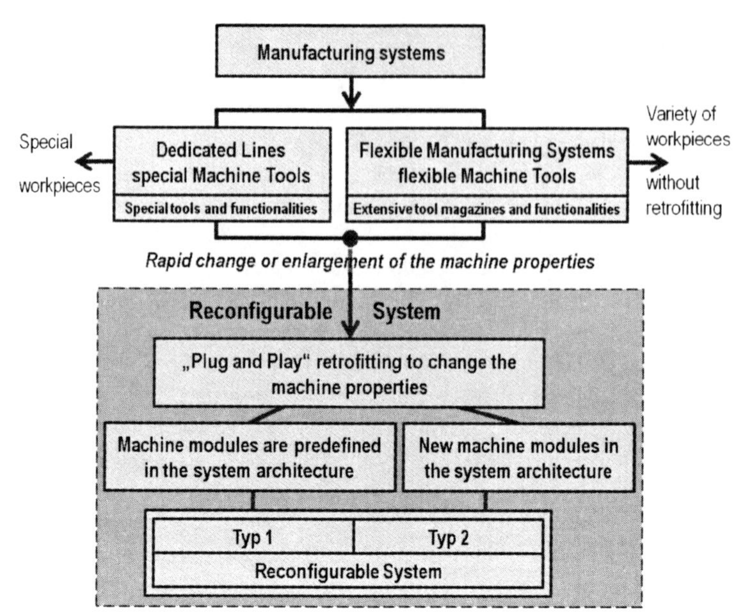

Typ1: - Reconfiguration and parameterisation of the controller automatically possible
- Calibration tools necessary

Typ 2: - Add-on software for automatical reconfiguration of the controller necessary
- Undefined system behaviour: Calibration tools and simulation software necessary

Fig. 4.21 Types of configurable and reconfigurable systems

Therefore, two types of reconfigurable systems can be distinguished. Type 1 consists of machine modules, which are pre-defined in the system architecture, whereas Type 2 has machine modules, which are not designed for the system architecture (Fig. 4.21). For the last one, an automatic reconfiguration process is not possible.

4.7 Summary and Conclusions

The concept of reconfigurable machine tools and robots requires that the interfaces of the components be kept to a minimum in order to enable fast modification. With self-sustaining mechatronic configurable modules that contain all components needed for a satisfactory function, the number of interfaces is reduced to a bus system for communication (set point values, feedback signals, etc.) and a bus system for the energy supply. The mechanical interface between components must have a system-compatible rigidity and the geometric accuracy should be easily adjustable. Conventional machine tools and robots meet these demands at the most for one interface, in order to adapt the technology to the requirements of the users, for example with different powered tools, measurement equipment or laser cutters. To date, no such modules are available in the area of motion generation.

The required controller components as well as a suitable architecture can be generated easily using the current know-how of the 'state-of-the-art'. Self-adapting Control systems which are based on an open architecture modular real-time platform with a defined application programming interface (API) allowing the exchange of software components, can automatically adapt themselves according to a mechanical machine structure based on mechatronic modules. For this, the self-adapting control systems have a mechanism for monitoring and identification of mechanical modules and a configuration run-time system responsible for the automatic adaptation.

The necessary steps for the reconfiguration of the control system, the set of needed software components, their interconnection and their sequence of initialization are deduced from an information model, which reproduces the expert's knowledge about the reconfiguration process. It describes the machine modules, the control software components and the dependencies between them in a formal manner so that algorithms can process them. Object-orientated descriptions of the mechanical, electrical and control software components and configuration patterns are used to reproduce the expert's knowledge about the generation of an executable control system. A model-based approach is used to describe the reconfiguration process.

All these approaches have been investigated in separate ways and projects, but an integrated and completely universal approach for reconfiguration and consequently its realization in regard to the equipment of machine tools and robots is still missing, because such machine concepts have not yet established themselves in the market.

It is a different picture in the case of reconfigurable machining systems, because here only known self-sustaining machine units are exchanged that are part of the state-of-the-art and not active or passive machine components. The communication network of the machines is limited to synchronous information when the work-piece is passed on and to feedback information about the process state. Therefore, the required control functions are 'state of the art' in the area of flexible manufacturing.

References

Daniel C., 1996, Dynamisches Konfigurieren von Steuerungssoftware für offene Systeme. ISW Forschung und Praxis vol 112. Springer-Verlag, Berlin-Tokio, Dissertation

Emmerich W., Aoyama M., Sventek J., 2007, The impact of research on middleware technology. SIGSOFT Software Engineering Notes 32/1:21–46

ISO/IEC 7498-1, 1994, Information technology – Open Systems Interconnection – Basic Reference Model: The Basic Model

Koren Y., 2005, Reconfigurable Manufacturing and Beyond. Keynote Speech of CIRP05 3rd International Conference on Reconfigurable Manufacturing, Ann Arbor, Michigan, USA, May 10–12

Koren Y., Heisel U., Jovane F., Moriwaki T., Pritschow G., Ulsoy G., van Brussel H., 1999, Reconfigurable Manufacturing Systems. Annals of the CIRP 48/2:527–540

Lutz P., 1999, Offenheit als oberste Maxime – OSACA/HÜMNOS: Trotz PC-Welt wird weiter an der offenen Steuerung gebaut. Fertigung 27/9:40–42

Meo F., 2008, Ocean Consortium – Ocean-Project-Homepage http://www.fidia.it/english/research ocean fr.htm (14.01.)

Müller J., 1997, Objektorientierte Softwareentwicklung für offene numerische Steuerungen. ISW Forschung und Praxis vol 120. Springer Verlag, Berlin-Tokio, Dissertation

N.N.: Unified Modeling Language (UML) Notation Guide, Version 1.4, www.uml.org

Object Management Group, Inc, 2004, Common Object Request Broker Architecture: Core Specification Version 3.0.3.

Projektgemeinschaft MoWiMa, 1998, Abschlußbericht des BMBF-Verbundprojekts MoWiMa. VDMA Maschinenbau Institut GmbH, Frankfurt/Main

Pritschow G., Rogers G., Bauer G., Kremer M., 2003, Open Controller Enabled by an Advanced Real-Time Network, CIRP 2nd Int. Conf. on Reconfigurable Manufacturing (Ann Arbor)

Pritschow G., Wurst K.-H., Seyfarth M., Bürger T., 2003, Requirements for Controllers in Reconfigurable Machining Systems, CIRP 2nd International Conference on Reconfigurable Manufacturing (Ann Arbor)

Pritschow G., Weck M., Bauer G., Kahmen A., Kremer M., 2004, OCEAN – Open Controller Enabled by an Advanced Real-Timer Network. Production Engineering XI/1:171–174

Pritschow G., Kircher C., Kremer M., Seyfarth M., 2006, Control Systems for RMS and Methods of their Reconfiguration. In: Dashenko A.I. (ed) Reconfigurable Manufacturing Systems and Transformable Factories. Springer Verlag, Berlin-Heidelberg, Chapter 10

Schmidt D.C., Kuhns F., 2000, An Overview of the Real-Time CORBA Specification. Computer 33/6:56–63

Wurst K-H., 1991, Flexible Robotersysteme – Konzeption und Realisierung modularer Roboterkomponenten. ISW Forschung und Praxis vol 85. Springer-Verlag, Berlin

Wurst K-H., Heisel U., Kircher C., 2006, (Re)konfigurierbare Werkzeugmaschinen-notwendige Grundlage für eine flexible Produktion. wt Werkstattstechnik online 96:H5, S. 257–265 http://www.werkstattstechnik.de. Düsseldorf: Springer-VDI-Verlag

Wurst K.-H., Kircher C., Seyfarth M., 2004, Conception of Reconfigurable Machining Systems – Design of Components and Controllers, 5th International Conference on International Design and Manufacturing in Mechanical Engineering (Bath, UK)

Chapter 5
Reconfigurable Machine Tools for a Flexible Manufacturing System

M. Mori and M. Fujishima[1]

Abstract A flexible manufacturing system with CNC machining centers is becoming decidedly appealing to automotive industry. Such production systems are required to have minimal cycle times and exhibit high flexibility. Consequently, the development and supply of machine tool systems that can fulfill the utmost important requirements such as flexibility, reliability, and productivity for mass production is necessary. This chapter will discuss newly developed CNC machine tools with reconfigurable features. The design concept, machine tool configuration, and application examples of the machines are addressed.

Keywords CNC Machine Tools, Machining Center, Reconfiguration, Flexibility

5.1 Introduction

These days, CNC machine tool systems with high flexibility and versatility to deal with dynamic changes in production volume and part variation are demanded. Traditional dedicated production lines, such as flexible transfer lines, cannot efficiently adapt to the nature of changing parts and fluctuating lot sizes. Manufacturing systems with variable machining centers and turning centers are gradually replacing dedicated systems for medium lot size production. This requires the production system's basic element, the machine tool, to exhibit high speed, precision, and to be reconfigurable, compatible, and convertible to create economic benefits for the customers. Technological breakthroughs and the need for expanded functionality in reduced package sizes are trending towards greater complexity of designed parts. In a multi-machine production line, this complexity is best addressed by dividing production into a series of interchangeable machining units of differing configurations, such as a CNC horizontal machining spindle-based or a vertical machining spindle-

[1] Mori Seiki Company, Japan

based machine, and a turning machine. Rather than achieve finished product status by multiple or manual setup on one or a few machines, each is assigned to a unique module. Addition of modules increases the available complexity of the part without sacrificing cycle time resulting in an efficiently optimized line system (Koren, et al. 1999, Koren, et al. 2002).

Additionally, frequent work piece design revisions and production lot size changes require retooling or re-modularizing the production line. Machining center units must be capable of insertion and replacement anywhere in the production line. This allows greater utilization of factory resources. Practically speaking, this implies that different machine types must be able to replace others of different configurations without disturbing the line. Therefore, machines must be designed around a common platform while maintaining vastly different machining capabilities. Making the interchange uncomplicated and straightforward allows maximum productivity amid work piece design and lot size changes.

In this chapter, a series of newly developed machine tool systems with reconfigurable features, intended for the automotive parts industry, is presented. The development concept of this machine series features a horizontal machining, vertical machining and turning center. All have been designed on a common platform and all form a vital building block for an ultimate transfer line. The requirement for this series is to deliver robust mass production performance along with scalability and flexibility to reconfigure the line as production requirements transform.

5.2 Reconfigurable Machine Tools Development

In order to address the requirements mentioned above, the machine tools should be designed for a compact and interchangeable footprint without compromise to the cutting performance. To perform required machining operations on a complex part, three machine types (horizontal, vertical, and turning) should be simultaneously available with identical floor footprints and cover heights. The workpiece should be accessible from the front, side and top, in order to accommodate various types of work handling devices, such as robot and swing arm, APC (Automatic Pallet Changer) transferring, gantry loader, etc. Maintaining a minimum non-cutting time (especially important for mass production) requires a higher acceleration speed for all drive axes and a shorter spindle acceleration time. Another important requirement is easy to conduct maintenance or replacement for the moving parts. The challenges to meeting or achieving these requirements are: designing an extremely compact tool magazine, keeping the machine width as narrow as possible, and achieving effortless maintenance in a small and packed space. Furthermore, using common parts among these machines should be given great consideration. For example, all three can share the same spindle, three-axis unit and bed.

Three machine configurations (horizontal, vertical and turning) have been designed simultaneously. The machine specifications for the horizontal case

(NXH2000DCG) are shown in Table 5.1. X, Y, and Z-axis strokes are 150 mm, 150 mm, and 220 mm, respectively. The maximum work piece size has a 100 mm height with a ϕ 140 mm swing diameter. The table size is 200 mm × 200 mm with a 1000 mm loading height for ease of setup. The rotation speed of the spindle is 12 000/min (optional speed: 20 000 rpm) with 11 Nm of torque maximum output (high torque option available). The acceleration time to 12 000/min is designed to be 0.5 sec. An HSK A40 tool holder was chosen. The maximum tool diameter is 50 mm with a maximum length of 180 mm. The rapid traverse speed is 50 m/min for all axes. The acceleration speed for each of axis is X: 1.2 g, Y: 2.7 g and Z: 3.5 g. Overall NXH2000DCG machine dimensions are 680 mm (width) × 1982 mm (height) × 2295 mm (length). The vertical and turning centers have identical machine width, but slightly different machine heights. The 11 Nm spindle torque and axial thrust of 3500 N were developed to provide generous machining capability despite the small size. For aluminum, this allows effortless drilling and tapping of up to 14 mm holes and use of an 80 mm diameter face mill.

Figure 5.1 shows the configuration for the horizontal and vertical machine centers. In order to meet the most severe operating conditions, DCG (Driven at the

Table 5.1 Specification of developed machine

Item		NXH2000DCG
Stroke	X-axis stroke (mm)	150
	Y-axis stroke (mm)	150
	Z-axis stroke (mm)	150
Work	Cylindrical	φ140×100mm (h)
	Table Size	200×200
	Table Height (mm)	1000
Spindle	Rot speed (min-1)	12000×20000
	Acc time (s)	0.5
	Torque (Nm)	11
Tools	Type	HSK A40
	Max Diameter (mm)	50
	Max length (mm)	180
Rapid Traverse	X-axis (m/min)	50
	Y-axis (m/min)	50
	Z-axis (m/min)	50
Acceleration	X-axis (G)	1.2 G
	Y-axis (G)	2.7 G
	Z-axis (G)	3.5 G
Magazine	Number of Tools	13
	Chip to Chip (s)	2.0 (MAS)
Machine Size	W × H × L (mm)	680×1982×2295

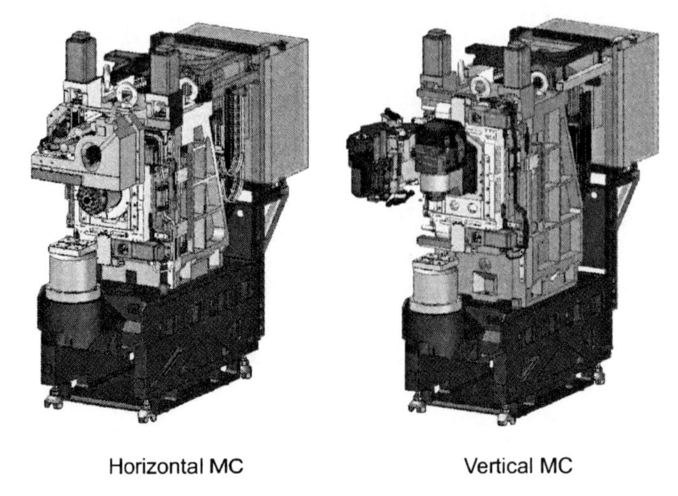

Fig. 5.1 Horizontal and vertical configurations

Center of Gravity) and Box-in-Box construction technologies are implemented. Mechanical common sense indicates that for stable motion, a component put into motion should be driven at its center of gravity. However, that this was never considered an overriding factor can be clearly seen through a survey of current machine tool structures. The fact is, finding design examples that willingly raise costs or sacrifice other elements to drive components at the center of gravity are difficult. A major design theme for this machine was the application of center of gravity drive to optimize stability during high speed/high acceleration motions. However, the true importance of pushing at the center of gravity was not fully grasped until nearly complete design models were evaluated using finite element analysis methods. Even up to the final design stages, the plan had been to use a single ball screw mounted at one side of the spindle to drive the vertical axis. The reason for this was cost. However, the results of a machine dynamics analysis, shown in Fig. 5.2, was convincing in showing the merits of center of gravity drive.A comparison between vibration amplitude when axial motion is stopped using a single ball screw and the vibration amplitude

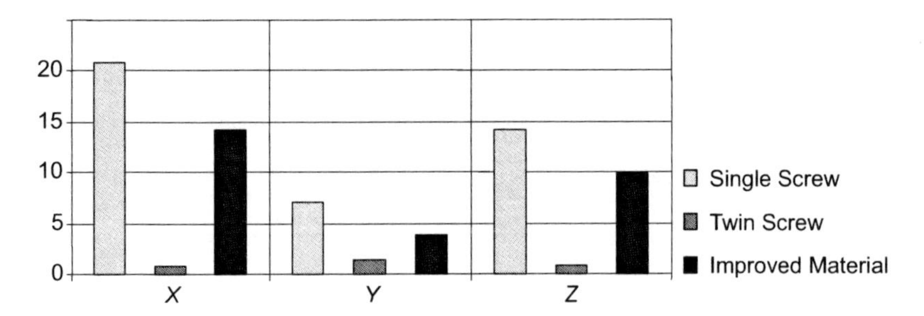

Fig. 5.2 Vibration suppression by the DCG configuration

for two ball screws in a center of gravity drive configuration was performed. The result showed an order of magnitude vibration reduction when pushing at the center of gravity with two ball screws.

These two technologies allow for an extreme acceleration rate of 3.5 g in the Z-axis while keeping minimal vibration and high accuracy. Improvements in servo motor technology and an ample ball screw lead provide 50 m/min federates in all three axes.

Cartridge type spindles were developed for both machines to enable fast and easy spindle changes. This feature is found to be critical when dealing with a production line because stopping one machine causes the entire line to stop. Other similar features allowing for easy replacement of consumable parts were also considered and implemented across the board.

With all three linear axes on the tool side, the table is fixed and open at the front and sides. Efficient chip handling and flexible work handling is the result of such a design. Chips can be routed out of the machine either by the rear or the front by way of a chip conveyor, shooter, or common pit. With the table configuration, chips are given a minimum number of catch points and tend to easily transfer into the evacuation system. Work can be transferred from the side, front, or top by a variety of loading systems giving the user flexibility to meet their requirements and type of work piece. Furthermore, gantry loaders, robot arms, APC, side shuttles, etc. can all be accommodated.

In order to achieve cutting performance at par with large machines, design and analysis was carried out simultaneously. Component and system level FEM analysis to the main structure, shown in Fig. 5.3, was performed through all design iterations. Various static and dynamic analyzes of other subsystems were carried out to ensure an optimized and working machine in all aspects. The best static and dynamic behavior was thoroughly investigated with particular attention to residual vibration reduction commonly found on high acceleration machines.

Figure 5.4 shows the modular features of the machine components. The column, bed, Y-axis, and X-axis saddles are identical for three types of machines. Installation of a vertical spindle and spindle head along with the corresponding ATC unit

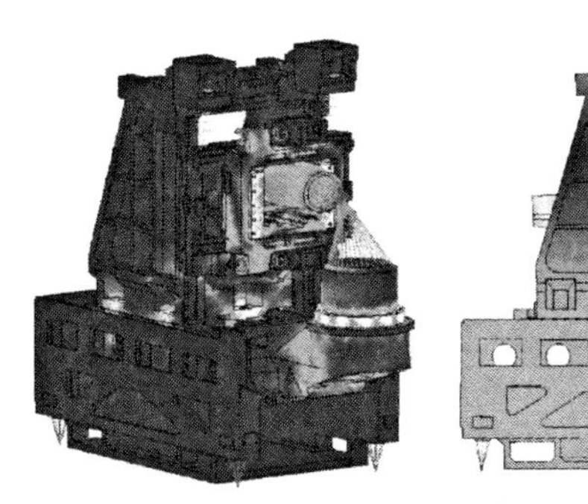

Fig. 5.3 Results of static and dynamic finite element (FE) analysis

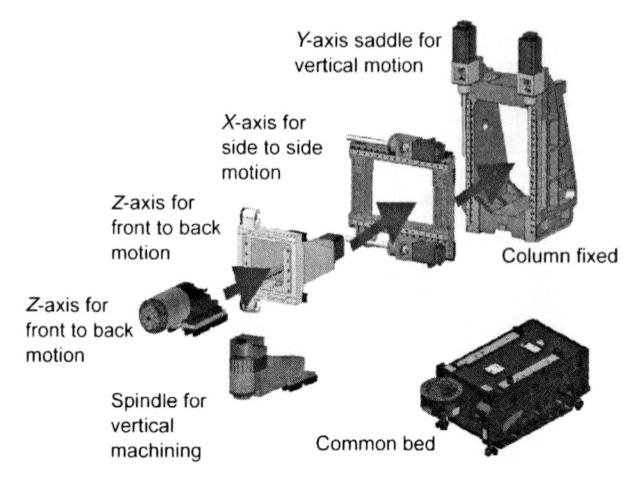

Fig. 5.4 Reconfigurable machine components

converts the system to either a horizontal or vertical machining center. A turning center is achieved by replacing the table unit with a turning spindle for the vertical configuration machine.

The flexibilities of production lines built with these machines are also found in Figs. 5.5 and 5.6. In Fig. 5.5, a 3-axis machine can be easily configured to a 2-axis machine by removing the x-axis saddle according to the application requirement.

Figure 5.6 shows different combinations with different types of machines and work handling devices, as well as chip management methods. The Gantry loader type work handling method is shown in the upper left figure while the transferring conveyer type is shown in the upper right figure. Coolant supply and chip management can be handled from the front of the machine by a centralized system, as shown in the upper right figure. The common chip trough runs on ground level below the work holding table and across the machines and moves chips and coolant to the

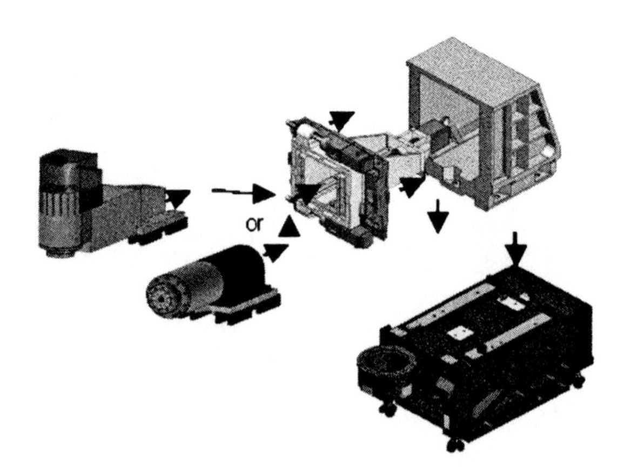

Fig. 5.5 Possible machine reconfiguration – horizontal to vertical or 3 to 2 axis

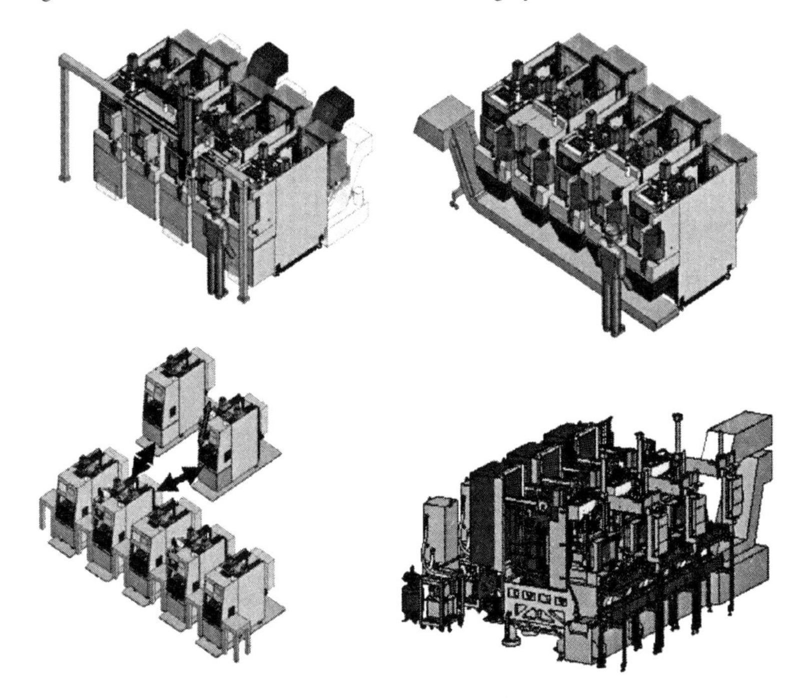

Fig. 5.6 FMS examples constructed by the developed machines

central filtration unit. The narrow width of the machine reduces the overall length of the chip trough, minimizing post-machining chip handling problems. Individual machine chip conveyers also handle chips and coolant from the rear, as shown in the upper left figure.

With the change of a machined part or machining operations, the machine can be replaced simply as shown in lower left figure, which is supportive to help customers reconstruct the product line within a reasonable time frame. The lower right figure shows a work handling method by pallet transferring system in front of machine. A swing arm is integrated with each machine to transport the part/pallet from table to the transfer conveyor and verse versa. It has been shown that the reconfigurable features of the machines make all these four different working handling methods possible. These different working handling methods are applied in various applications based on the part size, complexity and required production rate.

5.3 Application Examples

Proof of a design concept is achieved by acceptance of the intended users. Three application examples will be illustrated from tier one automotive parts suppliers representing a variety of part types and manufacturing processes. In each case, the design concept proved to perform at or above expectations, showing that this type of machine tool design concept is both practical and desirable.

An automotive piston line, shown in Fig. 5.7, was constructed consisting of three (3) horizontal machines, three (3) vertical machines, three (3) turning machines, and one (1) special purpose drilling machine. The piston to be machined was made of material AC4H and had roughing, boring, drilling, milling, turning, and grooving done to it. It was machined directly from a cast part. A water based coolant washes the chips into a central pit. With the use of a shuttle feeder, the total cycle time achieved, including machining time and transport time, was 13 seconds. Assuming a minimum line up-time of 90%, 6000 pieces are produced each day. Other equipment adapted to this line are scale, parts washer, loader, multi-axis robot and part stockers.

The ABS line was comprised of 11 horizontal machines and 3 vertical machines, as illustrated in Fig. 5.8. Chip and coolant discharge is handled with a centralized multi-machine conveyor. Work is transported via an APC and pallet transfer/pallet return system.

Fig. 5.7 Manufacturing system for piston production

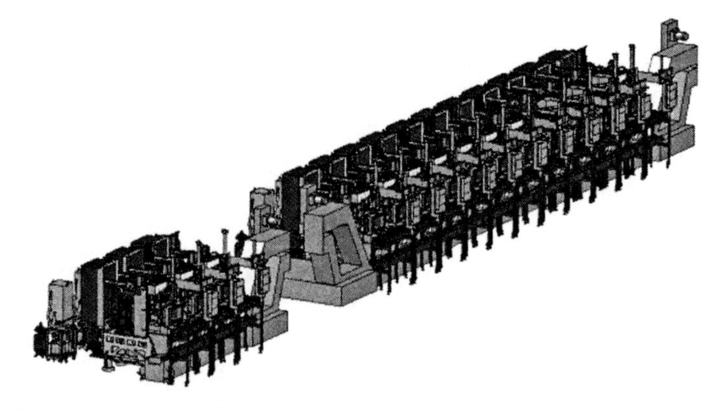

Fig. 5.8 Production line for ABS part production

The material is also AC4H. This part, starting from a cast surface, has extensive facing, drilling, boring, etc. done on six (6) faces so that a part reorientation is necessary. Cycle time has been reduced to less than 50 seconds per part. 1 500 finished pieces are produced daily. The line is able to hold tight accuracy and repeatability on several critical bore diameters on the part.

The throttle body line consists of only four (4) horizontal machines and was introduced to reduce the space required to machine this relatively small part, as only one process is performed. The concept of fitting a machine to the work size is used to design an efficient and cost effective layout. A single central multi-axis robot feeds each of the four machines with a total cycle time of 80 seconds, producing over 3800 pieces daily. The throttle body is made using ADC12-A and the chips are flushed to the rear of the machine with four (4) individual chip troughs.

5.4 Summary

A series of machines with identical widths and cover height has been developed for mass production. These machines have been designed with maximum flexibility and reconfigurability for constructing flexible manufacturing systems. Design concept validity has been proven in the industrial environment as evident by the utilization of these systems by automotive parts suppliers.

References

Koren Y., Heisel U., Jovane F., Moriwaki T., Pritchow G., Van Brussel H., Ulsoy A.G., 1999, Reconfigurable Manufacturing Systems, CIRP Annals, 48/2, (keynote paper)

Koren Y., Ulsoy A.G., 2002, Vision, Principles and Impact of Reconfigurable Manufacturing Systems, Powertrain International, pp 14–21

Chapter 6
Reconfigurable Machine Tools and Equipment

E. Abele and A. Wörn[1]

Abstract Reconfigurable Manufacturing Systems (RMS) are characterized by their quick adaptation to un-scheduled and un-predictable changes in production requirements. Trends, like the reduction of product life cycles, high products diversity at small lot sizes, as well as the fast development and implementation of new production technologies, call for new approaches in the design of flexible and life cycle overlapping machine tools. The flexibility of RMS comprises changes in machining technology, production capacity, machine structure and function as well as in work piece spectrum and material property. The presented design of RMS is based on a construction kit principle, which enables it to adjust to new production requirements by substitution, addition or removal of machine systems. A new trend in the area of RMS is the complete machining of work pieces by using different machining technologies in one machine workspace (Abele, Wörn, 2004). This paper describes a method that considers the constructive particularities of the Reconfigurable Multi-technology Machine tool (RMM) taking flexibility aspects into consideration.

Keywords Reconfigurable Machine Tool, Flexibility, Equipment

6.1 Introduction

In the area of investment goods, the trend of mass production changes towards customized production. Characteristics of this fundamental change in the production philosophy are the continuously shortening product life cycles, rapid quantity fluctuations, small batch sizes up to make-to-order production and at the same time a large number of variants and a high degree of alteration in the part spectrum (Fig. 6.1). Moreover industry and, in particular, automotive component suppliers are exposed to an enormous cost pressure and competition caused by the demands of the OEM

[1] Technische Universität Darmstadt, Germany

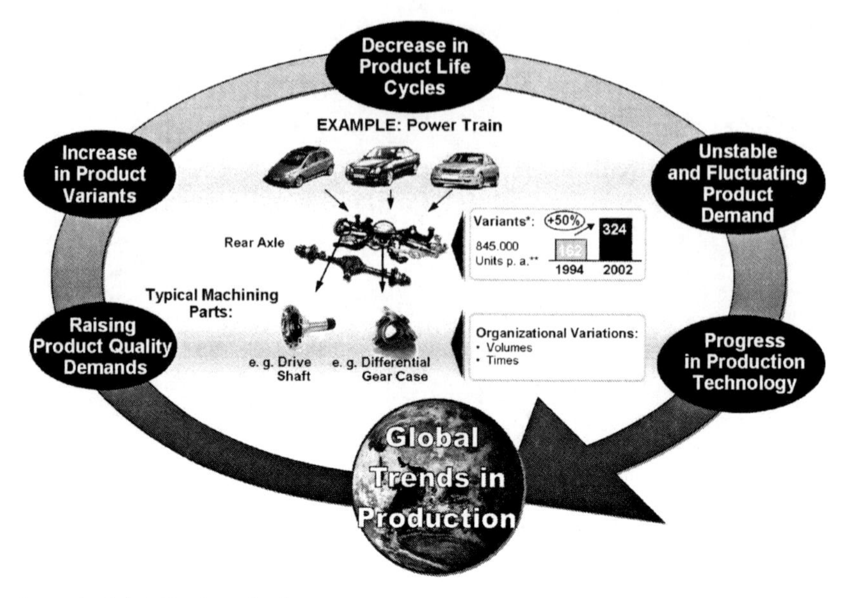

Fig. 6.1 Global trends in production

for continuous cuts in the products' prices. These demands can only be absorbed by higher productivity and efficiency in production processes. The innovation push and engineering progress in the field of production technology, power train engineering and materials science lead to a significant potential in production and to a continuous change of production processes. In addition, with the integration of different manufacturing technologies into one machine tool workspace leads to a tendency of increasing work piece quality. Thus, the requirements of future manufacturing systems are modified by the trends mentioned.

Present manufacturing machine tools are mostly designed with respect to the process and for the specific task they are about to execute. The design is not normally adjustable to variable process conditions. Amortization times for manufacturing tools are up to 15 years and require machine tool concepts, which can be efficiently utilized in the long term (Schuh, Wernhöhner, Kampker, 2004). Future manufacturing machine tools however must be designed with a view to the possibility of internal and external structure changes. Furthermore, they need to be economically adaptable towards changed production functions. Production life cycle overlapping usage, technical update ability, high productivity and integration of multiple production technologies are demands for those future machine tools (Koren, et al., 1999).

The Reconfigurable Multi technology Machine tool (RMM) design is the required solution. The concept of RMM is based on high productivity, flexibility and

What kind and how much flexibility?
- What kind of flexibility (variants, technology, products) is required?
- How can flexibility be measured and quantified?
- How much flexibility is really necessary? Short term, long term?

How to realize flexibility?
- Technical solutions
- Organizational solutions
- Overcapacity

How to optimize?
- Extra costs versus extra benefit
- Productivity versus flexibility

Standardization versus manufacturers' knowledge?

Fig. 6.2 Critical reflections about "reconfigurability"

expanded useful economic life. The RMM universality is hereby not achieved by an architecture that covers all options. Adaptability is carried out by a consistent architecture of the machine tool that adopts a construction kit design concept. Thus, nowadays, the structure of devices can be altered by the user where short ramp-up times and fast production adaptations can be managed. The economic useful lifetime of RMM is extended by the conversion and updating possibility in contrast with the conventional machine tool concepts.

The economic efficiency of RMM is indicated by the manifold usage of the machine components beyond multiple production life cycles. Thus, the RMM enables investment saving potentials compared with conventional, Product-Specific Machine Tools (PSMT). The RMM concept assures success when the appropriate necessary economic and technical flexibility is achieved.

At the beginning of the productive planning process the following basic questions should be answered, as depicted in Fig. 6.2.

6.2 Flexibility Requirements

Present investment decisions for machine tools concentrate primarily on three characteristics: costs, times and quality (Abele, Liebeck, Wörn, 2006). Future decision making requires another criterion, the aspect of flexibility by ever-changing production structures, engineering progress, and, for the economic long-term, usage of production facilities (Chryssolouris, 1996). Flexibility is the capability of a machine tool to adapt to new production requirements. Production facilities, which have a high degree of flexibility, can be rapidly changed-over at a low expenditure according to the characteristics of a particular production task (Koren et al.). This feature is important especially in make-to-order and medium serial production.

In order to cover a wide range of the production requirements, production facilities must be flexible in several fields, as illustrated in Fig. 6.3. Essential spheres of flexibility can be classified in the following ways (Schäffer, 1995, Milberg, 1996):

- Process flexibility describes the ability to install different manufacturing technologies in the machine tools such as turning, milling, laser-welding, polishing etc.
- Failure flexibility refers to the ability of production facilities to handle breakdowns and thus ensuring continuation of manufacturing. The criterion for evaluating failure flexibility is the machine productivity.
- Volume flexibility describes the ability of production facilities to operate profitably at different production volumes. The criterion here is the economic batch size.
- Expansion flexibility describes the capability to expand machine tools' capacity with minimal efforts. The criterion for its evaluation is the expansion ability.
- Product flexibility refers to the ability to change-over production facilities to produce a new set of products economically. Adaptation ability is the criterion used here.

The capability of production facilities to change-over is not only an important criterion in terms of production life cycle spanning usage; it refers also to different adaptation requirements during the production life cycle of a manufacturing product. Thereby, the product life cycle displays the customer's demand throughout the production life cycle, i.e. the period between the first and the last production of

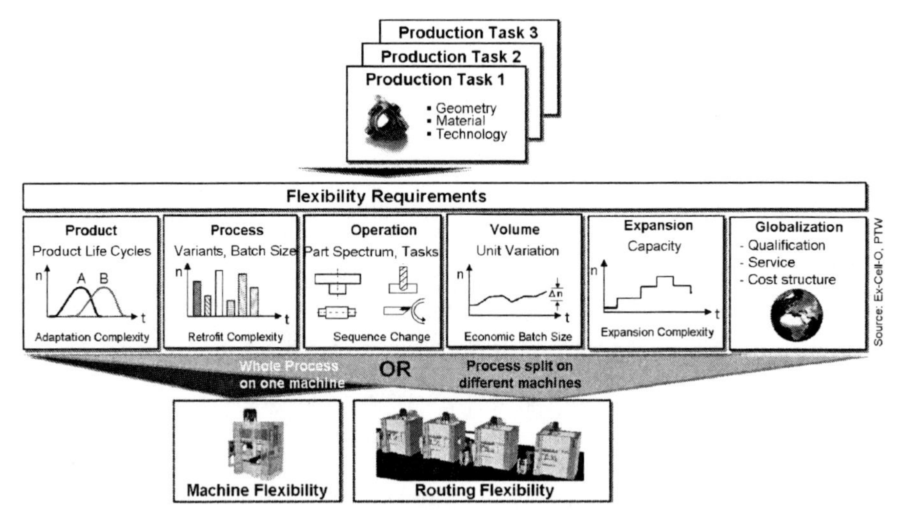

Fig. 6.3 Essential spheres of production flexibility

a specific product. In that period, several life cycle phases with different production characteristics are passed through. Those characteristics have different flexibility demands and require appropriate adjustment steps to the production facilities (Schuh, Wemhöhner, Kampker, 2004):

- The product introduction stage is characterized by test runs, functional prototyping of pilot products, and high diversity of parts at low quantities as well as a high degree of modification in terms of process parameters, production sequence and part design. At this stage production facilities must allow the adjustment regarding functional and technological changes in accordance with product and process.
- During the growth stage, product quantities rise steadily. Therefore, the flexibility targets are expansion and volume flexibility. Productivity and capacity of the production facilities have to increase continuously in accordance with product demand.
- The saturation stage requires a high production capacity at a low product modification rate. Breakdowns of production facilities have serious consequences on productivity and capacity. Therefore, failure flexibility of production facilities is important in this phase.
- The degeneration stage is characterized by steadily declining product quantities up to the termination of production. At this stage, surplus production facilities can be prepared for new production tasks, which require product flexibility. Alternatively, discrete devices and modules of the production facilities can be reused as spare parts.

The degree of flexibility depends on the attributes of the specific production facility. There are different approaches to a flexible design. Flexible machine tools can conform to several production requirements. In general, they are over dimensioned and less productive compared to customized production facilities. Flexible machine tools possess a large spectrum of potentially applicable components and functions that cope with the flexibility demands. As a result, investment costs of these machine tools rise along with the attained flexibility (Metternich, Würsching, 2000).

The concept of Reconfigurable Manufacturing Systems (RMS) provides a solution for the long-term adaptability of the machine tools (Koren, et al., 1999, Pouget, 2000). The ability of reconfiguration ensures the customized adjustment of the machine structure to the required production functions (Metternich, Würsching, 2000). A new approach in the field of RMS is the Reconfigurable Multi technology Machine tool (RMM) concept that enables the integration of multiple production functions in one machine workspace. The structure of RMM has a convertible design and consists of basic elements, which are common for every machine configuration and reconfigurable elements, which can be economically adapted through addition, substitution or structural change. Thus not only is the adaptation of the RMM within predefined solutions possible, but so is the adaptation towards unpredictable production requirements (Abele, Wörn, 2004).

6.3 Reconfigurable Multi-Technology Machine Tool (RMM)

6.3.1 Machine Tool Design

A RMM is considered as a system in a hierarchy of different distinct systems according to Fig. 6.4. In general, devices are separable from other devices and surrounding are regarded as systems (Ropol, 1975). Systems again are composed of sub-modules which are divided into different downstream hierarchic System Levels (SL) (Wiendahl, Heger, 2003).

The RMM structure is realized by a platform-based construction kit, consisting of defined systems at different hierarchic SL with assigned functions (Metternich, Würsching, 2000, Gunnar, Yxkull, 1996, Pahl, Beitz, 1996). The complex structure of RMM can be regarded as an organizational unit, which refers to a production planning process. This organizational unit can be sub-divided into smaller, manageable sub-modules at System Level SL −1 respectively SL −2 with the purpose to create an uncoupled agile machine tool structure with increased sub-modules autonomy and exchange-ability between these sub-modules. Thus, the fast adaptation of the RMM system to new requirements can be managed with a change in the system connections or by modifying the assembly structure on several system levels (SL) during its usage.

The RMM can operate as an independent machine tool or can be integrated into a superior Production System at a higher level (SL +1). RMM can thereby be arranged either sequentially or in parallel. In a sequential arrangement the production steps complete one another. Capacities can be extended in order to resume identical operations in parallel-arranged RMM. A combination of sequential and parallel machine tool arrangements is also possible.

The machine system modules are treated as "black boxes" (Milberg 1997). The functional description of the discrete system modules is done by means of attributes. These techniques facilitate the integration process of sub-systems into the RMM structure (Denkena, Drabow, 2003). Attributes are quantitative and qualitative characteristics of a system (e.g. the spindle power of a spindle system) as well as input or output quantities respectively (e.g. energy) between other systems and the system surrounding. Relations (do you mean Exchange of information) between systems occur during the interconnection of system input and output functions like energy, information, material and forces/moments. The integration of RMM subsystems requires designing effective system attributes such as the interfaces between modules that are located at system boundary lines (Gu, Hashemian, Sosale, 1997, 2003).

RMM design adopts system sharing-modularity architecture, which simplifies the integration of modules in RMM, where various modules share a common platform. Modules are autonomous process-oriented agile system units, with well-defined interfaces, that deal with a specific spectrum of process functions (Ropol, 1975). The goal is to assign delimitable process functions to independent RMM

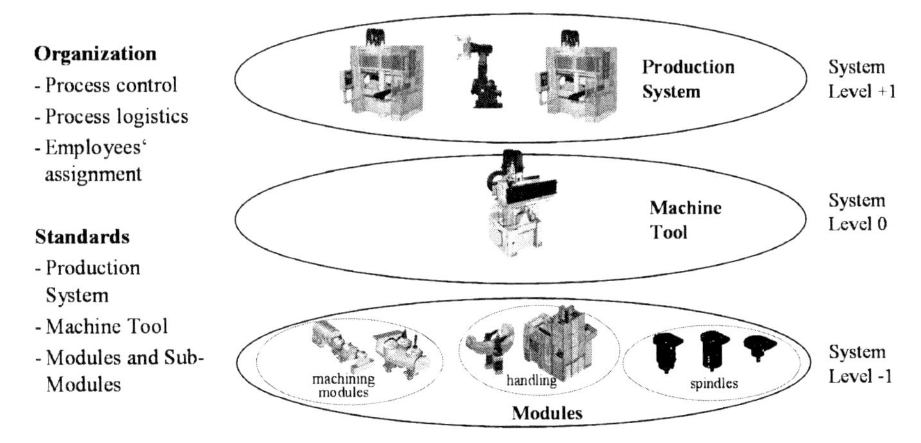

Fig. 6.4 Mutability in production

sub-systems. Modules feature a high degree of conversion flexibility, as again, they are composed of changeable sub-modules (e.g. spindle systems or tools) at SL −1. A modification in all relevant processes can be carried out. It ensures the composition and integration of the machine tools in compliance with product and process requirements (Fig. 6.4). This enables the introduction of new future machining technologies (Frick, 2003).

6.3.2 Modules

The possibility of creating intelligent connections and arrangements of different manufacturing modules in a RMM workspace offers various advantages (Figs. 6.6 and 6.7). The complete manufacturing of complex parts and part families is a suitable example (Fig. 6.5) (Frick, 2003). Complete manufacturing of work pieces is advantageous since part re-clamping and further handling operations can be eliminated, hence, increasing the work piece quality by up to 30%. Therefore, a positive result is gained from the re-clamping errors minimization. Thus, manufacturing tolerances specifications can be reduced and costly finishing processes, like grinding, can be reduced or avoided. Furthermore, the prevention of interruption of machining and the elimination of additional work piece handling and the re-clamping processes bring forth increased productivity. This leads to a reduction of the unit costs.

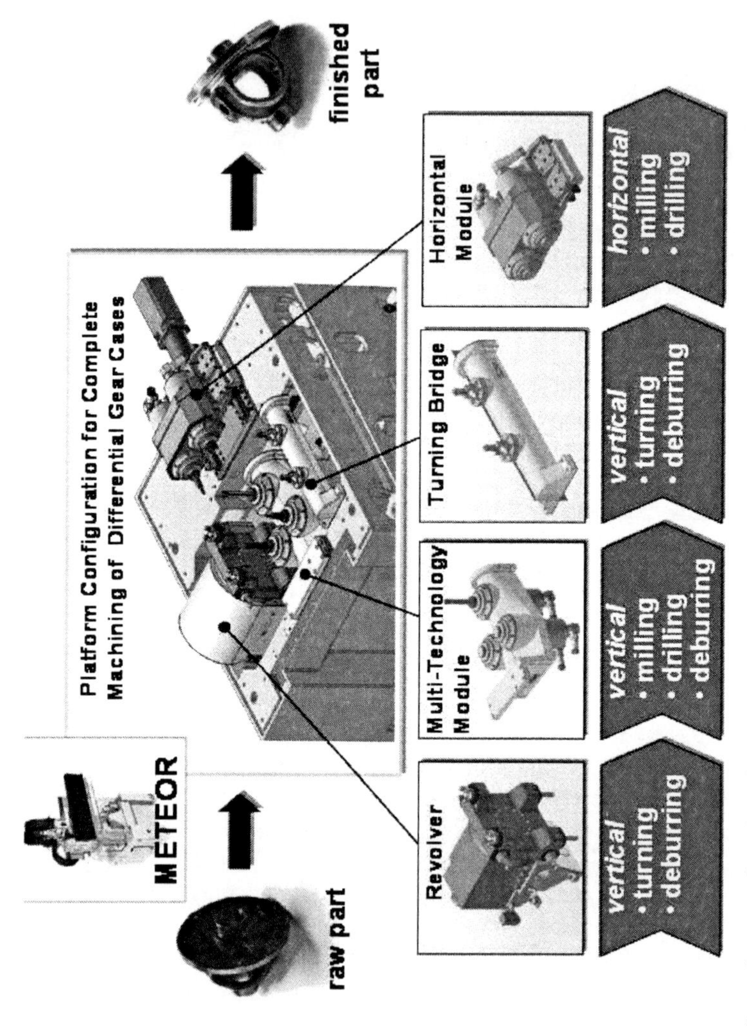

Fig. 6.5 Example of complete manufacturing of parts

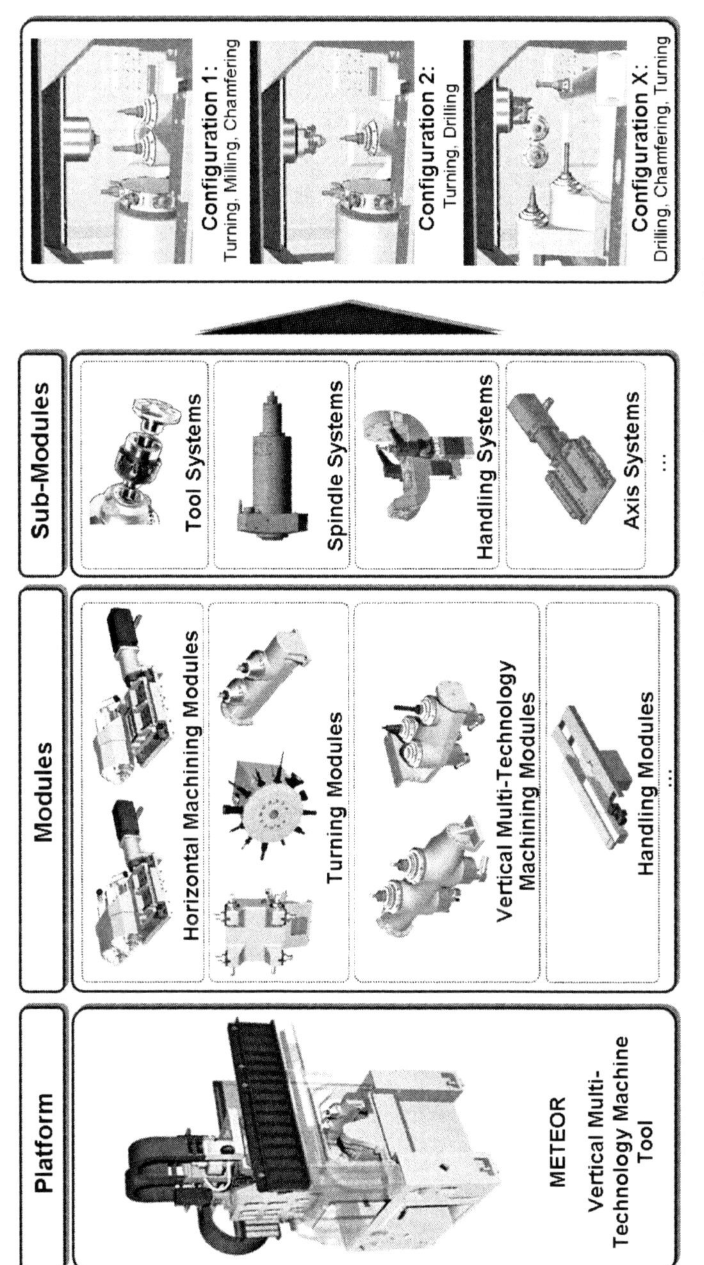

Fig. 6.6 Machine tool modular construction kit and examples of workspace configurations (Abele, Liebeck, Wörn, 2006)

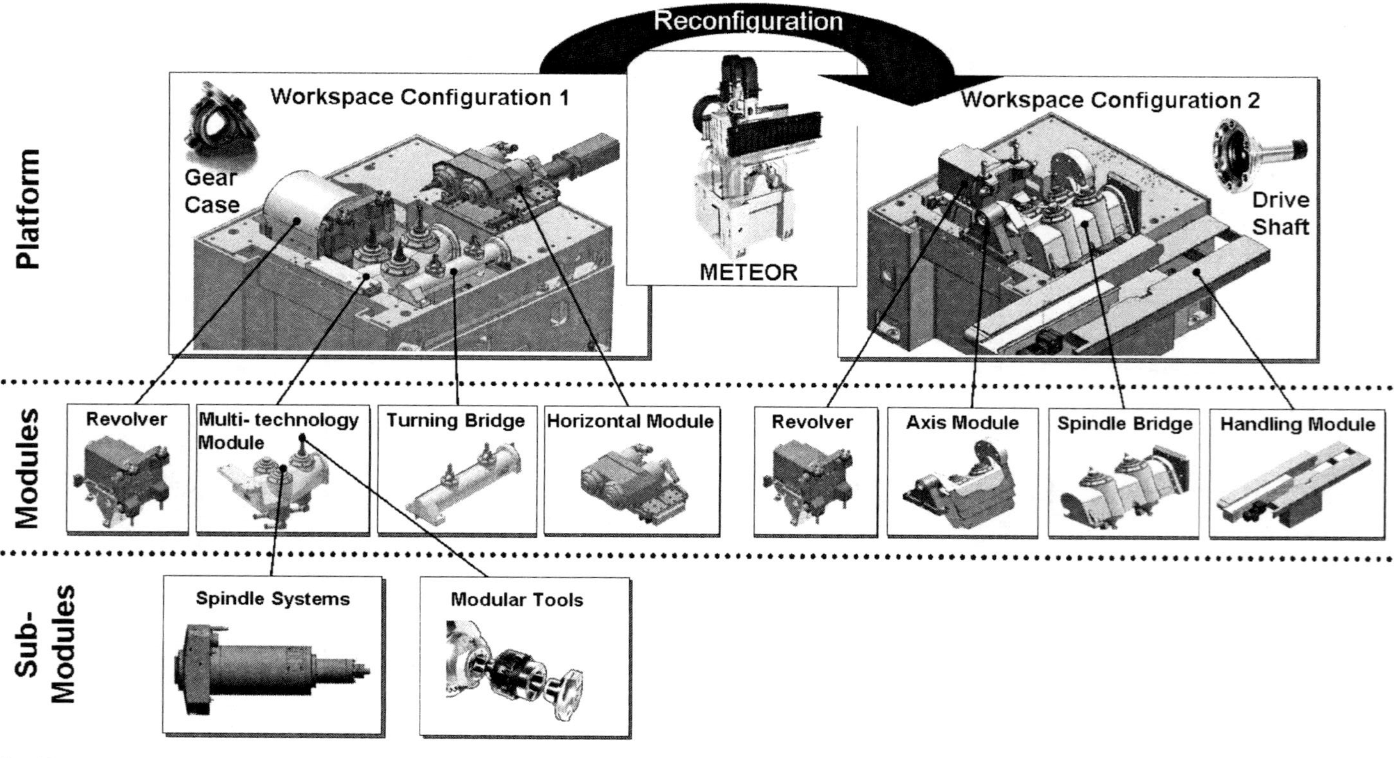

Fig. 6.7 Modularity provides adjustable RMT functionality (Abele, Liebeck, Wörn, 2006)

6.3.3 System Interfaces

System interfaces are auxiliary components that provide compatibility and enable the integration of sub-systems into the RMM. The reconfiguration scale of a module is highly dependent on the characteristic design and the specification of its interfaces. Therefore, standardized interfaces must be defined in the range of generality and divergence (modules and its sub-modules).

Interfaces can be classified into mechanical interfaces and functional interfaces for data, energy and auxiliary material transmission. Mechanical interfaces transmit forces and torques and perform locking and alignment functions (Abele, et al. 2006). Interfaces for parts like tool and tool-holding fixtures are standardized in general. The physical connections between modules must be specifically designed to facilitate the functional interactions and simplify assembly and dis-assembly operations.

The Multi-Coupling (MC) concept offers an integrated sub-system interface arrangement based on "Plug and Produce" functionality (Figs. 6.8, 6.9).

MC implement a combination of the following features: transfer of needed functions across the parting planes, locking and release mechanisms and alignment features, which include locating and positioning, safety and intelligence features. such as sensing, diagnosis and self-configuration capabilities. MC consist of plug-in and receiver devices which are linked manually or automatically. In many cases, exchangeable modules and sub-modules of the same type (e.g. spindle systems), have the same transfer functions and can be equipped with identical MC for connection and external functional supply. Further connections can be arranged through external mono-linking coupling devices.

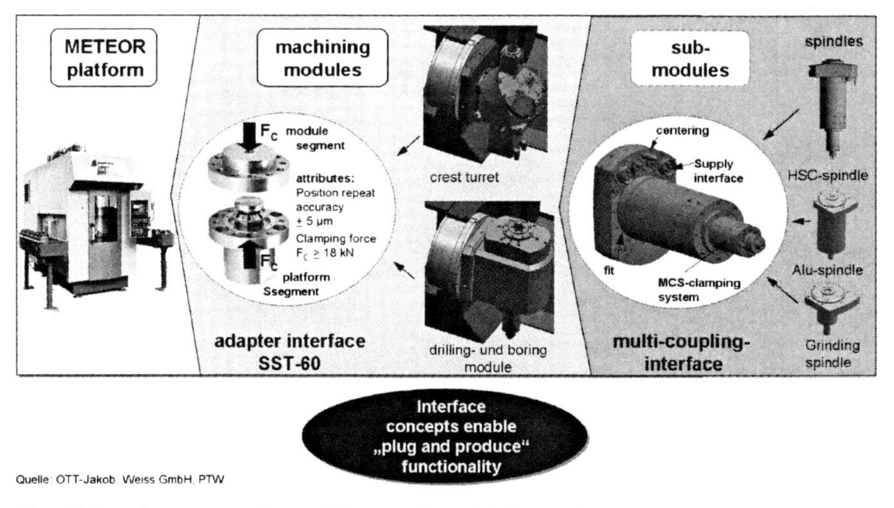

Fig. 6.8 Interface concepts for module and sub-module integration

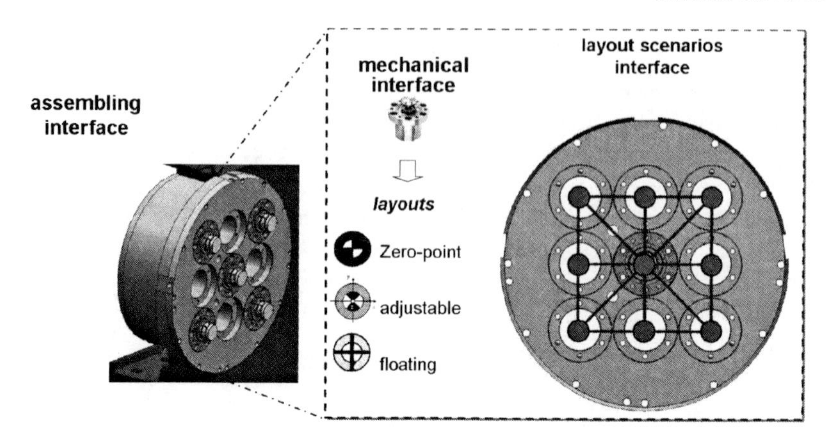

Fig. 6.9 Mechanical module interface design

MC dimensions are designed according to the maximum system requirements as long as the functions are located within a definite function tolerance zone. The functional supply (e.g. media, pressure, power) can be carried out by externally installed central supply modules. They regulate the functional flow taking account of the required function values of the module. The application of MC in the place of conventional irresolvable connections leads to extra costs. These additional investment costs must be compensated for with the savings potential as a consequence of the reduction of the maintenance and reconfiguration costs.

6.3.4 Expert Tool for System Configuration

The RMM construction kit system contains a given number of pre-defined subsystems. The combination of the sub-systems generates the RMM system configuration. Such an RMM configuration depends on the specific machining tasks. Different arrangement solutions can be accomplished for the same superior production function. For a given RMM construction kit, a configuration expert tool to configure a suitable set of modules would be required to support the planning of the RMM. The two main tasks of this tool would be the plausibility and collision check, as well as the visualization of the readily configured machine tool. As the user of this tool could not be a machine tool designer, one of the main requirements to the software is usability. As a direct consequence, the expert tool should be suitable to run on systems without high level hardware requirements. The user could be a technology planner, a service engineer or a salesman using a laptop on the shop floor for example. As a result, the following development application platforms could be suitable: The planning module could be developed in C++ to generate programs executable

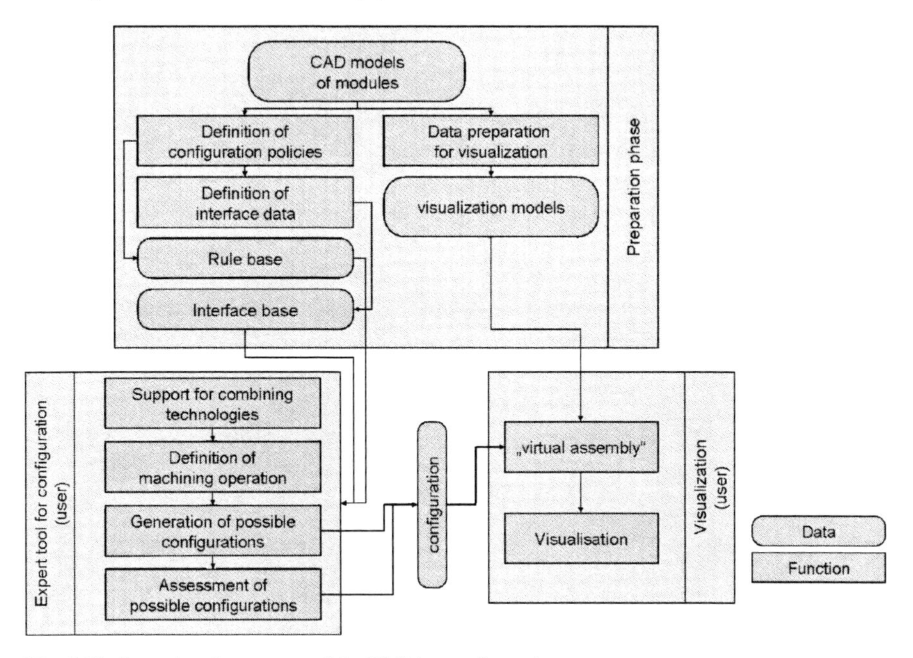

Fig. 6.10 Operational sequence of the RMM reconfiguration process

on PC based systems; the visualization could be based on Virtual Reality Modeling Language (VRML).

The virtual configuration of the RMM would be split up into three phases: (1) preparation, (2) configuration and (3) visualization phases according to Fig. 6.10. During the preparation phase, policies for possible configurations and interface information would be defined. These preparation steps would provide the rules and the interface basis. In addition, the visualization models would be prepared in this phase. Therefore, the CAD models of the systems could be reduced and exported into single VRML.

In the configuration phase, an expert tool for the configuration would be used to support the users in their selection of the available technologies and the definition of the machining operations. When using the defined machining operation list, the tool would generate the possible configurations according to the knowledge rule base developed in the preparation phase. Then, the user would be able to select the possible alternatives and obtain a 3D visualization of the configured machine tool. Consequently a "virtual assembly" would be performed using the VRML models generated in the preparation phase and visualized using a VRML viewer tool. This provides the opportunity to consider and select the desired configuration, i.e. the combination of the different modules and sub-modules of the machine tool.

6.4 Summary

This paper provided a detailed overview of a new development in the RMS field, the Reconfigurable Multi-technology Machine tools (RMM). It included details for the general flexibility required for adaptable machine tools and the essential machine tool flexibility needs during the production life cycle. The conceptual design of RMM, according to the presented construction kit approach and outlined configuration procedure via an expert system, was described. The presented configuration strategy matches the requirements of the production task and product life cycle overlapping usage. The RMM concept enables industry to cope with the challenges of future changing production demands.

Acknowledgements This RMM approach is a result of the research and development "METEOR" project, which is sponsored by the German Federal Ministry of Education and Research (BMBF) as part of the "Research for the Production of Tomorrow" frame concept and overseen by the Project Execution Organization (PFT) of the BMBF for production and manufacturing technology, the Research Center, Karlsruhe.

References

Abele E., Liebeck T., Wörn A., 2006, Measuring Flexibility in Investment Decisions for Manufacturing Systems. Annals of the CIRP 55/1:433–436

Abele E., Wörn A., 2004, Chamäleon im Werkzeugmaschinenbau. ZWF 99/4:152–156, Chamaeleon in machine tool manufacture

Abele E., Wörn A., Stroh C., Elzenheimer J., 2005, Multi Machining Technology Integration in RMS, 3rd Conference on Reconfigurable Manufacturing, University of Michigan, Ann Arbor, MI, USA

Abele E., Wörn A., Martin P., Klöpper R., 2006, Performance Evaluation Methods for Mechanical Interfaces in Reconfigurable Machine Tools, International Symposium on Flexible Automation, Osaka, Japan

Chryssolouris G., 1996, Flexibility and Its Measurements. Annals of the CIRP 45/2:581–587

Denkena B., Drabow G., 2003, Modular Factory Structures: Increasing Manufacturing System Changeability, Proceedings of the 2nd CIRP International Conference on Reconfigurable Manufacturing, Ann Arbor, MI, USA

Frick W., 2003, Dreh-Bohr-Fräszentren – Multifunktionale Maschinen zur Komplettbearbeitung komplexer Werkstücke. Verlag Moderne Industrie, Landsberg, Turning drilling milling centers – multi-functional machines to the complete processing of complex workpieces

Gu P., Hashemian M., Sosale S., 1997, An integrated design methodology for life cycle engineering. Annals of CIRP 46/1:71–74

Gu P., Slevinsky M., 2003, Mechanical Bus for Modular Product Design. Annals of the CIRP 52/1:113–116

Gunnar E., Yxkull A., Arnström A., 1996, Modularity – the Basis for Product and Factory Reengineering. Annals of the CIRP 45/1:1–6

Koren Y., Heisel U., Jovane F., Moriwaki T., Pritschow G., Ulsoy G., Van Brussel H., 1999, Reconfigurable Manufacturing Systems. Annals of the CIRP 48/2:527–540

Metternich J., Würsching B., 2000, Plattformkonzepte im Werkzeugmaschinenbau. wb Werkstatt und Betrieb 133/6:22–29, Platform concepts in the machine tool manufacture

Milberg J., 1992, Werkzeugmaschinen – Grundlagen, Zerspantechnik, Dynamik, Baugruppen und Steuerungen. Springer, Berlin, Machine tools – basics, chipping technology, dynamics, components and controls

Pahl G., Beitz W., 1996, Engineering Design – A Systematic Approach. Springer, London, pp 149–161

Pouget P.M., 2000, Ganzheitliches Konzept für rekonfigurierbare Produktionssysteme auf Basis autonomer Produktionsmodule. VDI-Fortschrittsbericht vol 537. VDI-Verlag, Düsseldorf, pp 92–152, Holistic concept for reconfigurable production systems on basis of autonomous production modules

Ropol G., 1975, Systemtechnik – Grundlagen und Anwendung. Carl Hanser, München, pp 23–71, System engineering – basics and application

Schäffer G., 1995, Systematische Integration adaptiver Produktionssysteme. (Systematic integration of adaptive production systems). Dissertation Technische Universität München, pp 42–44

Schuh G., Wemhöhner N., Kampker A., 2004, Lebenszyklusbewertung flexibler Produktionssysteme. wt Werkstatttechnik online 94/4:116–121, Life cycle evaluation of flexible production systems

Wiendahl H.-P., Heger C.L., 2003, Justifying Changeability – A Methodical Approach to Achieving Cost Effectiveness, 2nd CIRP International Conference on Reconfigurable Manufacturing, Ann Arbor, MI, USA

Chapter 7
Changeable and Reconfigurable Assembly Systems

B. Lotter[1] **and H-P. Wiendahl**[2]

Abstract Industrial assembly is subject to quick product changes, increasing numbers of variants and short planning spans of the customer. Because of the relatively high percentage of manual work the cost pressures from low wage countries is especially high. These challenges can be effectively met, however, through a comprehensive rationalization approach to the assembly, highly flexible assembly technology and qualified personnel. Depending on the product complexity, variant diversity and output rate there are a number of concepts available, which are situated between the competing demands of productivity and flexibility. This chapter describes the main features of manual, automated and hybrid assembly with a special attention to flexibility and reconfigurability.

Keywords Manual assembly, automated assembly, hybrid assembly, set-wise assembly, one-piece flow.

7.1 Introduction

The assembly of industrially produced products requires anywhere from 15 to 70% of the total manufacturing time. Efficient assembly technology is therefore, absolutely critical. Controlling the strongly fluctuating volume as well as the constantly growing number of variants is especially difficult. Due to the shorter life of the products on the market, the product and variant dependent components for an assembly system have to represent as small as possible portion of investment and to facilitate efficient operating with a large degree of variability in the size of production runs.

The industry uses variety of concepts to meet these demands. Figure 7.1 illustrates the areas of utilization for the three most important assembly systems: manual

[1] Assembly Consultant, Germany

[2] University of Hannover, Germany

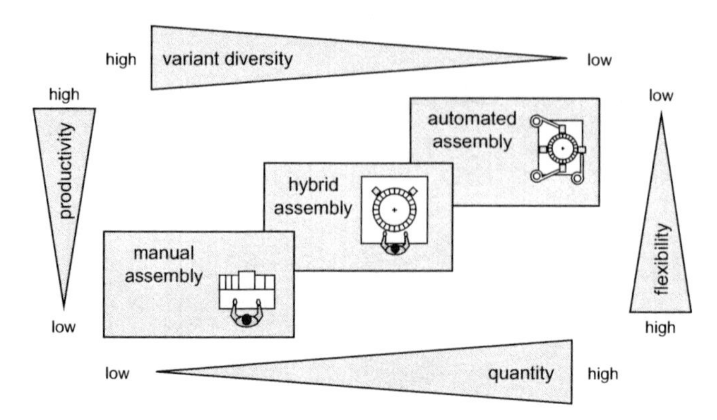

Fig. 7.1 Utilization areas for manual, hybrid and automated assembly concepts

assembly, hybrid assembly and automated assembly. It can be seen that with increasing the degree of automation the productivity increases; however the flexibility and control of the variant diversity sharply decreases.

Both the product and the market demand are critical when choosing an assembly concept. First of all, the dimensions, weight and structure as well as the variants and number of items of the product define the technical boundary conditions. Furthermore, the assembly system has to comply with a number of requirements including: anticipated sales duration, expected number of pieces per unit time as well as the market's demand that all variants be delivered in small lots and just-in-time.

It can generally be said that assembly systems were already constructed quite early with modular components. This is much easier than with manufacturing systems, because the necessary forces to connect the parts are usually much less than those required in machining and forming processes. This applies to the precision and rigidity of the machines frames and guides. In assembly systems the parts that need to be supplied and moved tend to be more diverse and more sensitive than in other manufacturing system.

In the course of developing industrial assembly systems, six basic forms have been created in order to meet the diversity of the resulting jobs. Figure 7.2 classifies these concepts into rigid (predominantly automated) and flexible (predominantly manual) assembly systems. The two groups are then classified further according to their output (pieces/hour) and the complexity of the product (expressed in number of assembled pieces).

The fundamental difference between the two groups is that with the rigid automated systems, the focus is on the technical design, whereas with the flexible manual systems, the layout of the employees' work dominates. It can be clearly seen in Fig. 7.2 that the limit of the output rate for a manual system is approximately 720 pieces per hour, which corresponds to a cycle time of five seconds. In comparison with an automated system (with the exception of so-called high speed automats)

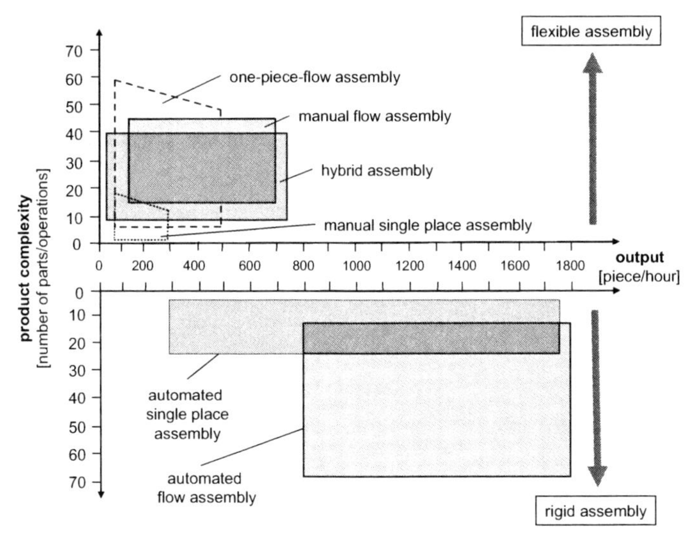

Fig. 7.2 Classification of assembly systems according to output and complexity

the limit is 1800 pieces/hour and thus a cycle time of two seconds. The next factor to be considered regarding the systems is their flexibility.

7.2 Flexible Manual Assembly Systems

The center point for manual assemblies is the operator. By using their hands, dexterity, senses and intelligence along with aids, such as tools, jigs and fixtures, gauges etc., the worker carries out the assembly operations. The output rate of the worker is dependent on a number of factors, such as the ergonomic design of the workstation and surroundings, e.g., room, light etc. Moreover, the layout of the workspace plays a significant role. The assembly work should be conducted within the operator's field of vision without them having to move their head and at a height lower than the heart. Repetitive movements, such as bending and straightening up should be avoided. Assembling within the ergonomically correct workspace enables a high level of efficiency through easy access. Furthermore, it prevents fatigue and assembly errors and is appropriate for employees who are older and/or have reduced performance.

However, it has to be taken into consideration that an ergonomically correct workstation has a relatively small workspace and is therefore only appropriate for assembling small products with minimal complexity. Complex products have to be assembled either by dividing the work content into a number of linked workstations or by completing it on a workstation with so-called set-wise assembly flows.

7.2.1 Single Station Assembly with Set-Wise Assembly Flow

Usually items are assembled manually piece by piece, that is; a product is completely assembled step-by-step before the next one is begun. In contrast, with set-wise assembly flow, the first part is assembled for the entire product run (e.g., 8, 12 or 16) one after the other, then the second part and so on up until the last part.

In order to facilitate this repetition of movement, a workstation layout with the assistance of two turntables as shown in Fig. 7.3 is recommended. Turntable 1 contains the run of the product's base part each in its own holder (in this case 18), whereas turntable 2 carries the parts that are to be mounted onto the base part in flexible bulk ware bins (here 6 parts P1 to P6). Each of the parts that are to be assembled can thus be rotated into the most ergonomically favorable position (the so called joint position) with regards to the base part.

An advantage of this arrangement is the forced repetitiveness of the movement. Further advantages include the short and for all parts similar distance to grasp the parts. The time required to access and return the tools is distributed among the number of base parts found on turntable 1, because they only have to be grasped once per product run. Furthermore, the individual parts are supplied corresponding exactly to the assembly sequence and thus the quality of the assembly also increases. Compared to one-piece assemblies, the complete assembly time can be decreased by approximately 30–50%. Nonetheless, the limits of this concept are reached when, due to the size, number or parts variants, the required table becomes too large to manage.

This concept focuses on having as efficient as possible assembly while preventing unnecessary worker movements. Flexibility of product variants can be achieved through previously setup turntables, which can be quickly interchanged. The vol-

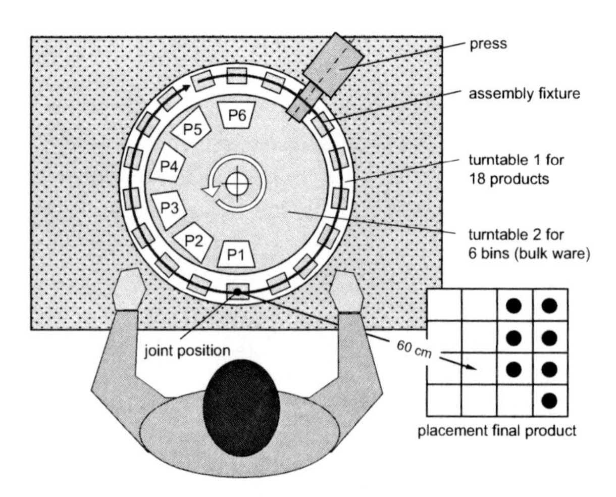

Fig. 7.3 Assembly station with set-wise sequencing for an electrical componentry

ume flexibility is also easily controlled by extending or shortening the daily operating schedule. Finally it is noteworthy here that a large proportion of the assembly workplaces components can be reused. In order to setup for another product, only new assembly fixtures need to be interchanged.

7.2.2 Single Station Assembly According to the One-Piece-Flow Principle

In order to assemble products with a large number of parts and product variants in small lot sizes, it is necessary to have numerous different components and tools readily available. If the assembly takes place at one workstation this leads to space problems. An efficient solution for this type of assembly job is single station assembly, according to the One-Piece-Flow Principle. In this case, the operator moves together with a part carrier along a series of supply bins and completes the base part piece-by-piece to the final product. Thus, as the name suggests, one product at a time flows through the individual assembly stations.

As an example of this type of workstation, Fig. 7.4 illustrates a design using a ball guide rail with a semi-circle layout. The actual assembly platform is a so-called assembly sled, which – when set on the ball guide rail – can easily be manually moved back and forth around the entire semi-circle. On the assembly sled sits a turntable, which holds the part carrier and allows the operator optimal access to both the carrier and the product. Components and tools are provided from the outer side of the system either by re-filling or changing the bulk ware bins. The layout of the bins corresponds to the assembly sequence. Depending on the size of the components and/or containers, up to 50 bins can be placed on the outer side of the workstation; all of them located below the ergonomically significant heart level of the operator.

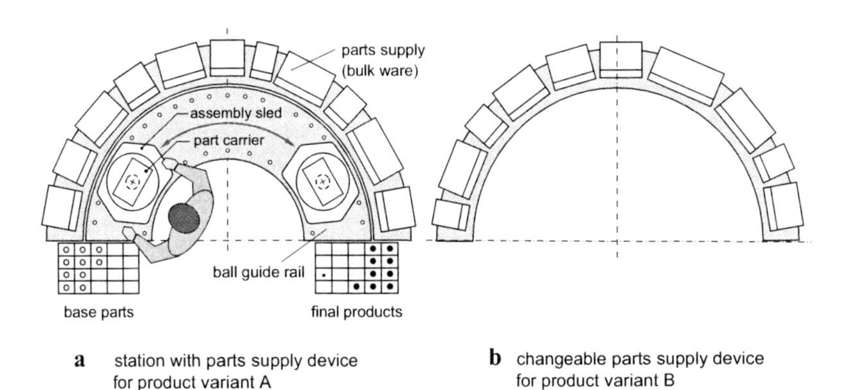

Fig. 7.4a,b One-piece-flow assembly station with change supply device for parts sets

All that is needed in order to change work positions is a minimal body movement whereby the work area remains within the field of vision of the worker.

The changeability of this workstation concept with regards to the product variants is achieved by removing the entire part supply device from the ball guide rail and replacing it with another one, which has already been setup. The required changeover time from one product to a new one is thus extremely short. However, the volume flexibility is not any greater than with set-wise assembly.

7.2.3 Multi-Station Assembly According to the One-Piece-Flow Principle

If the product variants continue to increase and product quantities rise up to approximately 100 000 pieces/year and delivery lots of 1 to 100 pieces are required, then a solution such as those displayed in Fig. 7.4 is no longer sufficient. The variant specific parts especially have to then be supplied exactly in accordance with the demanded lot sizes. In such cases, the obvious solution is to separate the provision of variant dependent items from variant independent items as illustrated in the following example.

The product to be assembled is a control circuit device. With a base area of 80 mm × 160 mm and a weight of approximately 2.5 kg, it consists of 48 different parts. Eight of the components or pre-assembled units determine the variant designs and 40 of them are variant neutral parts. The lot sizes fluctuate between 10 and 50 pieces.

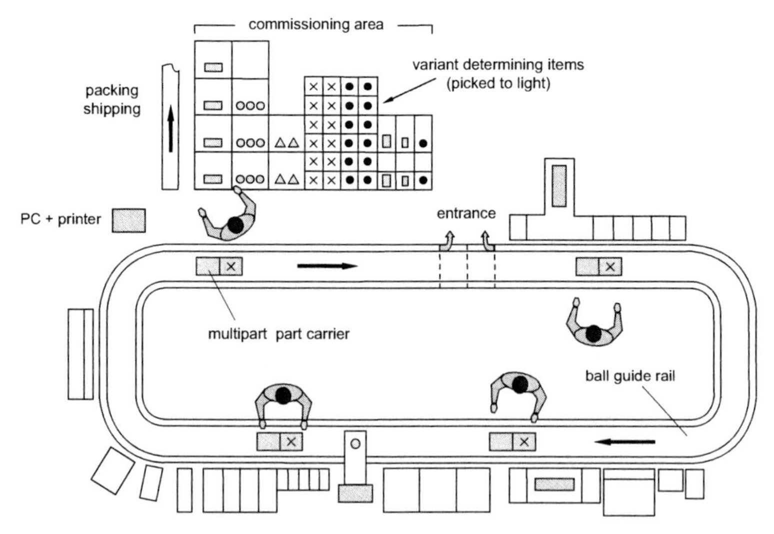

Fig. 7.5 Assembly system for highly variable products

The assembly systems designed for this (Fig. 7.5) consists of two parts: the commissioning area and the actual assembly line. The variant dependent components are stored in the commissioning area, which is organized like a supermarket, while the variant neutral parts are provided along the assembly line according to use.

The work flow is completed in the following steps: The commissioning area is equipped with a PC and a printer. The worker removes the completely assembled and tested product from the returning carrier and sets it onto the packing conveyor belt. He then calls up the next order on the PC, printing in turn an identification label, which is then placed on the part carrier at a previously determined place. In calling up the order, the so-called "pick to light" system is also activated. It consists of light displays on each box indicating to the worker, which parts need to be picked and placed on the empty part carrier for the next variant.

In order to obtain the necessary space for the parts, the carrier is constructed as a so-called tandem or compound part carrier as depicted in Fig. 7.6. The carrier, which the base part or assembly fixture is placed on, is shown in Fig. 7.6a while the expansion of the tandem carrier is illustrated in Fig. 7.6b. The coupled carrier plates hold the variant dependent items in a kind of inlet and in this case there are enough partitions for the product. If the product is more complex and has more variant specific parts that need to be mounted, the part carrier can be further supplemented with a second carrier plate as shown in Fig. 7.6c.

The worker now places the base part in the holder of the part carrier and the additional parts in the pockets on the tandem plate. As a part is removed from the corresponding box in the commissioning area, the light display goes out. When all of

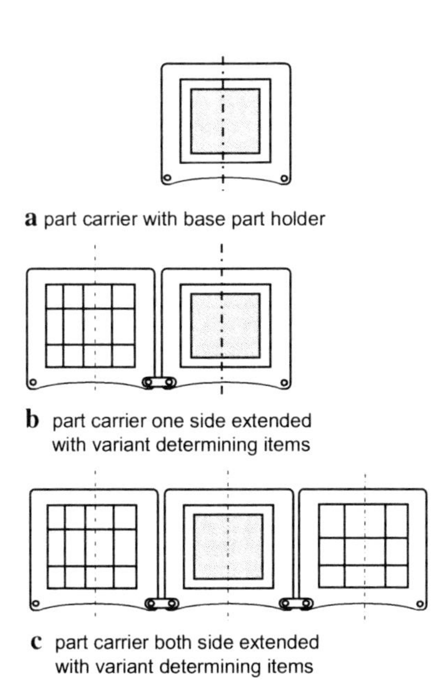

Fig. 7.6a–c Part carrier system for versatile products assembly

the lights are off, the part carrier, equipped with the variant specific items is released into the assembly line.

Subsequently, the worker leads the tandem carrier manually on a ball roll guide to the variant neutral parts and the corresponding assembly fixtures, adding them to the product piece by piece.

This concept has an equally high flexibility both with regards to the number of pieces as well as the number of variants. The implemented components are also extremely re-useable. Furthermore, it is easily reconfigured, because the individual elements are modular and mobile.

In this concept changeability regarding the variants is achieved by changing just the inlets of the part carriers. The production rate can be changed by attaching the work content to one, two or three workers in the inner circle of the system.

7.3 Flexible Automated Systems

In Fig. 7.2, it can clearly be seen that an output rate above 720 pieces/hour requires an automated systems. Linear transfer assembly lines are well suited for assembling products with a surface area up to approximately 300 × 400 mm and weighing up to approximately 20 kg.

Almost all linear transfer assembly lines use standardized basic modules. These serve as a platform for the so-called process modules, which conduct operations such as screwing, welding or testing. Linked to the basic modules, these form the automated stations. The process modules are in turn comprised of product neutral basic platforms and customer or procedure specific ones. The process modules are inserted into the automated stations manually using a loading platform, whereas data and energy is transferred via plug-in connections. Due to the mobility of the process modules modifying the system can be completed in much less than an hour, sometimes requiring as little as a few minutes. In contrast, converting a rigid system can take anywhere from a number of days up to a week. Further fundamental components include the manual modules, which can also be integrated into the system.

Figure 7.7 illustrates an example of a system consisting of basic, process and transfer modules; manual and automated stations are thus combinable.

Due to the comparatively high capital costs, customers stipulate that the systems be able to grow or shrink with the varying demand during the life-cycle. Thus, the ramp-up phase of a new product can start off with a relatively small system, consisting of one to two manual workstations and an automated cell. This keeps the necessary capital expenditures for resources within manageable limits. If the production numbers develop positively during the following period, the assembly system can be adapted to the growing demands in a number of extension stages as depicted in Fig. 7.8.

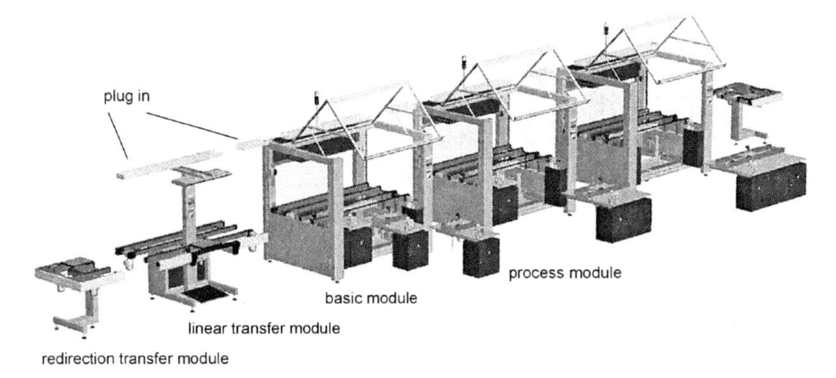

Fig. 7.7 Modular system for linear transfer assembly lines (Courtesy Teamtechnik)

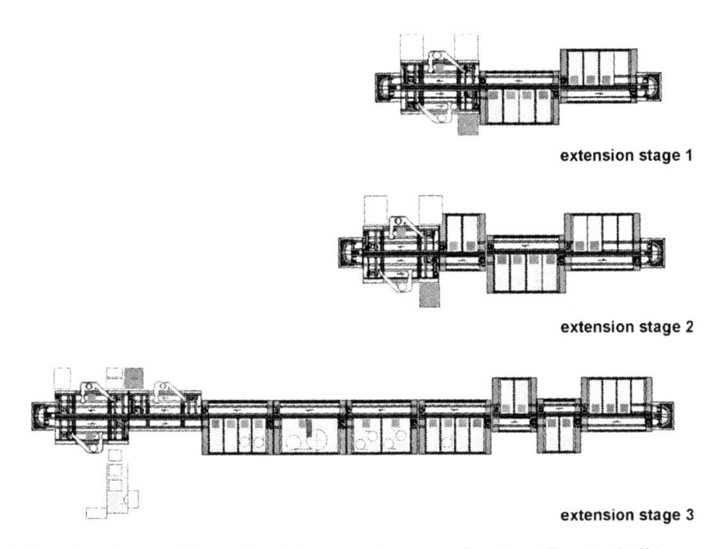

Fig. 7.8 Extension stages of linear transfer assembly lines (Courtesy Teamtechnik)

On the other hand, the size of the system can also be reduced in the same way when the number of products decreases during the end phase of its life-cycle. Due to the standardized construction of the modules, the system components that then become free are suitable for assembling the product's next generation.

Changeability with respect to the production rate is ensured in this concept by changing from manual to automatic stations and/or adding or removing stations. The disadvantage lies in the relatively high costs for the process modules if they are not needed any more. Also if the lifetime of the product is much shorter than the technical life of the modules the reuse of the modules becomes problematic.

7.4 Hybrid Assembly Systems

7.4.1 Characteristics

As previously explained, with an increasing number of pieces, purely manual workstations often do not have sufficient capacities. Before starting to consider a fully automated solution, it makes sense to consider the concept of a mixed manual-automated assembly. These are referred to as ‘hybrid assemblies’. Such hybrid assembly systems are facilities for assembling units and/or products, in which automated workstations are combined with manual workstations, With regards to the number of pieces, variant diversity, productivity and flexibility they are positioned between manual and automated assembly systems, see Fig. 7.1.

The starting point for planning a hybrid assembly system is a pure manual assembly. From there, the most favorable ratio between the automated and manual tasks will be determined by adjusting the degree of automation for individual assembly operations to the respective assembly jobs. By changing the number of assembly workers on the manual workstations, there is a high level of flexibility regarding the number of pieces. Sudden changes in the demand can thus be easily accommodated. For example an assembly system with four manual pick and place stations can also be manned with only two operators instead of four, thus leading to a 50% reduction in the production rate. When, for example, during the holidays there are only a few orders, the system could even be operated with only one person. The resulting product output would then be approximately 25% of the system’s maximum possible output rate.

A further advantage of hybrid systems is that the initial degree of automation can be adapted to changes in the product rate during the entire service life using a number of extension stages. Only when the entire potential of one stage is exhausted, is the next level of extension implemented based on the actual sales numbers.

Finally if the demand of products continues to increase, it is possible to supplement the system with further hybrid cells. The danger of a bad investment, especially during the ramp-up stage is thus decreased.

With a hybrid system, it is important to always try and construct it using product neutral components, hence, increasing the proportion of system modules that can be reused after the end of the products’ life-cycle. Because the manual process and not the automatic processes sets the pace for the stations cycle time the stochastic variability of the manual stations cycle time creates no problem.

Next, based on concrete examples, hybrid assembly systems will be further discussed.

7.4.2 Example of a Hybrid Assembly System

A clamping tool is to be assembled with a production run of approximately 2.5 million/year. It consists of 12 different parts, which can be combined into five different variants. The following planning tasks are identified:

- The system's operating time is 230 days/year with two shifts, each seven hours long.
- The assembly costs per unit can be at most 0.30 €.
- The time for return on investment must not exceed 5 years.
- Both an automated assembly system and a hybrid one are to be planned and compared with one another.

For the automated assembly, an 80% degree of utilization, a capital investment of 1.6 million € and a unit assembly cost of 0.29 € with 2.5 million pieces/year was determined. The plan for the hybrid assembly technology included implementing two systems. The total capital investment was calculated at 445 000 €, the degree of utilization at 90%, and the unit at 0.245 €.

Figure 7.9 illustrates the solution that was found for a hybrid system and the three possible extension stages including the output rate data. Figure 7.10 explains the work flow on one of the workstations in detail.

The work content of the first extension stage is split between workstation 'A' and workstation 'B'. Assembly occurs according to the set-wise flow principle, on a sliding tandem carrier with 12 assembly fixtures, which is synchronized in the station. The base part is supplied in a palette magazine.

On workstation 'A', 8 to 10 different parts are manually inserted depending on the product variant. After-wards, the tandem carrier is manually transferred to work-station 'B' by the moving operator. Here, the worker manually inserts the rest of

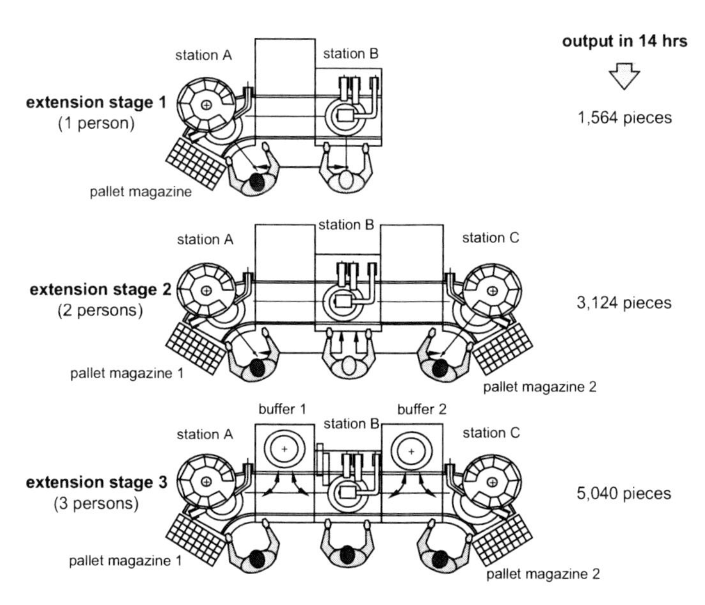

Fig. 7.9 Extension stages of a hybrid assembly system (example clamping tool)

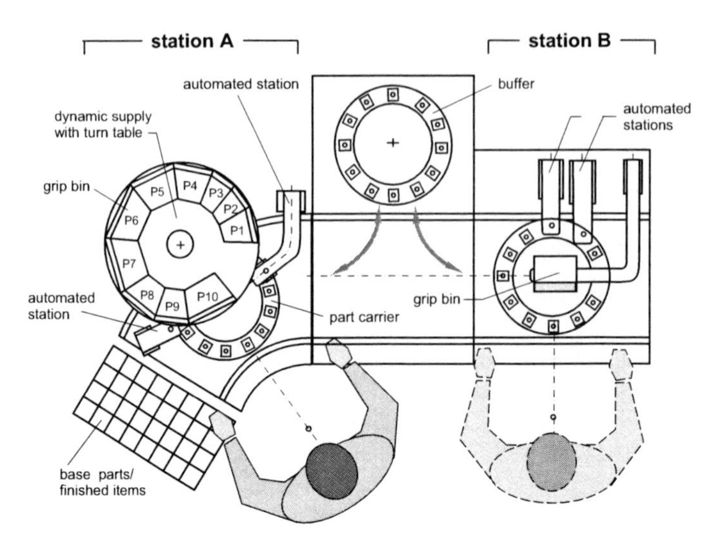

Fig. 7.10 Hybrid assembly system – extension stage 1

the parts and at the same time runs the automated processes. After completion, the worker transfers the carrier back to workstation 'A' and unloads the 12 finished products. The assembly cycle then re-starts.

The output rate of the first extension stage in a two shifts mode in a total of 14 hours per workday is 1 564 pieces. Approximately 70% of the total work content is conducted on workstation 'A' and approximately 30% on workstation 'B'.

In the second extension stage (see Fig. 7.10) the system is expanded through an additional workstation 'C' (a duplicate of workstation 'A'). As a result two employees who alternatively use workstation 'B' are required. With two shifts, the output rate increases to 3 124 pieces per day.

In the third extension stage, certain operations from workstation 'A' and 'C' along with an additional worker are assigned to workstation 'B'. The system's output rate, with two shifts increases to 5 040 pieces per day. In order to decouple workstations 'A' and 'B' and/or 'B' and 'C' buffers are planned. These enable the exchange of part carriers between the workstations. With increasing capacity demands, a second hybrid assembly system can also be likewise implemented in the different extension stages.

The capital expenditures are also made in stages in parallel with the implementation of extension stages. Thus, for a system in the first extension stage € 127 000 is required, for the second extension stage € 54 000 and for the third extension stage € 41 500. For a second system, the capital expenditures can be expected to behave similarly.

In Fig. 7.11, the trend of the assembly costs as a function of production output, for both the fully automated system and the system with three hybrid cells are compared.

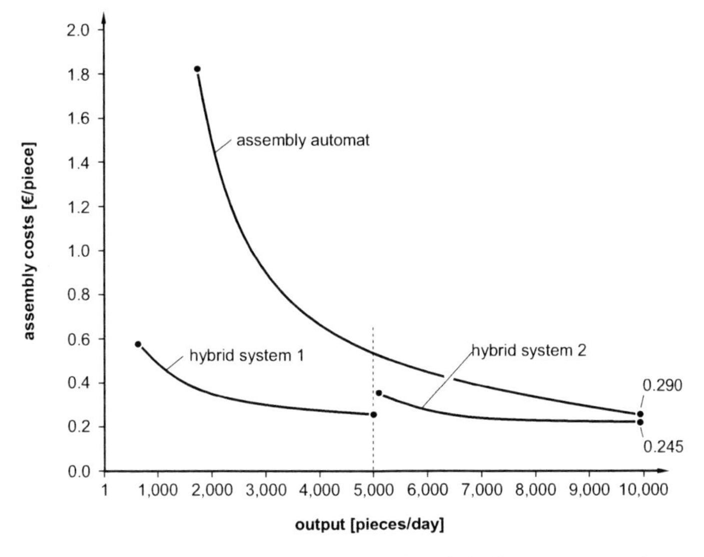

Fig. 7.11 Comparison of assembly costs for automated vs. hybrid system (example)

The basic advantages of the hybrid solution are that: the calculated assembly costs of € 0.245 is consistently lower than the automated system, the capital expenditures are clearly lower, and in the (frequently observed) case where the production rate is not large enough the capital risk is noticeably lower.

Numerous, similar analyzes of other assembly items have shown that this example is no exception. Data for seven examples were used in order to compare the results of automated and hybrid assembly. The planning data are indicated in Fig. 7.12. Using standardized values, both the personnel costs as well as the fixed and variable costs were determined through work center costing. Only the daily operating periods vary.

example	item name	parts [–]	planned output [pieces / year]	working schedule [hrs / day]
1	electrical switch	12	2,500,000	15
2	electric motor	7	2,200,000	15
3	drill chuck (final assembly)	8–10	2,850,000	14
4	water boiler	12	500,000	14
5	water faucet	17	1,500,000	22.5
6	clamping tool	12	2,300,000	14
7	planetary gear	25	950,000	15

Fig. 7.12 Product samples automated vs. hybrid assembly

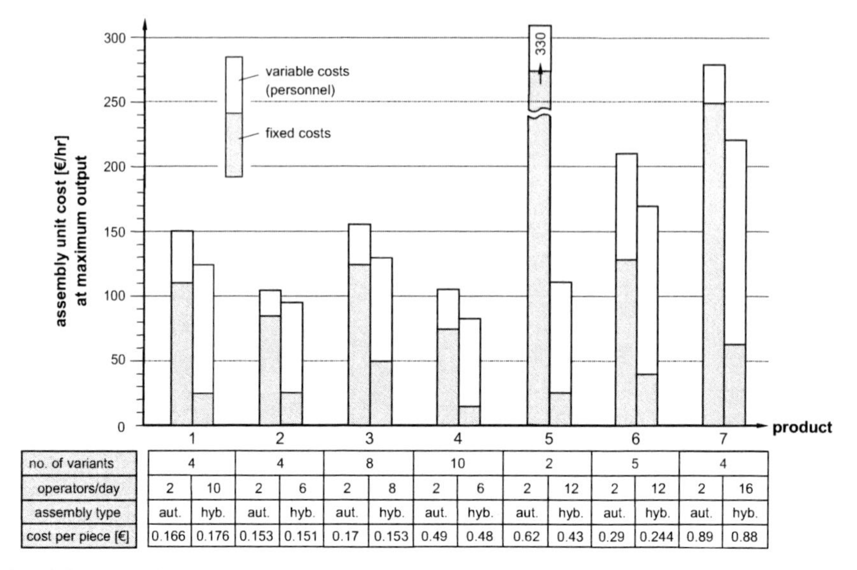

product	1		2		3		4		5		6		7	
no. of variants	4		4		8		10		2		5		4	
operators/day	2	10	2	6	2	8	2	6	2	12	2	12	2	16
assembly type	aut.	hyb.	aut.	hyb.	aut.	hyb.	aut.	hyb.	aut.	hyb.	aut.	hyb.	aut.	hyb.
cost per piece [€]	0.166	0.176	0.153	0.151	0.17	0.153	0.49	0.48	0.62	0.43	0.29	0.244	0.89	0.88

Fig. 7.13 Comparison of assembly costs for seven different products

Figure 7.13 shows the calculated assembly costs of the seven examples, split into fixed and variable costs. The costs of the automated system (in the left column) are compared to the costs of the hybrid solutions (in the right column). The assembly unit costs and the number of employees in the final capacity stages are also shown.

7.4.3 Analysis of the Results for Automated and Hybrid Assemblies

The basic results can be summarized in the following statements:

- For the automated assembly systems, the assembly unit costs are predominantly determined by the fixed costs.
- For the hybrid systems, the assembly unit costs are mostly determined by the variable costs.
- Hybrid systems distinguish themselves through their employment friendliness despite a high level of rationalization.
- The assembly unit costs of the hybrid assembly technology are competitive in comparison to the automated solution for the entire range of production runs.
- The hybrid systems achieve the target assembly unit costs even with a relatively small production rate.
- Implementing hybrid systems requires a comparatively low capital expenditure. Combined with the possibility of extending the system in stages, the risk of a bad investment is decreased.

- The low degree of automation in hybrid systems increases the possibility of re-using modules in a new system, following the end of a products life-cycle.
- Changeability with respect to output rate of automated as well as of hybrid assembly can be achieved by either gradually increasing the degree of automation and/or adding or removing of modules. The difference between these options lies in the amount of capital investment.

Naturally, hybrid assembly technology also has its limitations. The following aspects should be considered though when making a decision:

- Certainty of the sales prognosis in relation to the product life, number of variants, lot sizes, yearly production runs, ramp-up and its unique selling proposition.
- An automated assembly is more efficient when a high production output rate with a cycle time of less than three seconds is required and sales are certain.
- When the decision is not clear, it is recommended that both possible solutions be evaluated and compared.

7.5 Conclusion

Assembly systems are easier to design for high changeability compared to manufacturing systems for mechanical parts. In general they are of modular layout not only regarding the processes but also the transfer and buffer functions as well as the control system. The product variants can be handled by insertion of removable part fixtures into trays and work piece carriers. The volume variation is usually taken care of by the possibility to replace manual operations by automatic modules and/or duplication of stations. The general trend is to make as many components of the system as possible independent of the specific shape of the parts to be handled. Future business models aim for using modules leasing concepts in order to reduce the capital tied in fixed assets.

References

Al-Kashroum O., 1998 Einfluss des Arbeitsablaufs bei der Montage (The Impact of Work Flow in Assemblies). Dissertation TU Chemnitz

Kinkel S., (ed)2004, Erfolgsfaktor Standortplanung (Success Factors for Location Planning). Springer Verlag, Berlin/Heidelberg

Lotter B. et al., 2006, Primär-Sekundär-Analyse (Primary Secondary Analysis). Expert-Verlag Remmingen 2002 Lotter, B; Wiendahl H.-P. (eds.) Montage in der industriellen Produktion. Ein Handbuch für die Praxis (Assembly in Industrial Production – A Practical Handbook). Springer Verlag Berlin/Heidelberg

Lotter E., 2006, Hybride Montagesysteme (Hybrid Assembly Systems). In: Lotter B., Wiendahl H.-P. (eds) Montage in der industriellen Produktion – Ein Handbuch für die Praxis. Springer Verlag, Berlin/Heidelberg, Assembly in Industrial Production – A Practical Handbook

Rosskopf M., Reinisch H., 2004, Prozessmodulare Gestaltung von Produktionssystemen (Design of Modular Process Production Systems). In: Wiendahl H.-P., Gerst D., Keunecke L. (eds) Variantenbeherrschung in der Montage – Konzept und Praxis der flexiblen Produktionsendstufe (Mastering Variants in Assembly – Concept and Practices of the Flexible Production Final Segment). Springer Verlag, Berlin/Heidelberg, pp 231–260

Part III
Logical Enablers

Reconfigurable Control Systems for Robots

Chapter 8
Unified Dynamic and Control Models for Reconfigurable Robots

A.M. Djuric and W.H. ElMaraghy[1]

Abstract A highly reconfigurable control system that intelligently unifies reconfiguration and manages the interaction of individual robotic control systems within a Reconfigurable Manufacturing System (RMS), is presented. A Reconfigurable Plant Model (RPM) representing different robotic systems was developed to perform any reconfigurable control process. The RPM has seven reconfigurable modules: Reconfigurable Puma-Fanuc (RPF) model, Unified Kinematic Modeler and Solver (UKMS), Reconfigurable Puma-Fanuc Jacobian Matrix (RPFJM), Reconfigurable Puma-Fanuc Singularity Matrix (RPFSM), Reconfigurable Robot Workspace (RRW), Reconfigurable Puma-Fanuc Dynamic Model (RPFDM), Reconfigurable Puma-Fanuc Dynamic Model Plus actuators (RPFDM+). The Reconfigurable Control Platform (RCP) was developed for the Reconfigurable Plant Model using MATLAB/Simulink® software. The PUMA 560 robot was selected for the case study. Using information of the kinematic and dynamic parameters for PUMA 560 robot and its DC motors parameters, the reconfigurable "PI" controller was designed in a function of the motor parameter. The system response exhibits a very good performance. The reverse modeling of the reconfigurable modules can be used for developing a new Reconfigurable Robot Meta Model.

Keywords Industrial Robots, Dynamics, Control, Reconfigurable Modules

8.1 Design of Reconfigurable Modules for the Reconfigurable Robotics, Automation and Intelligent Systems Industry

The globalization of industry means that customers and production will commonly exist worldwide. To support this new trend, there is a need to develop new machines and software packages, which are quickly and easily changeable and adapt-

[1] IMS Center, University of Windsor, Canada

able to new customer needs. This goal can be achieved by applying a reconfigurable approach in designing new machines and software. To be able to use machines in a future reconfigurable industry, there is a need to treat them as reconfigurable automated machines. To complement modeling and creation of new reconfigurable systems, modular robots and machines, we can define the existing systems, robots and machines as reconfigurable, according to their reconfigurable aspects. These techniques can be extended to incorporate future reconfigurable elements.

For creating the reconfigurable modules, the unification of different automated machines was achieved by using the "power of comparison" between different automated systems. The reconfigurable kinematic and dynamic modules represent the reconfigurable plant model (RPM) prepared for the reconfigurable controller design.

This process can be reversed. Using the reconfigurable kinematic, dynamic and control modules the new reconfigurable machines can be produced. The first part of this methodology uses current automated machines for future reconfigurable industry by representing them as reconfigurable modules. The second part of this methodology is using developed modules for designing new reconfigurable automated machines, such that they perfectly match in a new environment. This methodology represents a bridge between current industrial practices and future reconfigurable trends.

This approach was applied to 6R (six rotational joints) industrial robots, and the reconfigurable kinematic and dynamic models were developed by Djuric and ElMaraghy (2006, 2007).

8.1.1 Description of a Robot Model

The most important information for modeling robots is their kinematics and dynamics properties. The properties of kinematics are: number of joints, type of joints, positive direction for each joint, base frame, tool frame, links' lengths and links' offsets. The dynamics properties are: links' masses, radial distance to the center of each link and moment of inertia about a center of mass of each link (Djuric and ElMaraghy, 2007).

8.1.2 Reconfigurable Aspects of Industrial Robotic Systems

In designing reconfigurable modules for the existing robots, it is crucial to know what is reconfigurable and what the robot's reconfigurable aspects are. A classification is presented in Fig. 8.1 based on the analysis of many industrial robots. It has four reconfiguration levels and each level has its own groups of robots:

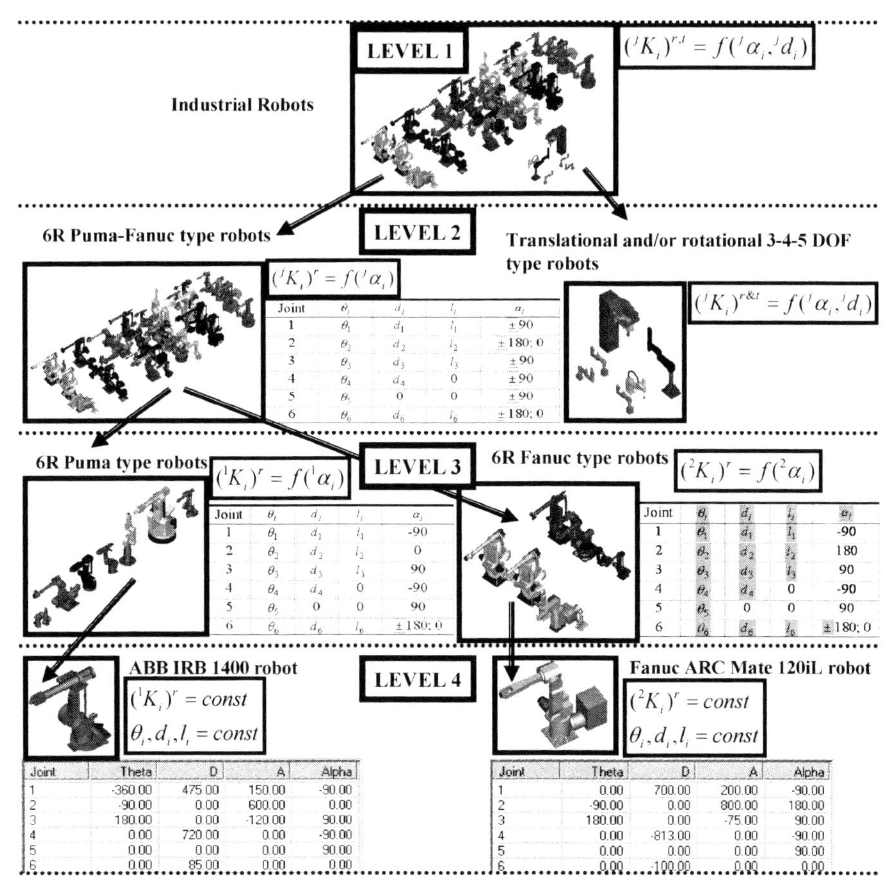

Fig. 8.1 Classification of industrial robots

- Level 1: All Industrial robots and they belong to one group;
- Level 2: 6-R robots that belong to one group and robots with combination of rotational and translational joints that are in a second group;
- Level 3: First group of robot are 6R Puma type robots and second group is 6R Fanuc type robots;
- Level 4: Many single robots from the two groups of robots. These single robots represent their own groups.

8.1.3 Reconfigurable Kinematic and Dynamic Modules

By analyzing the similarities between different robotic systems, a Reconfigurable Puma-Fanuc (RPF) model was developed (Djuric and ElMaraghy, 2006). The RPF model has six rotational joints. Each joint can have either left or right positive directions. All combinations of their orientations were exhausted in order to produce the relationship between each coordinate system, as shown in Fig. 8.2. Their D-H parameters are expressed in Table 8.1.

The main purpose of having the RPF kinematic model is to have both directions for each joint (Fig. 8.2) available to solve a generic inverse kinematic problem with the aid of the D-H roles in Table 8.1. This was achieved by defining reconfigurable parameters, which connect all possible joint directions. Each joint direction depends on the twist angle α_i, which has different values for each joint. By calculating sine and cosine of the twist angle the reconfigurable parameters are shown in Eq. 8.1:

$$\begin{aligned} &K_1 = \sin\alpha_1 \quad K_2 = \cos\alpha_2 \quad K_3 = \sin\alpha_3 \\ &K_4 = \sin\alpha_4 \quad K_5 = \cos\alpha_5 \quad K_6 = \sin\alpha_6 \,. \end{aligned} \tag{8.1}$$

The direct and inverse kinematic for the RPF model was solved and the results are a function of reconfigurable parameters from Eq. 8.1. The process of unified kine-

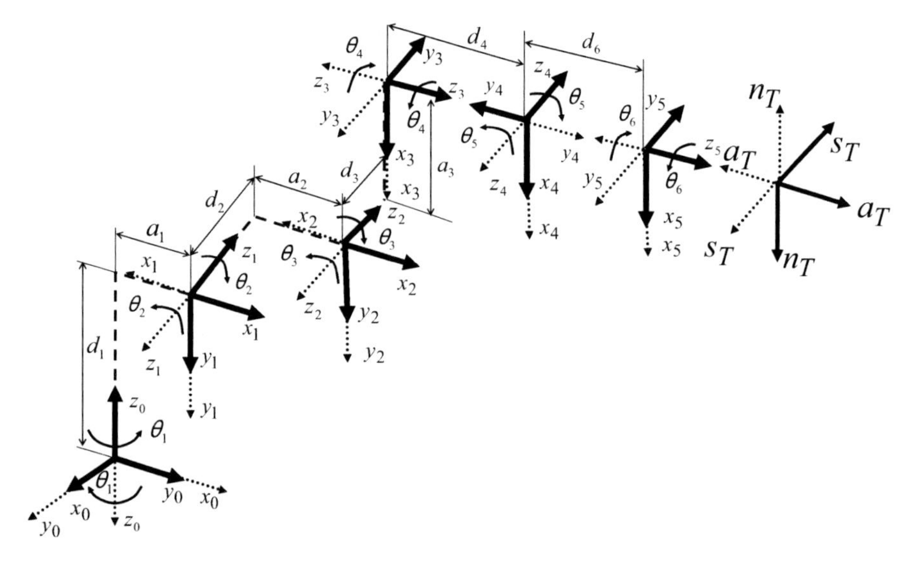

Fig. 8.2 Reconfigurable puma-fanuc kinematic model

Table 8.1 D-H parameters of RPF model

Joint	θ_i	d_i	a_i	α_i
1	θ_1	d_1	a_1	$\pm 90°$
2	θ_2	d_2	a_2	$\pm 180°; 0$
3	θ_3	d_3	a_3	± 90
4	θ_4	d_4	0	± 90
5	θ_5	0	0	± 90
6	θ_6	d_6	a_6	$\pm 180°; 0$

matic modeling and solving was automated using the developed software, "Unified Kinematic Modeler and Solver (UKMS)", (Djuric and ElMaraghy, 2006).

Four more reconfigurable modules were developed for the RPF model. For the singularity analysis, the Reconfigurable Puma-Fanuc Jacobian Matrix (RPFJM) was developed in a function of the same reconfigurable parameters and the general solution is presented in Eq. 8.2; (Djuric and ElMaraghy, 2007).

$$^{6}(^{0}J) = \begin{bmatrix} J_{11} & J_{12} & J_{13} & J_{14} & J_{15} & 0 \\ J_{21} & J_{22} & J_{23} & J_{24} & J_{25} & J_{26} \\ J_{31} & J_{32} & J_{33} & J_{34} & J_{35} & 0 \\ J_{41} & J_{42} & J_{43} & J_{44} & J_{45} & 0 \\ J_{51} & J_{52} & J_{53} & J_{54} & J_{55} & 0 \\ J_{61} & J_{62} & J_{63} & J_{64} & 0 & K_{6} \end{bmatrix} \quad (8.2)$$

For the simple singularity analysis, the Reconfigurable Puma-Fanuc Singularity Matrix (RPFSM) was calculated using the previously developed Jacobian matrix, (Djuric and ElMaraghy, 2007).

The next important reconfigurable module was a solution from the Workspace of RPF model, using the first three joints for calculation. Varying their joint limits from minimum to maximum, the complete 3-D reachable workspace was described.

The 3-D envelope is called the "1-2-3 envelope". Excluding Joint1, the 2-D envelope was calculated and named "2-3 envelope". A unique control algorithm was developed after analyzing the motion of Joints 2 and 3. Applying the FBP (Filtering Boundary Points) Method (Djuric, 2007), the Reconfigurable Robot Workspace (RRW) was calculated. Using the UKMS software the "2-3 envelope" and "1-2-3 envelope" for the robot ABB IRB 1400 were created and are shown in Fig. 8.3.

The RPFDM represents the dynamics of the robot links, Eq. 8.3, and the RPFDM+ represents coupled dynamics of robot links and actuators dynamics, Eq. 8.4, (Djuric, 2007). For computing the dynamics of the RPF model the recursive Newton-Euler algorithm and Automatic Separation Method (ASM) (Djuric, 2007),

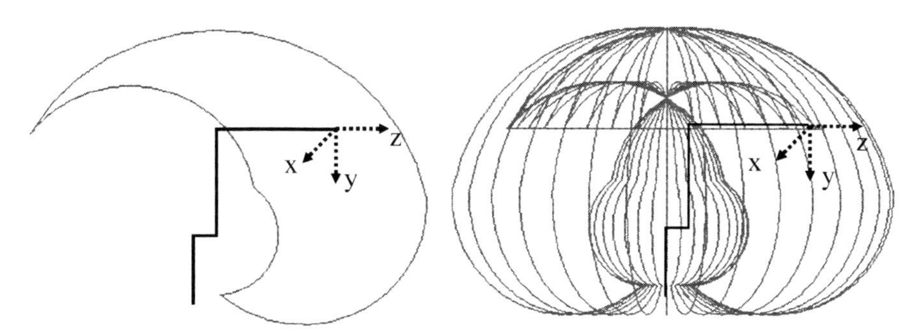

Fig. 8.3 "2-3 envelope" and "1-2-3 envelope" for the robot ABB IRB 1400

were developed. The ASM procedure, an innovation in this research, was used for automatic simplification and organization of the dynamic equations.

$$A(q)\ddot{q} + B(q)[\dot{q}\dot{q}] + C(q)[\dot{q}^2] + G(q) = \tau \tag{8.3}$$

$$\hat{C}V = \hat{A}\ddot{q} + \hat{B}\dot{q} + A(q)\ddot{q} + B(q)[\dot{q}\dot{q}] + C(q)[\dot{q}^2] + G(q)\,. \tag{8.4}$$

where the following matrices are: $\hat{A} = \mathrm{diag}\left(\frac{J_{mi}}{n_i^2}\right)$, $\hat{B} = \mathrm{diag}\left(\frac{B_{mi}}{n_i^2} + \frac{K_{ti}K_{bi}}{n_i^2 R_{mi}}\right)$ and $\hat{C} = \mathrm{diag}\left(\frac{K_{ti}}{n_i^2 R_{mi}}\right)$.

All seven reconfigurable modules represent the Reconfigurable Plant Model (RPM).

8.2 Design of Reconfigurable Control Platform (RCP)

The RPF model was prepared for the reconfigurable controller design. The controller must be designed such that it depends on the model parameters. The Reconfigurable Control Platform (RCP) was developed using MATLAB/Simulink® software. As a case study with the PUMA 560 robot was selected. Using the kinematic and dynamic parameters for the PUMA 560 robot and its DC motors parameters, the reconfigurable PI controller was designed as a function of the motor parameter. From the literature review, (Corke, 1998), (Corke and Armstrong, 1994), (Corke, 1994), (Corke and Armstrong, 1995), (Armstrong, et al., 1986), and (Leahy, et al., 1986) the information for the PUMA 560 is presented by Djuric (2007), and used in the following calculation.

The control design for the electro-mechanical dynamic model was done in two steps: First, the PI (proportional integral) controller was designed for each DC motor. Second, these controllers were tuned and used for the complete motors and links dynamics for each robot joint.

8.2.1 DC Motor Reconfigurable Position Control Design

To design a position controller for each robot actuator, we need to analyze all six DC motors according to the given parameters. Combining the electrical and mechanical equations of each motor and applying the Laplace transformations, the transfer function for the DC motor was developed and shown in Eq. 8.5:

$$\frac{\theta(s)}{V(s)} = \frac{K_{\mathrm{t}}}{(J_{\mathrm{m}}s^2 + B_{\mathrm{m}}s)(L_{\mathrm{m}}s + R_{\mathrm{m}}) + K_{\mathrm{t}}K_{\mathrm{b}}s}\,. \tag{8.5}$$

The block diagram for the DC motor is shown schematically in Fig. 8.4.

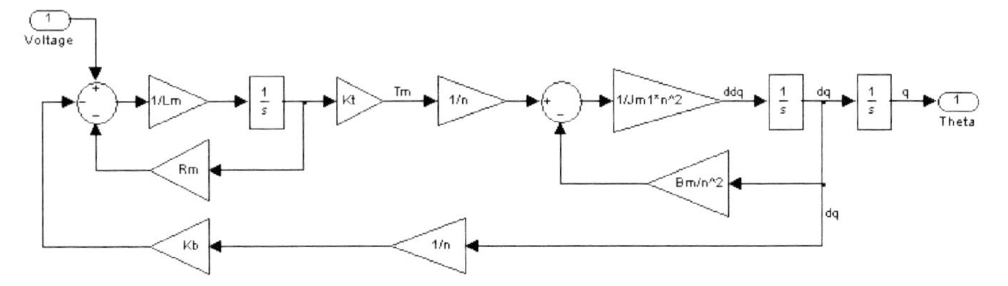

Fig. 8.4 DC motor block diagram

For the angular position of the motor shaft, the initial PI controller was designed via the Root Locus method. This procedure has been done for all six DC motors (Djuric, 2007). The analyzes show that the control parameters and gain can be expressed as a function of the each motor parameters (Eqs. 8.6 and 8.7).

$$K_{\mathrm{P}i} = R_{\mathrm{m}i}\,, \quad K_{\mathrm{I}i} = L_{\mathrm{m}i}\,, \qquad i = 1,2\ldots 6 \tag{8.6}$$

$$\mathrm{Gain}_i = \frac{V_{\max}}{60/2\pi}\left(\frac{K_{\mathrm{t}i}}{B_{\mathrm{m}i}R_{\mathrm{m}i}} + K_{\mathrm{t}i}K_{\mathrm{b}i}\right)\,, \qquad i = 1,2\ldots 6\,. \tag{8.7}$$

The control scheme for the DC motor including the gear ratio is presented in Fig. 8.5.

The MATLAB/Simulink scheme was designed such that the electrical and mechanical parts are separate. The electrical scheme output is $\frac{K_{ti}i_{mi}}{n_i}$, $i = 1,2\ldots 6$, and is input into the mechanical part of the scheme. Each link is treated separately and their equations are as follows:

$$a_{i1}\ddot{q}_1 + \ldots + a_{i6}\ddot{q}_6 + b_{i12}\dot{q}_1\dot{q}_2 + \ldots + b_{i56}\dot{q}_5\dot{q}_6 + c_{i1}\dot{q}_1 + \ldots + c_{i6}\dot{q}_6 + G_i = \tau_i\,. \tag{8.8}$$

Generalized joint coordinates, velocities and acceleration are expressed by $q_i, \dot{q}_i, \ddot{q}_i$. The PI controller for each link is connected to the electrical part of the motor. The output of the electrical part of the motor is the motor's torque. The mechanical part of the motor is combined with link dynamics and is schematically presented in Fig. 8.6. Also, the link dynamics from Fig. 8.6 is shown in detail in Fig. 8.7.

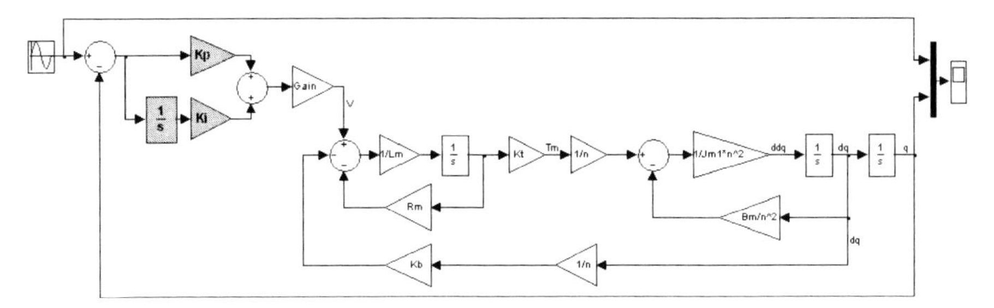

Fig. 8.5 PI controller for the DC motor

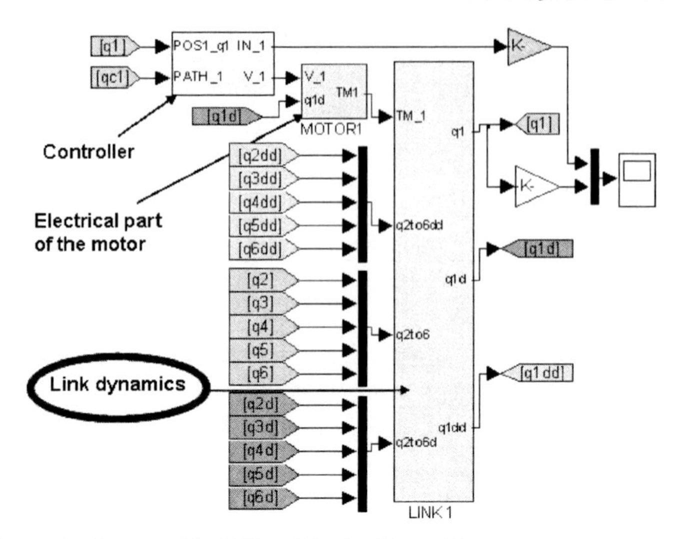

Fig. 8.6 Schematic diagram of the PUMA 560 robot First Link

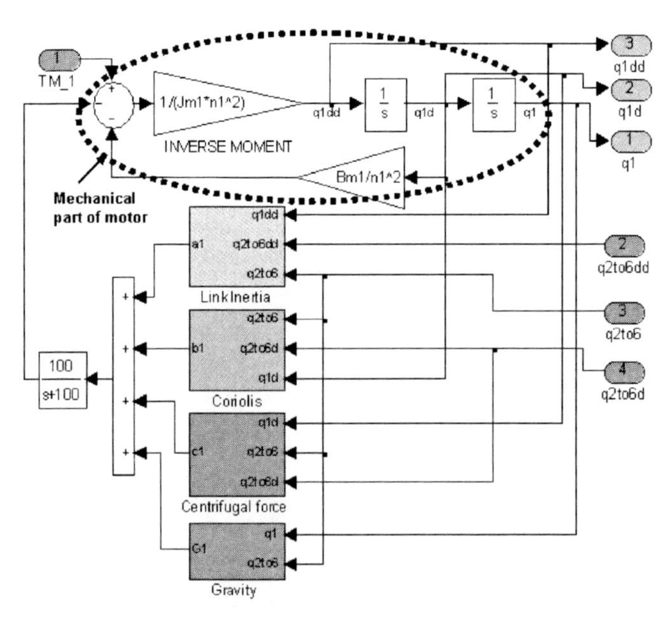

Fig. 8.7 Schematic diagram of the first link dynamics and first motor dynamics

The reconfigurable control block from Fig. 8.6 is shown in detail in Fig. 8.8. It has two main blocks: PI control block and the command wave form block.

Figure 8.9 details the electrical part of the DC motor block from Fig. 8.6.

According to the DH parameters from Djuric (2007), the PUMA 560 kinematic model and the motion trajectory was created using UKMS software. The path contains ten points (1-2-3-4-5-6-7-8-9-1-HOME) (see Fig. 8.10).

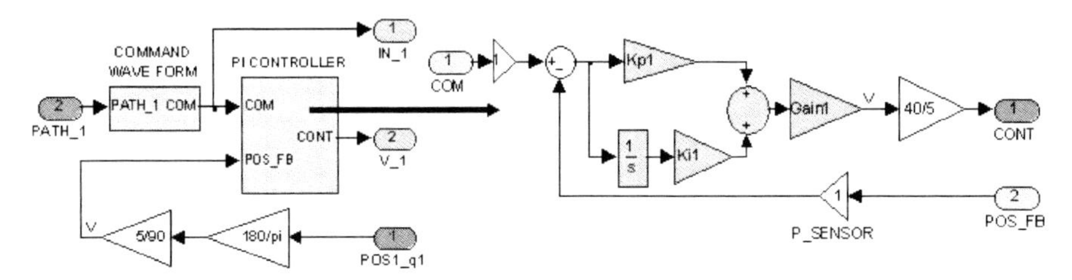

Fig. 8.8 Schematic diagram of the reconfigurable "PI" controller with input command

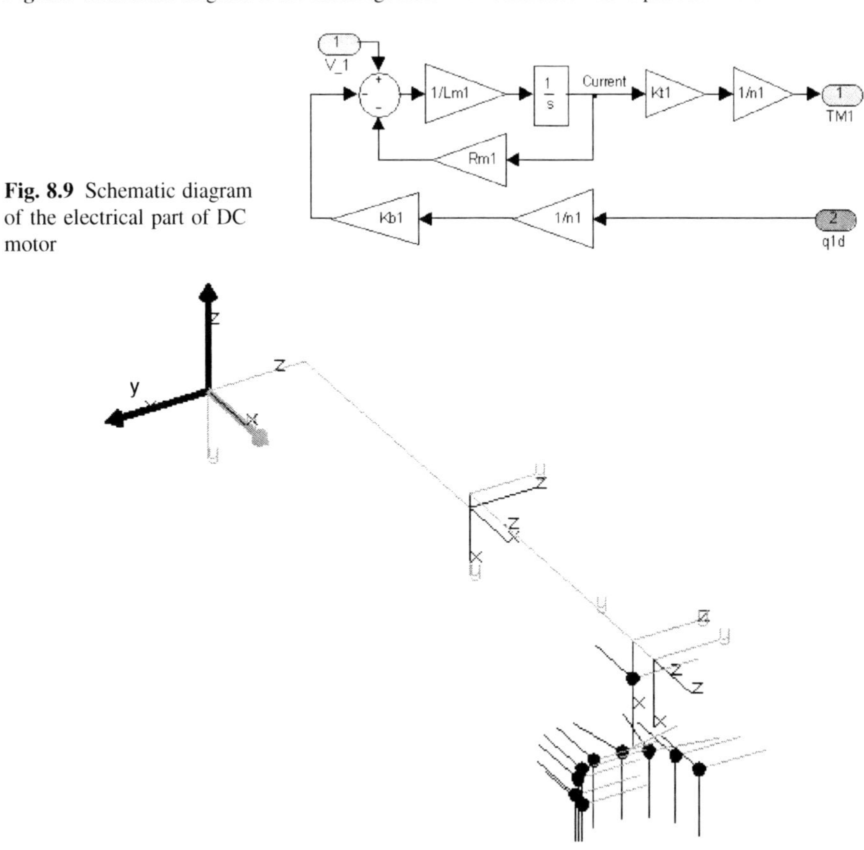

Fig. 8.9 Schematic diagram of the electrical part of DC motor

Fig. 8.10 PUMA 560 robot path in UKMS software

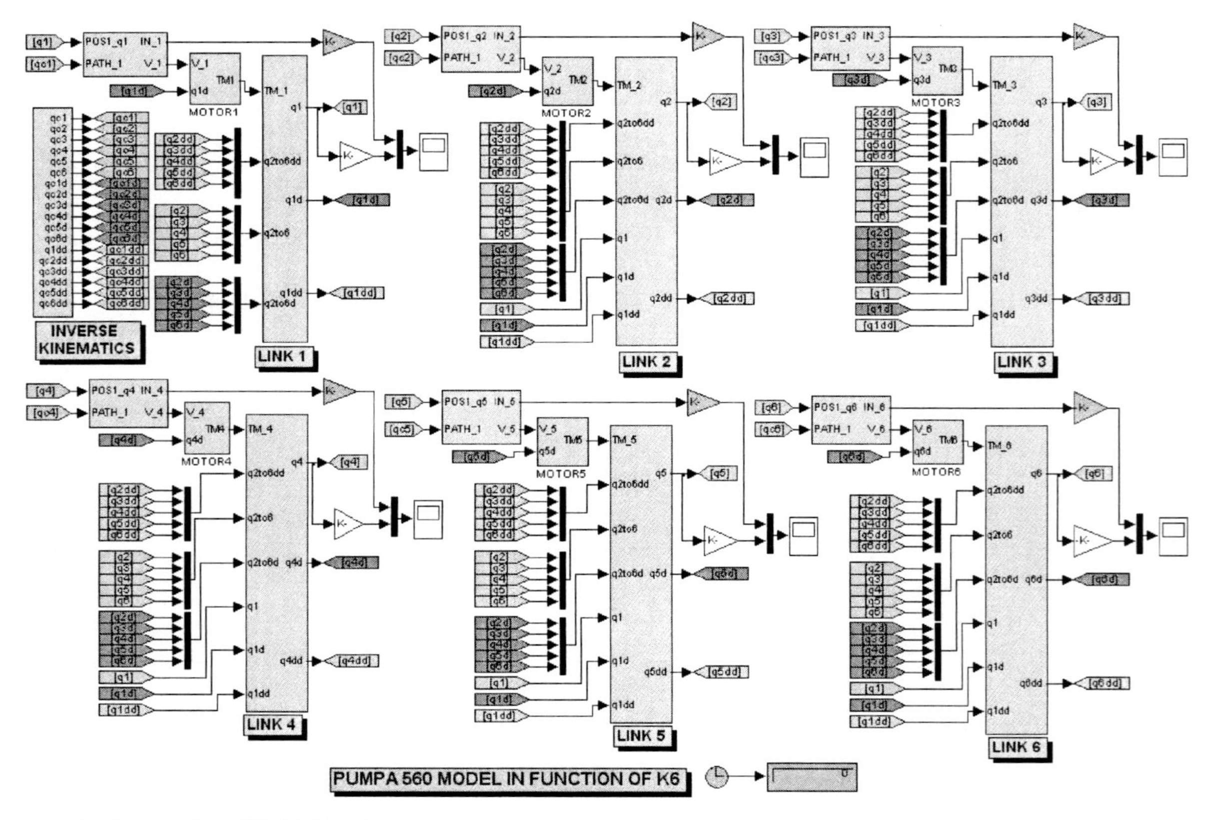

Fig. 8.11 Schematic diagram of the PUMA 560 robot

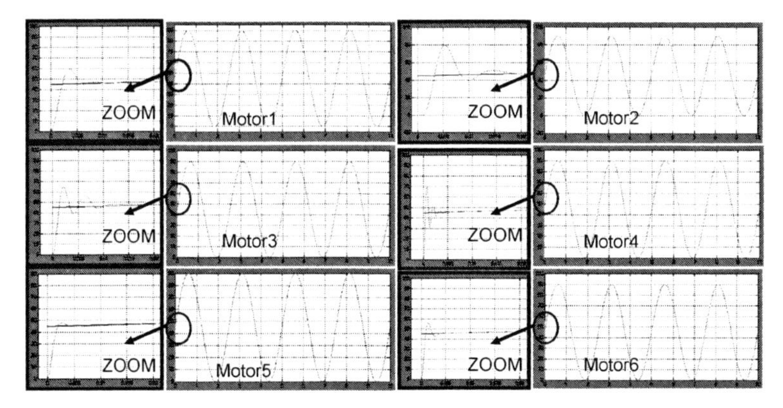

Fig. 8.12 Schematic diagram of the PUMA 560 robot

The complete schematic diagram of the PUMA 560 is presented in Fig. 8.11. This diagram has six links, six motors, six reconfigurable controllers, and a block with a robot path (inverse kinematic block).

The response of the PUMA 560 robot using reconfigurable PI controller for each joint is presented in Fig. 8.12. It is very clear that the system has a fast response with some overshoot.

8.3 Design of Reconfigurable Robot Platform (RRP)

The plant model of the selected robot and its controller can be auto-generated using the predefined reconfigurable model parameters. There are six steps for defining robot kinematics, dynamic and control using all reconfigurable modules. These are:

1. Check if selected robot belongs to the RPF group of robots.
2. Using UKMS software (Djuric and ElMaraghy, 2006) for modeling robots, select joint coordinate systems, link lengths, link offsets, and joint limits from the selected robot manufacturer. This will produce a robot kinematic model, which can be used for creating different motion trajectories for specific applications. These motion trajectories will be used in step six for the position control simulation.
3. The kinematic model contains information used for automatic generation of its Jacobian matrix using (RPFJM), singularity matrix using (RPFSM) and workspace (RRW).
4. By adding a mass of each link and knowing all kinematic information (D-H parameters), the robot dynamic model was calculated using the RPFDM module.
5. Using RPFDM+ and required motor information, the complete dynamic model of the selected robot was calculated.

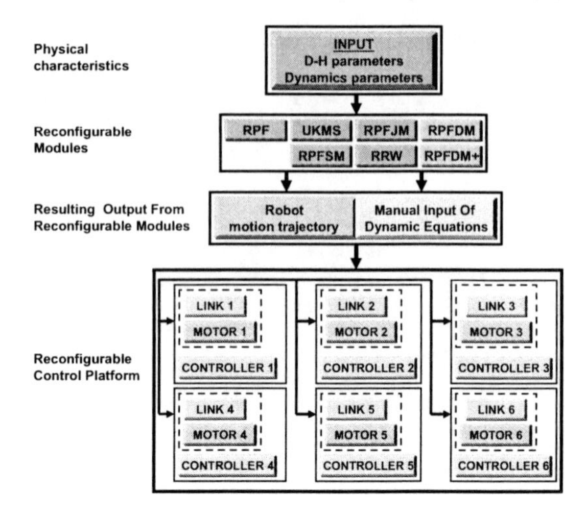

Fig. 8.13 Reconfigurable robot platform (RRP)

6. Using the Reconfigurable Robot Platform (RRP), changing expressions for each link dynamic are possible. In a separate *.m (Matlab) file, all motors' parameters were inputted, and by running the simulation all six PI controllers, were automatically generated.

This procedure is graphically presented in Fig. 8.13.

8.4 Reverse Modeling of Reconfigurable Robot Meta-Model

The presented methodology can be used to design a new truly reconfigurable and unified robot platform. Using the developed reconfigurable kinematic and dynamic modules, the new reconfigurable machines can be produced, so that they perfectly match in a new environment. The Meta-model of a reconfigurable robot can be

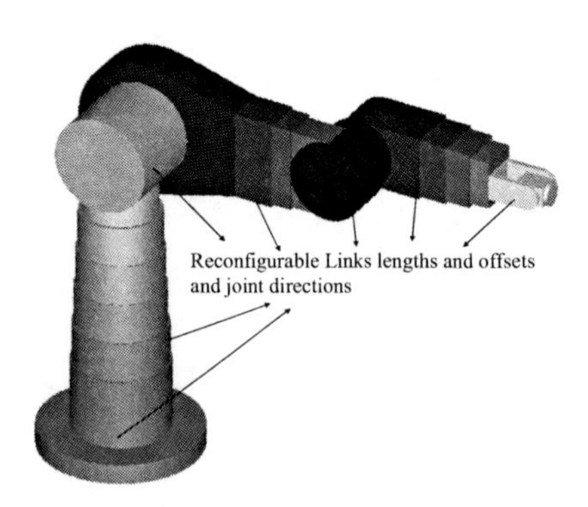

Fig. 8.14 Meta model of a reconfigurable robot

built such that it satisfies all kinematic characteristics of the previously developed unified model (see Fig. 8.14). This means that each joint has the ability to easily change positive directions and to increase or decrease link lengths and offsets. This methodology is a bridge between current industrial practices and future global trends.

8.5 Conclusions

The RPF (Reconfigurable Puma-Fanuc) model and its seven solutions/modules: Unified Kinematic Modeler and Solver (UKMS), Reconfigurable Puma-Fanuc Jacobian Matrix (RPFJM), Reconfigurable Puma-Fanuc Singularity Matrix (RPFSM), Reconfigurable Robot Workspace (RRW), Reconfigurable Puma-Fanuc Dynamic Model (RPFDM), Reconfigurable Puma-Fanuc Dynamic Model Plus (RPFDM+) and Reconfigurable Control Platform (RCP) are graphically shown in Fig. 8.15.

The reconfigurable modules can be used for a current manufacturing system and for the future reconfigurable industry. The presented methodology can be applied to any automated machine, to develop reconfigurable modules and reconfigurable machines. Furthermore, by unifying another group of automated machines a new Meta-Model can be developed.

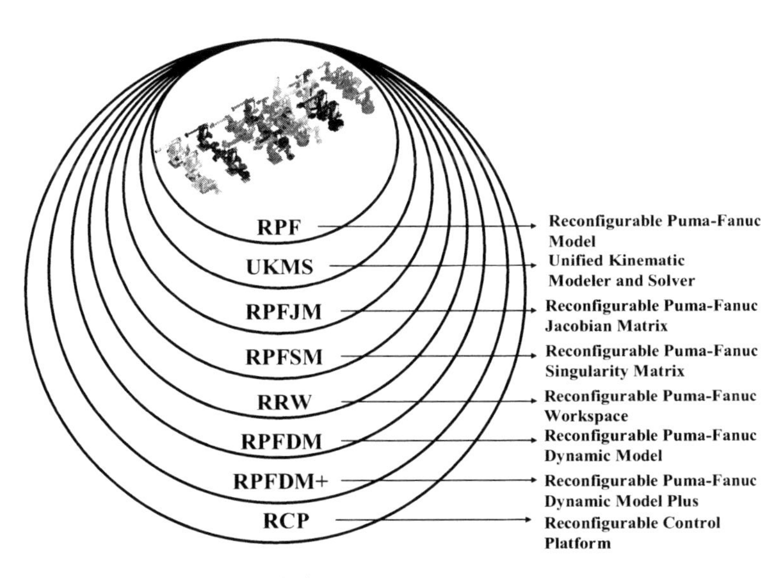

Fig. 8.15 RPF model and its five solutions

Nomenclature

Abbreviations

ASM:	Automatic Separation Method.
D-H:	Denavit – Hartenberg.
FBP:	Filtering Boundary Points
RMS:	Reconfigurable Manufacturing System
RPM:	Reconfigurable Plant Model
RPF:	Reconfigurable Puma-Fanuc
UKMS:	Unified Kinematic Modeler and Solver
PI:	Proportional Integral
RRP:	Reconfigurable Robot Platform
RPFSM:	Reconfigurable Puma-Fanuc Singularity Matrix
RRW:	Reconfigurable Robot Workspace
RPFDM:	Reconfigurable Puma-Fanuc Dynamic Model
RPFDM+:	Reconfigurable Puma-Fanuc Dynamic Model Plus actuators
RCP:	Reconfigurable Control Platform

Symbols

a_{T} :	Approach vector
a_i :	i^{th} Link offset
A :	6×6 Inertia matrix
B :	6×15 Coriolis torques matrix
B_{m} :	Motor damping coefficient
C :	6×6 Centrifugal torques matrix
d_i :	i^{th} Link length.
G :	6×1 Gravity torques vector
i_m :	Armature current
J_m :	Moment of Inertia of the motor referred to the motor shaft
${}^{6}({}^{0}J)$:	Jacobian matrix for 6R robots
K_i :	i^{th} Configuration parameter of the ith twist angel α_i
K_b :	Voltage constant
K_t :	Torque constant
L_m :	Armature inductance
n :	Gear ratio
n_T :	Normal vector
q :	Vector of generalized joint coordinates
$\dot{q}$:	Vector of joint velocities
$\ddot{q}$:	Vector of joint acceleration
R_m :	Armature resistance
s_T :	Sliding vector
T_m :	Motor generated torque
V :	Armature voltage
$V_{\max}$:	Maximum voltage
α_i :	Twist angle
θ_i :	Joint angle
θ_m :	Angular position of armature
τ :	Load torque

References

Armstrong B., Oussama K., Burdick J., 1986, The Explicit Dynamic and Inertia Parameters of the PUMA 560 Arm. International Conference on Robotics and Automation, 510–518

Corke P.I., 1994, The Unimation Puma servo system, MTM-. 226 report, CSIRO Division of Manufacturing Technology, Australia

Corke P.I., 1998, A Symbolic and Numeric Procedure for Manipulator Rigid-Body Dynamic Significance Analysis and Simplification. Robotica 16/5:589–594

Corke P.I., Armstrong B., 1994, A Search for Consensus Among Model Parameters Reported for the PUMA 560 Robot. Proc. IEEE Conf. Robotics and Automation, 1608–1613

Corke P.I., Armstrong B., 1995, A meta-study of PUMA 560 dynamics: A critical appraisal of literature data. Robotica 13:253–258

Djuric A.M., 2007 Reconfigurable Kinematics, Dynamics and Control Process for Industrial Robots. Doctoral Dissertation, University of Windsor

Djuric A.M., ElMaraghy W.H., 2006, Generalized Reconfigurable 6-Joint Robot Modeling. Transactions of the CSME 30/4:533–565

Djuric A.M., ElMaraghy W.H., 2007, A Unified Reconfigurable Robots Jacobian. Proc. of the 2nd Int. Conf. on Changeable, Agile, Reconfigurable and Virtual Production 811–823 (CARV 2007)

Leahy Nugent M.B.M. Jr., Valavanis K.P., Saridis G.N., 1986, Efficient Dynamic for a PUMA-600. IEEE 3:519–524

Chapter 9
Reconfigurable Control of Constrained Flexible Joint Robots Interacting with Dynamic and Changeable Environment

Y. Cao[1], **H. ElMaraghy**[2], **W. ElMaraghy**[2]

Abstract This chapter deals with the effect of changes at the machine/robot physical level and new reconfigurable control strategies to enable such change. Joint flexibility constitutes the major source of compliance in most industrial robots. It is important to account for joint flexibility when dealing with force control problems. In addition, the type of environment that the robot is in contact with, or the object that the robot works on, may be made of different materials. Hence, force control strategies suitable for both rigid and soft contact is needed corresponding to different parts of the object/surface while performing the task. A decoupling-based force/position control of flexible joint robot is first designed for rigid, stiff and dynamic environments. A reconfigurable force control scheme is proposed for when the robot's working trajectory covers different types of environments. Numerical simulation results are presented to demonstrate the effectiveness of the proposed decoupling approach and the reconfigurable force control scheme. The desired contact force can be obtained, whether it is rigid or soft environment, without stopping the robot, due to the active reconfiguration of the controller. This novel reconfigurable control scheme can be extended, by including other well-designed controllers, thus achieving more versatile control reconfiguration under changeable situations.

Keywords Robots, Control, Reconfiguration, Flexible Joints

9.1 Introduction

Common tasks carried out by industrial robots include assembly, deburring, grinding, scribing, writing, and chamfering. These tasks require control of both the position and velocity of the end-effector as well as the contact force between the end-

[1] Mechanical Engineering Department, University of British Columbia, Canada

[2] IMS Center, University of Windsor, Canada

H.A. ElMaraghy (ed.), *Changeable and Reconfigurable Manufacturing Systems*,

effector and the environment. In most literature, this problem is categorized as the control of constrained mechanical systems. There have been numerous publications dealing with three main issues: (1) type of control approaches; (2) type of robot (rigid or flexible); (3) type of environment (rigid, stiff, or dynamic).

The control approaches reported in the literatures can be further divided into two broad categories: impedance control and hybrid force/position control. Hogan (1985) first presented the impedance control to obtain a prescribed static or dynamic relationship between the force and position of the robot end-effector. However, following of the desired force profile exactly is impossible using impedance control strategies. Raibert and Craig (1981) proposed the basic hybrid position/force control scheme where the end-effector force is explicitly controlled in selected directions and its position is controlled in the remaining (complementary) directions. Adaptive approaches based on the hybrid controllers were used to estimate the robot plant parameters on-line and guarantee asymptotically exact force trajectory tracking (Whitcomb et al. 1997, Arimoto et al. 1993).

Hybrid methods consider the robot in contact with the environment as a system of dual vector spaces, or a 'dual system'. Dual system is a mathematical structure comprising two vector spaces of equal dimension and a scalar product that takes one argument from each space (Featherstone, 2003). When a robot is in contact with the environment, work (or power, virtual work) is produced by the scalar product of two different types of vectors namely force and velocity, which are objects of different physical and geometric natures. Based on the concept of duality (or reciprocity) the relation between force and velocity and exact feedback linearization were accomplished (Yoshikawa et al. 1988, Yoshikawa, et al. 1993). McClamroch and Wang (1988) explicitly utilized the duality relation and the constraints to decouple the dynamics of constrained mechanical systems and develop a stable hybrid position/force-control algorithm. In this work, this concept will be followed when designing the decoupled force/position controller for a flexible joint robot in contact with both rigid and compliant environments while executing a given task.

Application of hybrid control approaches are normally found in the following two situations: (1) robots in hard contact with a rigid environment, as in Raibert and Craig (1981) and Featherstone (2003); and (2) robots in soft contact with a compliant environment, as in Siciliano and Villani (1996), Villani et al. (1999) and McClamroch (1989). In these works, the environment is considered either infinitely rigid or purely stiff (with a spring-damper model), during the contact tasks, as shown in Fig. 9.1a and b. For stiff environments, the interaction force is a linear function of deflection of the contact surface. High stiffness environments can be considered as rigid. However, there are applications characterized by significant elasto-dynamics of the environment for which the simple spring/damper model would be inadequate (De Luca and Manes 1991, Vukobratovic 1998). Figure 9.1c shows an example of a robot interacting with a dynamic environment modeled by a mass-spring-damper system. This can represent a robot moving a tool over a surface, which behaves as a system with distributed parameters. A direct consequence of the presence of the dynamic environment is the coupling between the two sub-spaces of motion along

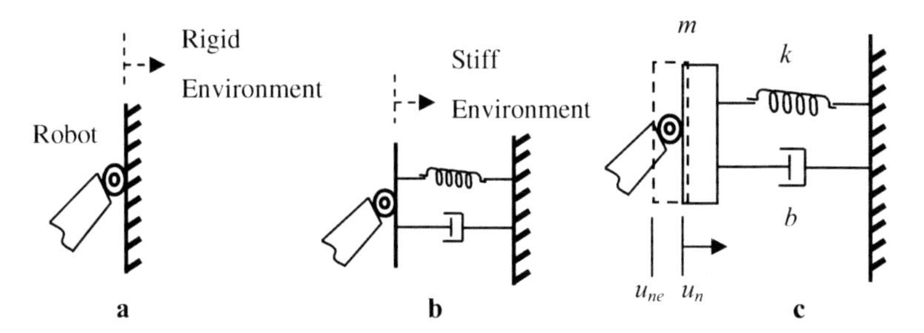

Fig. 9.1 a–c Robot interacting with various environment: **a** rigid; **b** stiff; **c** dynamic

the surface and the force normal to the surface. Vukobratovic and Ekalo (1996) proposed a model-based control scheme for simultaneous position-force control. Recently, Karan (2005) presented a robust position/force control scheme, which overcomes the problem of uncertainty in the parameters of a dynamic environment. Literature survey shows that the nonlinear decoupling technique based on force-control of flexible joint robots interacting with dynamic environments has received less attention. Early work by Jankowski and ElMaraghy (1991) proposed a decoupling approach in the joint space when the robot is in contact with a rigid surface.

The contribution of the approach presented in this chapter is twofold. First, based on the result from Jankowski and ElMaraghy (1991) and the duality relationship between force and motion, the nonlinear decoupling technique is applied in the task space to a general n-link flexible joint robot in contact with rigid or soft environment. Then feedback linearization is used to ensure asymptotically stable tracking of force and position. Second, a reconfigurable force control scheme based on the idea presented in Cao and de Silva (2006) is proposed for robots whose constrained motion brings them into contact with different types of environments. This is motivated by the practical issues encountered in finishing operations (e.g. grinding and deburring) by robots and changes to the parts material brought about by changes in their design. Systems designers typically program the robot to move its tool along a path defined by discrete points. However, along that path, the object that the robot works on may consist of parts of different hardness due to selective heat treating or different materials, or the object being supported by an elastic base. The success of the automated operations depends on the adaptability to the changing rigidity and consistency of controlling the contact force without stopping. Without actively reconfiguring the force control law, the tool tip may gouge the part, leave unwanted burrs on the edge, or, in a worst-case scenario, the tip may break requiring immediate repair and associated cost. Hence, more flexible force control is needed corresponding to difference situations. In this chapter, the proposed reconfigurable force control scheme is designed such that the robot can perform force control task on objects in dynamic and changeable environment and complete the task without stopping.

9.2 Dynamic Model of Flexible Joint Robot in Contact with Different Environment

The dynamic model of a flexible joint robot have n degrees of freedom and interacting with the environment can be written as:

$$M(q)\ddot{q}+V(q,\dot{q})+K(q-\alpha)=\tau_{\mathrm{F}} \tag{9.1}$$

$$I_{\mathrm{m}}\ddot{\alpha}+K(\alpha-q)=\tau_{\mathrm{m}}\ , \tag{9.2}$$

where

q: Generalized coordinates representing the angle of the robot links, $q \in \Re^n$;
α: Generalized coordinates representing the angle of the actuators' rotors (shafts) $\alpha \in \Re^n$;
$M(q)$: Inertia matrix associated with the rigid links, $M(q) \in \Re^{n\times n}$;
$V(q,\dot{q})$: Vector of Coriolis, centrifugal and gravity generalized force, $V(q,\dot{q}) \in \Re^{n\times 1}$;
I_{m}: diag$[I_{\mathrm{m}i}]$, the positive definite diagonal matrix of the moments of inertia of the motors, $I_{\mathrm{m}} \in \Re^{n\times n}$;
K: diag$[K_i]$, the positive definite diagonal matrix of stiffness of the rotors, $K \in \Re^{n\times n}$;
τ_{F}: Joint torque contributed from the contact force with environment, $\tau_{\mathrm{F}} \in \Re^n$;
τ_{m}: input torque from the motors, $\tau \in \Re^n$.

Note that $I_{\mathrm{m}i}$ and K_i $(i=1,2,\ldots,n)$ are the inertia and the stiffness of the ith joint. n is degrees of freedom of the robotic manipulator. The joint torque, τ_{F}, contributed by the contact force with the environment can be formulated depending on the type of contact surface. In order to derive the expression for the contact force, let's first define some variable.

u: position vector of the robot end-effector expressed in the task space or constraint frame, $u \in \Re^n$;
J_{T}: Task space Jacobian.

We can partition vector u into the following from, $u = \begin{bmatrix} u_t & u_n \end{bmatrix}^{\mathrm{T}}$ $(u_{\mathrm{t}} \in \Re^{n-m}$, $u_{\mathrm{n}} \in \Re^m)$, where u_{t} and u_{n} are vectors of coordinates in the tangent and normal space, respectively. Assume that the robot is subject to m holonomic, frictionless and deformable constraint surfaces characterized by $\psi(q) = \begin{bmatrix} \psi_1(q) & \cdots & \psi_{\mathrm{m}}(q) \end{bmatrix}^{\mathrm{T}} = u_{\mathrm{n}} - u_{\mathrm{ne}}$, $m \leqslant n$, where u_{ne} is the equilibrium position in the normal direction. The constraint force in the joint space, τ_{F}, can be expressed by $\tau_{\mathrm{F}} = J_{\mathrm{n}}^{\mathrm{T}} f$, where

$f \in \Re^m$ is m-dimensional vector of represents normal contact force components,
$J_n(q)$ is the jacobian of the holonomic constraints, i.e., $J_n = \frac{\partial \psi}{\partial q} \in \Re^{m\times n}$.

It is also found that matrix J_n is the m row of matrix J_{T}, or $J_n = [O\ I]J_{\mathrm{T}}$, where $I =$ unity matrix $\in \Re^{m\times m}$ and $O =$ zero matrix $\in \Re^{m\times(n-m)}$.

When the robot is in contact with stiff environment, contact force is generated as a result of environmental stiffness. If we assume that the direction of the contact force f is normal to the un-deformed environment, then the contact force in joint coordinates can be obtained as

$$f = k\,(u_{\mathrm{n}} - u_{\mathrm{ne}})\ , \tag{9.3}$$

where k represents the environment stiffness and is a constant positive definite matrix, $k \in \Re^{m\times m}$.

When the robot is interacting with the dynamic environment, the contact force expressed in the compliance frame can be given by the following model:

$$f = m\ddot{u}_n + b\dot{u}_n + k\,(u_{\mathrm{n}} - u_{\mathrm{ne}})\ , \tag{9.4}$$

where m, b, and k are $m\times m$ matrices of equivalent inertia, damping constant, and stiffness of the environment.

9.3 Decoupled Controller Design

Let $q_{\mathrm{d}}(t)$ be the desired trajectory of the link position, $\lambda_{\mathrm{d}}(t)$ the desired magnitude of the constraint force with a rigid surface and $f_{\mathrm{d}}(t)$ the desired contact force with a deformable surface. The control objective is to make sure that the manipulator's end-effector follows a desired trajectory and at the same time maintains a desired contact force between the end-effector and the surface even when the surface compliance changes along the trajectory.

When the robotic system is subject to m environmental constraints, the original n degrees of freedom will be left with only n–m degrees of freedom. Thus the systems motion is governed by a set of n–m independent equations (position/velocity relations) and a set of m dependent equations (force relations). Here, we are going to perform the coordinate reduction on the vector u in the constraint frame. First we need to obtain the explicit expression between u and torque input τ_{m}.

9.3.1 Contact with Rigid Surface

Differentiating the link Eq. 9.1 of the dynamic model twice yields,

$$Mq^{(4)} + 2\dot{M}q^{(3)} + \ddot{M}\ddot{q} + \ddot{V} + K(\ddot{q} - \ddot{\alpha}) = \ddot{J}_n^{\mathrm{T}} f + 2\dot{J}_n^{\mathrm{T}}\dot{f} + J_n^{\mathrm{T}}\ddot{f}\ . \tag{9.5}$$

Note the notion, $q^{(i)} = d^i q/dt^i$. Substituting $\ddot{\alpha}$ obtained from Eq. 9.2 into Eq. 9.5 gives

$$\begin{aligned} &Mq^{(4)} + 2\dot{M}q^{(3)} + \ddot{M}\ddot{q} + \ddot{V} + K\ddot{q} - KI_{\mathrm{m}}^{-1}\tau_{\mathrm{m}} + KI_{\mathrm{m}}^{-1}K(\alpha - q) \\ &\quad = \ddot{J}_n^{\mathrm{T}} f + 2\dot{J}_n^{\mathrm{T}}\dot{f} + J_n^{\mathrm{T}}\ddot{f} \,. \end{aligned} \tag{9.6}$$

The velocity of the end-effector in the constraint frame is related to the joint velocities by $\dot{u} = J_T\dot{q}$. The relationship between the fourth derivative of joint variables q and the end-effector position vector u (in the task space) is

$$q^{(4)} = J_{\mathrm{T}}^{-1}\left(u^{(4)} - 3(\dot{J}_{\mathrm{T}}q^{(3)} + \ddot{J}_{\mathrm{T}}\ddot{q}) - J_{\mathrm{T}}^{(3)}\dot{q}\right) \,. \tag{9.7}$$

Substituting Eq. 9.7 into Eq. 9.6 gives

$$\bar{M}u^{(4)} - \bar{M}\left(3(\dot{J}_{\mathrm{T}}q^{(3)} + \ddot{J}_{\mathrm{T}}\ddot{q}) + J_{\mathrm{T}}^{(3)}\dot{q}\right) + \bar{V} = KI_{\mathrm{m}}^{-1}\tau_{\mathrm{m}} + J_n^{\mathrm{T}}\ddot{f} \,, \tag{9.8}$$

where $\bar{M}(q) = MJ_{\mathrm{T}}^{-1}$ and

$$\bar{V}(q,\dot{q},\ddot{q},q^{(3)},\alpha,f,\dot{f}) = 2\dot{M}q^{(3)} + \ddot{M}\ddot{q} + \ddot{V} + K\ddot{q} + KI_{\mathrm{m}}^{-1}K(\alpha - q) - \ddot{J}_n^{\mathrm{T}} f - 2\dot{J}_n^{\mathrm{T}}\dot{f} \,.$$

Since $u = \begin{bmatrix} u_{\mathrm{t}} \; u_{\mathrm{n}} \end{bmatrix}^{\mathrm{T}}$ and further, the end-effector is in contact with the rigid environment, no motion is allowed in the directions that are normal to the constraint hyper-surfaces, hence, $u_n^{(4)} = 0$. With some manipulators, Eq. 9.8 becomes

$$M_{\mathrm{r}}\begin{bmatrix} u_{\mathrm{t}}^{(4)} \\ \ddot{f} \end{bmatrix} + V_{\mathrm{r}} = KI_{\mathrm{m}}^{-1}\tau_{\mathrm{m}} \,, \tag{9.9}$$

where $M_{\mathrm{r}} = \begin{bmatrix} \bar{M}\begin{bmatrix} I \\ 0 \end{bmatrix} & -J_n^{\mathrm{T}} \end{bmatrix}$; $V_{\mathrm{r}} = -\bar{M}\left(3(\dot{J}_{\mathrm{T}}q^{(3)} + \ddot{J}_{\mathrm{T}}\ddot{q}) + J_{\mathrm{T}}^{(3)}\dot{q}\right) + \bar{V}$.

A generalized computed torque control law can be chosen,

$$\tau_{\mathrm{m}} = I_{\mathrm{m}}K^{-1}\left(M_{\mathrm{r}}y_{\mathrm{r}} + V_{\mathrm{r}}\right) \,, \tag{9.10}$$

where $y_{\mathrm{r}} = \begin{bmatrix} y_{u_{\mathrm{t}}} \; y_{\mathrm{f}} \end{bmatrix}^{\mathrm{T}}$. This control law leads to two closed-loop systems $u_{\mathrm{t}}^{(4)} = y_{u_{\mathrm{t}}}$ and $\ddot{f} = y_{\mathrm{f}}$. Finally, controller design is completed on the linear side of the problem by choosing

$$\begin{aligned} y_{u_{\mathrm{t}}} &= u_{\mathrm{td}}^{(4)} + K_3\left(u_{\mathrm{td}}^{(3)} - u_{\mathrm{t}}^{(3)}\right) + K_2\left(\ddot{u}_{\mathrm{td}} - \ddot{u}_{\mathrm{t}}\right) \\ &\quad + K_1\left(\dot{u}_{\mathrm{td}} - \dot{u}_{\mathrm{t}}\right) + K_0\left(u_{\mathrm{td}} - u_{\mathrm{t}}\right) \\ y_{\mathrm{f}} &= \ddot{f}_{\mathrm{d}} + K_{\mathrm{v}}\left(\dot{f}_{\mathrm{d}} - \dot{f}\right) + K_{\mathrm{p}}\left(f_{\mathrm{d}} - f\right) \,, \end{aligned} \tag{9.11}$$

where K_3, K_2, K_1, K_0 and $K_{\mathrm{v}}, K_{\mathrm{p}}$ are constant diagonal feedback gain matrices, respectively.

9.3.2 Contact with Stiff Environment

Since the contact surface is rigid any more, motion is allowed in the normal direction. Hence, $\dot{u}_{\mathrm{n}} = J_{\mathrm{n}}\dot{q}$. Similarly, differentiating Eq. 9.1 twice and combining with the solution for $\ddot{\alpha}$ obtained from Eq. 9.2, one can obtain the following expression between u and torque input τ_{m}.

$$\begin{aligned} &MJ_{\mathrm{T}}^{-1}u^{(4)} - MJ_{\mathrm{T}}^{-1}\left(3(\dot{J}_{\mathrm{T}}q^{(3)} + \ddot{J}_{\mathrm{T}}\ddot{q}) + J_{\mathrm{T}}^{(3)}\dot{q}\right) + 2\dot{M}q^{(3)} + \ddot{M}\ddot{q} + \ddot{V} \\ &\quad + K\ddot{q} + KI_{\mathrm{m}}^{-1}K(\alpha - q) - \ddot{J}_{n}^{\mathrm{T}}f - 2\dot{J}_{n}^{\mathrm{T}}\dot{f} - J_{n}^{\mathrm{T}}\ddot{f} = KI_{\mathrm{m}}^{-1}\tau_{\mathrm{m}}\ . \end{aligned} \tag{9.12}$$

In brief, the above equation can be expressed as

$$M_{\mathrm{s}}u^{(4)} + V_{\mathrm{s}}(q,\dot{q},\ddot{q},q^{(3)},\alpha,f,\dot{f},\ddot{f}) = KI_{\mathrm{m}}^{-1}\tau_{\mathrm{m}}\ , \tag{9.13}$$

where M_{s} and V_{s} can easily be identified from Eq. 9.12. Now the corresponding feedback linearizing control for the dynamics given by Eq. 9.13 can be chosen as

$$\tau_{\mathrm{m}} = I_{\mathrm{m}}K^{-1}(M_{\mathrm{s}}y_{\mathrm{s}} + V_{\mathrm{s}})\ . \tag{9.14}$$

A globally linearized and decoupled equation of motion: $u^{(4)} = y_{\mathrm{s}}$ is obtained. Similarly linear position and force controllers can be designed for the corresponding task space variables u. Let $y_{\mathrm{s}} = \left[y_{u_{\mathrm{t}}}\ y_{u_{\mathrm{n}}}\right]^{\mathrm{T}}$, where y is partitioned corresponding to free and constrained motion. This leads to $u_{\mathrm{t}}^{(4)} = y_{u_{\mathrm{t}}}$ and $u_{\mathrm{n}}^{(4)} = y_{u_{\mathrm{n}}}$. To ensure asymptotical stability for the unconstrained direction of motion, $y_{u_{\mathrm{f}}}$ is chosen the same as in Eq. 9.11. Since the desired contact force is achieved by regulating the end-effector in the constrained direction u_{c} and $u_{n}^{(4)} = \frac{1}{k}f^{(4)}$, then $y_{u_{\mathrm{n}}}$ is designed as

$$\begin{aligned} y_{u_{\mathrm{n}}} = \frac{1}{k}\Big\{&f_{\mathrm{d}}^{(4)} + K_{\mathrm{c3}}\left(f_{\mathrm{d}}^{(3)} - f^{(3)}\right) + K_{\mathrm{c2}}\left(\ddot{f}_{\mathrm{d}} - \ddot{f}\right) \\ &+ K_{\mathrm{c1}}\left(\dot{f}_{\mathrm{d}} - \dot{f}\right) + K_{\mathrm{c0}}\left(f_{\mathrm{d}} - f\right)\Big\}\ . \end{aligned} \tag{9.15}$$

With proper choice of diagonal matrices for K_{c0}, K_{c1}, K_{c2}, K_{c3}, asymptotic force tracking in the constrained directions is guaranteed.

9.3.3 Contact with Dynamic Environment

Now the decoupled controller for a robot interacting with the dynamic environment may be developed. Substitute Eq. 9.4 into Eq. 9.1, we have

$$M(q)\ddot{q} + V(q,\dot{q}) + K(q - \alpha) = J_{n}^{\mathrm{T}}\left[mJ_{\mathrm{n}}\ddot{q} + m\dot{J}_{\mathrm{n}}\dot{q} + b\dot{u}_{\mathrm{n}} + k(u_{\mathrm{n}} - u_{\mathrm{ne}})\right]\ . \tag{9.16}$$

Rearrange the above equation and solve for $\ddot{q}$, yields

$$\ddot{q} = \left(M - J_n^{\mathrm{T}} m J_{\mathrm{n}}\right)^{-1} [J_n^{\mathrm{T}} \left(m\dot{J}_{\mathrm{n}}\dot{q} + b\dot{u}_{\mathrm{n}} + k(u_{\mathrm{n}} - u_{\mathrm{ne}})\right) - V(q,\dot{q}) - K(q-\alpha)] \,. \tag{9.17}$$

As can be seen from Eq. 9.17, the dynamics of the environment is included in the dynamics of the robot. Differentiate Eq. 9.1 twice to get

$$Mq^{(4)} + 2\dot{M}q^{(3)} + \ddot{M}\ddot{q} + \ddot{V} + K\ddot{q} + KI_{\mathrm{m}}^{-1}K(\alpha - q) - \ddot{J}_n^{\mathrm{T}} f - 2\dot{J}_n^{\mathrm{T}}\dot{f} - J_n^{\mathrm{T}}\ddot{f} = KI_{\mathrm{m}}^{-1}\tau_{\mathrm{m}} \,. \tag{9.18}$$

Obtain f, $\dot{f}$ and $\ddot{f}$ from Eq. 9.4, substitute into Eq. 9.18 and solve for $q^{(4)}$. Then using Eq. 9.7, the expression between the end-effector task space position vector and the motor input torque τ_{m} is obtained

$$M_{\mathrm{d}}u^{(4)} + V_{\mathrm{d}} = KI_{\mathrm{m}}^{-1}\tau_{\mathrm{m}} \,, \tag{9.19}$$

where $M_{\mathrm{d}} = \left(M - J_n^{\mathrm{T}} m J_{\mathrm{n}}\right) J_{\mathrm{T}}^{-1}$ and

$$\begin{aligned} V_{\mathrm{d}} = &-\left(M - J_n^{\mathrm{T}} m J_n\right) J_{\mathrm{T}}^{-1} \left[\left(3(\dot{J}_{\mathrm{T}}q^{(3)} + \ddot{J}_{\mathrm{T}}\ddot{q}) + J_{\mathrm{T}}^{(3)}\dot{q}\right)\right] + 2\dot{M}q^{(3)} + \ddot{M}\ddot{q} + \ddot{V} + K\ddot{q} \\ &+ KI_{\mathrm{m}}^{-1}K(\alpha - q) - J_n^{\mathrm{T}}\left[m\left(3(\dot{J}_n q^{(3)} + \ddot{J}_n\ddot{q}) + J_n^{(3)}\dot{q}\right) + bu_{\mathrm{n}}^{(3)} + k\ddot{u}_{\mathrm{n}}\right] \\ &- \ddot{J}_n^{\mathrm{T}} f - 2\dot{J}_n^{\mathrm{T}}\dot{f} \,. \end{aligned}$$

Partitioning u according to tangent and normal space, and making use of the expression for $u_{\mathrm{n}}^{(4)}$ obtained from differentiate twice of Eq. 9.4, which is $u_{\mathrm{n}}^{(4)} = m^{-1}\ddot{f} - m^{-1}\left(bu_{\mathrm{n}}^{(3)} + k\ddot{u}_{\mathrm{n}}\right)$, then Eq. 9.19 can be rewritten as

$$M_{\mathrm{d}} \begin{bmatrix} u_{\mathrm{t}}^{(4)} \\ m^{-1}\ddot{f} \end{bmatrix} - M_{\mathrm{d}} \begin{bmatrix} 0 \\ bu_n^{(3)} + k\ddot{u}_{\mathrm{n}} \end{bmatrix} + V_{\mathrm{d}} = KI_{\mathrm{m}}^{-1}\tau_{\mathrm{m}} \,. \tag{9.20}$$

A generalized computed torque control law can be chosen as follows,

$$\tau_{\mathrm{m}} = I_{\mathrm{m}}K^{-1}\left(M_{\mathrm{d}}y - M_{\mathrm{d}} \begin{bmatrix} 0 \\ bu_n^{(3)} + k\ddot{u}_{\mathrm{n}} \end{bmatrix} + V_{\mathrm{d}}\right) , \tag{9.21}$$

where $y = \left[y_{u_{\mathrm{t}}}\ y_{u_{\mathrm{n}}}\right]^{\mathrm{T}}$. This control law leads to two decoupled linear closed-loop systems, $u_{\mathrm{t}}^{(4)} = y_{u_{\mathrm{t}}}$ and $m^{-1}\ddot{f} = y_{u_{\mathrm{n}}}$. Finally, the controller design is completed on the linear side of the problem by choosing

$$\begin{aligned} y_{u_{\mathrm{t}}} &= u_{\mathrm{td}}^{(4)} + K_{\mathrm{t3}}\left(u_{\mathrm{td}}^{(3)} - u_{\mathrm{t}}^{(3)}\right) + K_{\mathrm{t2}}\left(\ddot{u}_{\mathrm{td}} - \ddot{u}_{\mathrm{t}}\right) + K_{\mathrm{t1}}\left(\dot{u}_{\mathrm{td}} - \dot{u}_{\mathrm{t}}\right) + K_{\mathrm{t0}}\left(u_{\mathrm{td}} - u_{\mathrm{t}}\right) \\ y_{u_{\mathrm{n}}} &= m^{-1}\left(\ddot{f}_{\mathrm{d}} + K_{\mathrm{n1}}\left(\dot{f}_{\mathrm{d}} - \dot{f}\right) + K_{\mathrm{n0}}\left(f_{\mathrm{d}} - f\right)\right) , \end{aligned} \tag{9.22}$$

where $K_{\mathrm{t3}}, K_{\mathrm{t2}}, K_{\mathrm{t1}}, K_{\mathrm{t0}}$ and $K_{\mathrm{n1}}, K_{\mathrm{n0}}$ are constant diagonal feedback gain matrices, respectively.

9.4 Reconfigurable Control Scheme

When the flexible joint robot is in contact with a surface containing both rigid, stiff and dynamic environment, it can be stated that the manipulator dynamics belong to either the invariant manifolds S_r (rigid surface), S_s (stiff surface), and S_d (dynamic surface). A switching control scheme is adopted for flexible joint robot moving in contact with surfaces of different compliance. Appropriate control law should be used depending on the type of contact surface. The following summarizes the switching logic.

If $q \in S_\mathrm{r}$, the motion of the robot remains on the manifold defined by S_r, then

$$\text{Control Law:} \qquad \tau_\mathrm{m} = I_\mathrm{m} K^{-1} (M_\mathrm{s} y_\mathrm{s} + V_\mathrm{s}) \, .$$

If $q \in S_\mathrm{s}$, the motion of the robot remains on the manifold defined by S_s, then

$$\text{Control Law:} \qquad \tau_\mathrm{m} = I_\mathrm{m} K^{-1} (M_\mathrm{s} y_\mathrm{s} + V_\mathrm{s}) \, .$$

If $q \in S_\mathrm{d}$, the motion of the robot remains on the manifold defined by S_d, then

$$\text{Control Law:} \qquad \tau_\mathrm{m} = I_\mathrm{m} K^{-1} \left(M_\mathrm{d} y - M_\mathrm{d} \begin{bmatrix} 0 \\ b u_n^{(3)} + k \ddot{u}_\mathrm{n} \end{bmatrix} + V_\mathrm{d} \right) \, .$$

A schematic diagram of this reconfigurable controller is shown in Fig. 9.2. C_r, C_s and C_d represent the position/force controller for rigid, stiff and dynamic environment, respectively. Supervisory module is the switching decision making module, which is triggered by the change of the type of the surface (or environment). Then the corresponding force controller is selected to continue the task. The supervisory module also monitors the performance of the system and it can possess learning and self-organizing capabilities as well. Detailed framework of this supervised control switching system can be found in Cao and de Silva (2006).

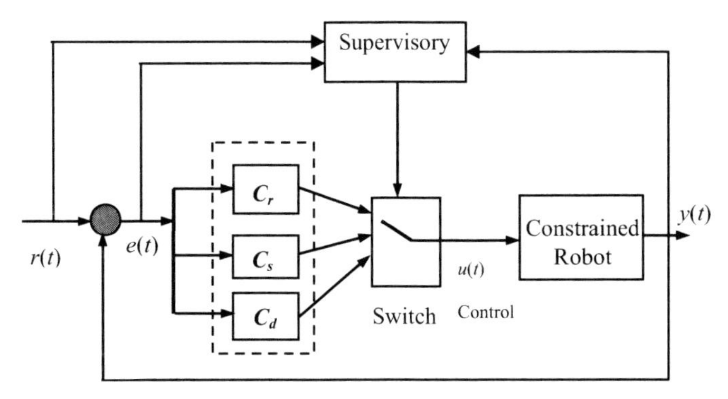

Fig. 9.2 Reconfigurable force control scheme

9.5 Simulation Study

A two-link flexible joint robot (Fig. 9.3) is chosen. The following system parameters, initial conditions, and desired values were used in the numerical simulation:

System parameters:

$$m_1 = 5\,\text{kg},\ m_2 = 5\,\text{kg},\ l_1 = 1\,\text{m},\ l_2 = 1\,\text{m},\ K_1 = K_2 = 500\,\text{N/m},\ k = 500\,\text{N/m}$$

Constraint Surface:

$$x + y = 2 \quad \text{or}$$
$$\psi(q) = l_1 \cos\theta_1 + l_2 \cos(\theta_1 + \theta_2) + l_1 \sin\theta_1 + l_2 \sin(\theta_1 + \theta_2) - 2\,.$$

Initial Conditions: Coordinate of point A in constrain frame is: $(\sqrt{2}, -\sqrt{2}+0.05)$.
Desired Maneuver:
Move from $\boldsymbol{A} \rightarrow \boldsymbol{B}$; with a sine-on-ramp profile, i.e.

$$u_{\text{nd}}(t) = \frac{u_{\text{nd}}(t_{\text{d}}) - u_{\text{n}}(0)}{t_{\text{d}}} \left\{ t - \frac{t_{\text{d}}}{2\pi} \sin\left(\frac{2\pi}{t_{\text{d}}} t\right) \right\} + u_{\text{n}}(0)\,,$$

where u_{nd} is the desired trajectory; t is time; and t_{d} is the time required for the maneuver. Note that $u_{\text{n}}(0)$ refers to the coordinate of initial location $\boldsymbol{A}$ in the tangent space, i.e. $u_{\text{n}}(0) = -\sqrt{2} + 0.05$. At the same time, it is also desired to maintain a constant normal force of 10 N in the task space.

The contact surface is assumed to be rigid when $u_{\text{n}} < 0$ and stiff when $u_{\text{n}} \geqslant 0$. Control law is switched at the boundary of the two types of surfaces. Values of the gains for the decoupled linear position controller are selected as $K_3 = 168$, $K_2 =$

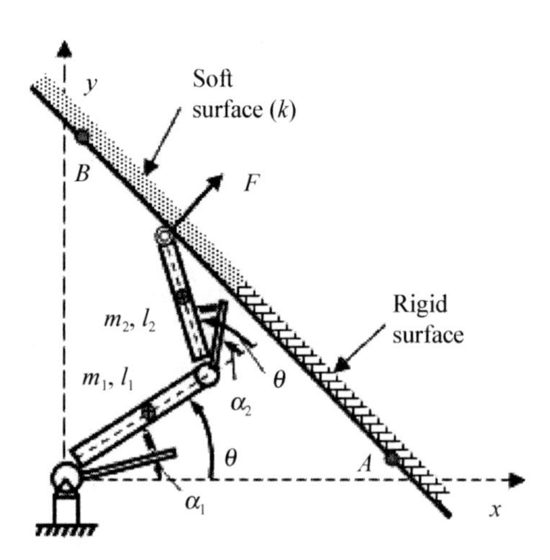

Fig. 9.3 Simulation example of the switched force control

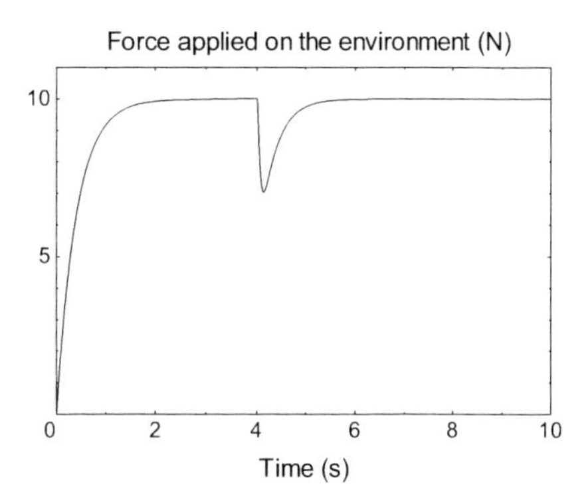

Fig. 9.4 Time history of the Contact force

7696, $K_1 = 53760$, and $K_0 = 102400$, which are essentially the coefficients of the fourth polynomial $(s+80)^2(s^2+8s+16)=0$. The roots of the second-order system $(s^2+8s+16)$ are dominant ones since they are less than one tenth of the third and fourth roots (-80). Hence, with this set of gains, it is expected that the end-effector's position exhibit critically damped response with natural frequency of $\omega_n = 4\,\text{rad/s}$ corresponding to a step-input. Values of the gains in the force controller for the rigid environment (Eq. 9.11) are selected as $K_v = 100$, $K_p = 250$. Values of the gains in the force controller for the stiff environment (Eq. 9.15) are selected as $K_{c0} = 409600$, $K_{c1} = 163840$, $K_{c2} = 10304$, $K_{c3} = 184$, which is essentially the coefficients of the fourth polynomial, $(s+80)^2(s^2+16s+64)=0$. Similarly, response of the contact force with the stiff environment is expected to be critically damped with natural frequency of $\omega_n = 8\,\text{rad/s}$ corresponding to a step-input.

The time history of the contact force is plotted in Fig. 9.4. Figure 9.5 presents the time history of the system variables including response of the links (θ_1, θ_2), tracking error of the links' angle (e_1, e_2) and the control inputs (τ_{m_1}, τ_{m_2}). With the decoupled nonlinear position/force controller, first, the desired trajectory tracking and contact force is achieved when the end-effector is on the rigid surface. Control switching happens at the same time when the end-effector is in contact with the soft surface. Switching of the control law happens at $t = 4\,\text{s}$. Tracking error occurs when the end-effector is in contact with the soft surface. This is necessary since the contact force is generated due to the deformation of the surface. There is also a short period of variation of the contact force from the desired value. This can be explained by the robotic dynamics at the time of the surface type change. When the end-effector moves from rigid surface to soft one, the end-effector dips into the soft surface resulting in a sudden change of the angular position of the links in order to maintain the 10 N desired contact force. Since the control law switches while the end-effector is still moving, dramatic change of the control inputs is observed. The maximum control input goes to around 400 Nm. This sudden change in the control inputs may not be acceptable in practice depending on the capacity of the motor. In order to prevent damage to the hardware, limits on the control inputs can be imposed.

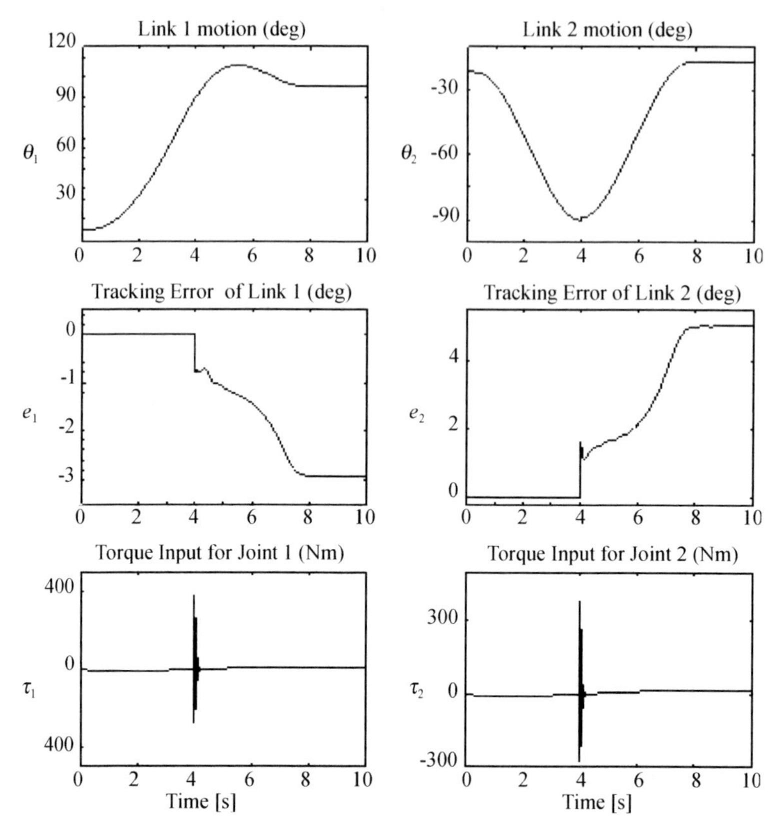

Fig. 9.5 Response of the system variables

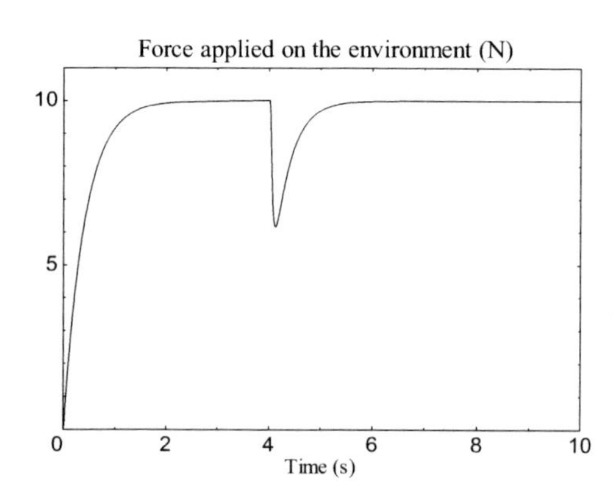

Fig. 9.6 Contact force with torque input limit ±50 Nm

The thresholds are set to −50 to 50 Nm. Figures 9.6 and 9.7 show the updated system response. No significant change can be observed in the response of either the tracking error or the contact force. At the time of control switching, the variation of

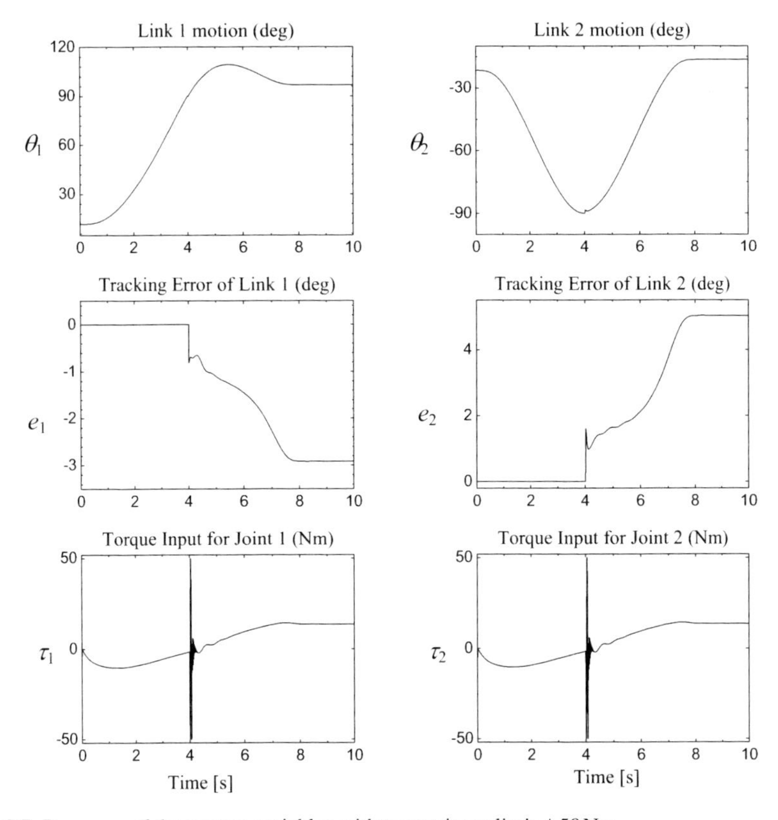

Fig. 9.7 Response of the system variables with torque input limit ± 50 Nm

the contact force is only slightly higher than the no threshold case. Setting limit on the torque input or voltage to the actuator is a very common practice in reality. However, the value of the threshold cannot be lower than what is needed to achieve the desired contact force. Results indicate that as long as such limits do not interfere with the minimum torque required to hold the system at desired location and desired contact force. As can be seen in Fig. 9.7, the control input for joint 1 is about 12 Nm at stable state. If the threshold is set to be lower than 12 Nm, the system is unstable.

Acknowledgements This research was conducted at the Intelligent Manufacturing Systems (IMS) Center where the first author held a Post Doctoral Fellowship. The research support from The Natural Science and Engineering Research Council (NSERC) of Canada is greatly appreciated.

References

Arimoto S., Liu Y., Naniwa T., 1993, Model-based adaptive hybrid control for geometrically constrained robots. Proc IEEE International Conference on Robotics and Automation. Atlanta GA USA 163–73.

Cao Y., de Silva C.W., 2006, Supervised switching control of a deployable manipulator system. International Journal of Control and Intelligent System 34/2:153–165

De Luca A., Manes C., 1991, Hybrid force/position control for robots in contact with dynamic environments. Proc of Robot Control SYROCO 377–382.

Featherstone R., 2003, A dynamic model of contact between a robot and an environment with unknown dynamics. In: Jarvis R.A., Zelinsky A. (eds) Robotics Research: The Tenth International Symposium. Springer, Berlin Heidelberg Bew York, pp 433–446

Hogan N., 1985, Impedence Control: An approach to manipulation: Part I – Theory; Part II – Implementation; Part III-Applications. J Dyn Syst Meas Control 107:1–24

Jankowski K.P., ElMaraghy H.A., 1991, Nonlinear decoupling for position and force control of constrained robots with flexible joints. Proc of IEEE Inter Conf on Robotics and Automation. Sacramento California USA 2:1226–1231

Karan B., 2005, Robust position-force control of robot manipulator in contact with linear dynamic environment. Robotica 23:799–803

McClamroch N.H., 1989, A singular perturbation approach to modeling and control of manipulators constrained by a stiff environment. Proc of the 28th IEEE Conference on Decision and Control. Tampa FL USA 2407–2411

McClamroch N.H., Wang D., 1988, Feedback stabilization and tracking of constrained robots. IEEE Trans Auto Control 33/5:419–426

Raibert M.H., Craig J.J., 1981, Hybrid position/force control of manipulators. J Dyn Syst Meas Control 102:126–133

Siciliano B., Villani L., 1996, A passivity-based approach to force regulation and motion control of robot manipulators. Automatica 32/3:443–447

Villani L., deWit C.C., Brogliato B., 1999, An exponentially stable adaptive control for force and position tracking of robot manipulators. IEEE Trans Rob Auto 44:778–802

Vukobratovic M., Stojic R., Ekalo Y., 1998, Contribution to the position/force control of manipulation robots interacting with dynamic environment- a generalization. Automatica 34/10:1219–1226

Vukobratovic M., Ekalo Y., 1996, New approach to control of robotic manipulators interacting with dynamic environment. Robotica 14:31–39

Whitcomb L.L., Arimoto S., Naniwa T., Ozaki F., 1997, Adaptive model based hybrid control of geometrically constrained arms. IEEE Trans Rob Auto 13:105–116

Yoshikawa T., Sugie T., Tanaka M., 1988, Dynamic hybrid position force control of robot manipulators: controller design and experiment. IEEE Trans Rob Auto 4/6:699–705

Yoshikawa T., Zheng X.Z., 1993, Coordinated Dynamic Hybrid Position/Force Control for Multiple Robot Manipulators Handling One Constrained Object. Int J Robot Res 12/3:219–230

Process Planning

Chapter 10
Reconfiguring Process Plans: A New Approach to Minimize Change

A. Azab, H. ElMaraghy and S.N. Samy[1]

Abstract In a customer driven market, the increasing number of product variants is a challenge most engineering companies face. Unpredictable changes in product design and associated engineering specifications trigger frequent changes in process plans, which often dictate costly and time consuming changes to jigs, fixtures and machinery. Process Planning should be further developed to cope with evolving parts and product families, increased mass customization and reduced-time-to-market. Agility and responsiveness to change is important in process planning. The current methods do not satisfactorily support this changeable manufacturing environment. They involve re-planning or pre-planning, where new process plans are generated from scratch every time change takes place, which results in production delays and high costs due to consequential changes and disruptions on the shop floor. The obvious cost, limitations and computational burden associated with the re-planning/pre-planning efforts are avoided by the developed methods. A novel process planning concept and a new mathematical programming model have been developed to genuinely reconfigure process plans to optimize the scope, extent and cost of reconfiguration and to overcome the complexity and flaws of existing models. Hence, process planning has been fundamentally changed from an act of sequencing to that of insertion. For the first time, the developed methods reconfigure process plans to account for changes in parts' features beyond the scope of original product families. A new criterion in process planning has been introduced to quantify the extent of resulting plan changes and their downstream implications. The presented method was shown to be cost effective, time saving, and conceptually and computationally superior. This was illustrated using two case studies in different engineering domains. The developed hypothesis and model have potential applications in other disciplines of engineering and sciences.

Keywords Process Planning, Mathematical Programming, Product Evolution, Metal Cutting, Assembly

---

[1] Intelligent Manufacturing Systems (IMS) Center, University of Windsor, Canada

10.1 Introduction

In a world dominated by advancements in telecommunication and transportation means, national economies are gradually being dissolved into a single global one through international trade, direct foreign investments, treaties such as GATT, and labor migration. In effect, the manufacturing globe is flattening through practices like supply chain, strategic alliances, subcontracting, outsourcing, leasing of manufacturing facilities and virtual corporations. As a result, competition has increased exponentially and markets have been characterized by turbulent patterns; demand is continuously varying with increasing uncertainty. Following customer needs and satisfaction have become important strategic goals for enterprises to survive and maintain their market shares.

One of the main policies leading manufacturing enterprises is adopting increasing product diversity and customization; economies of scale and mass production paradigms are gradually being shifted to economies of scope and mass customization. Market share per product model has decreased significantly. For most industries, a significant percentage of this change takes place after start of production; for example, 50 % of product variants in the automotive industry are generated after start up (Schuh and Eversheim, 2004). Change not only has to be realized, but it also has to done responsively and cost effectively by adapting agile and lean manufacturing and business paradigms. Changeability addresses these needs on the system level through flexible manufacturing and future reconfigurable manufacturing. On the production planning and control level, existing concepts and practices for the different support functions such as capacity planning, process planning and scheduling have to be further developed and evolved in order to meet these challenges.

In order to cope with this mandated continuous, iterative and rapid product design changes, instead of re-planning from scratch each time a change is demanded, it seems intuitive to modify or rather reconfigure existing plans according to the desired design changes and could prove to be a far better approach. Not only is computational complexity dramatically reduced, and hence computational time is saved, but also more efficient optimized solutions are obtained that could be implemented in minimal time and with less effort and cost. A new performance index has been introduced to quantify resulting changes in a process plan; a reconfiguration metric, which evaluates the extent of reconfiguration in the generated plans is formulated; the less the change in the resulting process plans, the better the plans. Change is costly; changes in a process plan immediately translate into changes in all related downstream activities on the shop floor including reconfiguring machine tools, changing existing manufacturing setups and tooling, re-training of labor on new plans, possible resulting quality issues, ramp-up time, opportunity cost, scheduling changes and the like. In this chapter, a new process planning approach is presented to minimize this change. Application and verification of the developed concepts, model and method have been carried out in more than one manufacturing domain.

10.2 Related Work

The witnessed evolution of Parts/Products Families has made obvious the need for "Evolvable and Reconfigurable Process Plans", which are capable of responding efficiently to both subtle and major changes in "Evolving Parts/Products Families" and changeable and Reconfigurable Manufacturing Systems (ElMaraghy, 2006 and ElMaraghy et al., 2008). Agility is necessary to produce individualized products for the constantly reconfiguring companies structures (Warnecke, 1993) and their production systems. There is a dearth of literature that addresses the problem of genuinely reconfiguring process plans. Existing methods may be classified as either pre-planning or re-planning efforts. The so-called non-linear process planning is an example of pre-planning scenarios, where alternate process plans are developed and stored ahead of time in anticipation of potential future changes. There is an obvious cost and computational burden involved in this approach for changes that may not materialize. Total re-planning, where for every product design change a whole new process plan is re-created, with limited benefit from available plans with their existing fixtures (setups) and tooling, also represents a major cost for manufacturers.

Process planning can be characterized as either variant or generative. Retrieval-type process planning techniques, based on a master template of a composite part, lend themselves to Reconfigurable Manufacturing Systems (RMS) that are predicated on a defined part/product family. Hetem (2003) discussed research, development and deployment of concepts and technologies to develop variant process planning systems for RMS. Bley and Zenner (2005) presented an overall integrated management concept based on variant planning by generating a generalized product model. Both papers presented a strictly variant type system, which did not support the introduction of new features into the part family caused by changing demands. Generative process planning is better able to handle products variety by generating process plans from scratch using rule- and knowledge-based systems, heuristics and problem specific algorithms (ElMaraghy, 1993). Pure generic generative process planning systems are not yet available. Azab and ElMaraghy (2007c) developed a hybrid sequential process planning approach, where both variant and generative planning are combined sequentially.

Generative mathematical modeling and programming are not generally used, but rather informal procedural methods are utilized in process planning (Azab, 2003) and solved using either non-traditional optimization methods or search heuristics. More specifically, generative process planning solutions for changeable and Reconfigurable Manufacturing Systems (RMS) is lacking. Xu et al. (2004) presented a clustering method for multi-part operations. Reconfigurable Machine Tools (RMT), tolerance-based and concurrent machining-based clustering methods for a single part were proposed. Shabaka and ElMaraghy (2005) developed an approach for selecting different types of machines and their appropriate configurations to produce different types of parts and features, according to the required machines capabilities. The structure of machine tools was represented by a kinematic chain. More than one minimum machine configuration for a single operation cluster was generated, hence, increasing the flexibility in machine tool selection and operations assignment (Shabaka and ElMaraghy, 2007). Shabaka and ElMaraghy (2006) also developed a Genetic Algorithms (GAs) method for operation selection and operation sequencing, where operations that have related tolerance or logical constraints, are clustered together and manufactured on the same machine. Azab et al. (2006) tailored a random-based heuristic, based on Simulated Annealing to solve the same problem. At the end of each solution iteration, an evolutionary mutation operator was applied to increase the search efficiency by enlarging the explored solution space. Jin et al. (2007) introduced a novel method of process route and layout design to accelerate and rationalize the reconfiguration process of an RMS. A directed network model based on graph theory was constructed. Azab and ElMaraghy (2007b) suggested a formulation based on the Quadratic Assignment Problem (QAP), which overcame the complexity of sub-tour elimination in the classical TSP formulations and modeled the fundamental precedence constraints for the first time in the literature. Song et al. (2007) presented a dynamic CAPP system structure to support RMS, where dynamic resource allocation using neural networks was suggested.

Hitherto, process planning is generally reviewed. Understandably, metal-cutting is the natural application whenever process planning is mentioned in the literature. First, the domain of machining requires extensive process planning; second, it is a fundamental manufacturing process that represents a significant percentage among others. For other processes, such as assembly, the developed general approach still holds. The proposed conceptual planning framework is applied and verified in assembly where the main objective is the same, i.e. to determine the feasible optimal assembly sequence that minimizes assembly handling time. Most reviewed literature also conceptually considered planning from scratch. At this macro planning level, which is the scope of this chapter, assembly planning remains a combinatorial optimization problem. Hence, in the last two decades, increased application of non-traditional optimization and graph-theoretic methods was observed. For example, Chen (1990) proposed Hopfield neural networks to solve a Traveling Salesperson Problem (TSP) formulation for automated assembly planning, where AND/OR precedence relationships were mapped into networks of neurons. Park and Chung (1991) graphically modeled the problem, where all possible planning alternatives were exhaustively enumerated. Parallelism of assembly tasks, whether by the use of multiple robots or workers, was taken into account. A graph-theoretic approach was employed by Laperriere and ElMaraghy (1994) to generate an optimal assembly sequence. Only the feasible search space was considered by including the precedence constraints when generating the search graph and an A^* algorithm was used. Huang and Wu (1995) recommended backward rather than forward search as being more efficient by considering the dis-assembly of the same product. Zhao and Masood (1999) employed a graph set technique for creating an assembly model.

Guan et al. (2002) presented a hierarchical evolution algorithm approach, where a compound chromosome encoding was constructed to represent the abundant assembly process information. Geometric reasoning was used to distinguish the geometric feasibility of the chromosomes. Del Valle et al. (2003) developed a model to support multi-robotic assembly environments to minimize the total assembly makespan by using Genetic and greedy algorithms. Galantucci et al. (2004) proposed a hybrid fuzzy logic Genetic Algorithms (GAs) method to plan the automatic and optimal assembly and dis-assembly operations. Tseng et al. (2007) considered the global logistic supply chain aspect by formulating a mathematical programming model to evaluate all the feasible multi-plant assembly sequences.

The developed new model for reconfiguring process plans at the macro (process sequence) level applies to many manufacturing processes and has been applied to both metal cutting and assembly. This novel approach, which genuinely reconfigures a process plan rather than re-generates it, has been to the author's knowledge introduced for the first time in literature by ElMaraghy, H. and Azab, A. (see reference list). The concept, its rationale and formulation and results of its applications (e.g. in metal cutting) are cohesively and concisely reported in this chapter. In addition, its application in assembly, in collaboration with S. Samy, is also described in Sect. 10.7.1.

10.3 Conceptual Basis

Reconfigurable Process Planning (RPP), a new term coined by (ElMaraghy, 2006), represents partial reconfiguration of process plans for new products some features/operations of which are not within the boundaries of an existing product family and its composite model and master plan. This means that the new part/product belongs to an evolving parts/products family where features are added or removed over time to satisfy customer demands and technological drivers. A novel approach has been presented to reconfigure the master plan or plans of existing parts/products on the workshop to meet the requirements of the new part/product and its features/operations, with the objective of minimizing the differences between the new and the old plans. Therefore, instead of generating the new plans from scratch, only new portions of the process plan, corresponding to the new additional features/operations (and their machining operations), are generated and optimally positioned within the overall process plan (Azab and ElMaraghy, 2007a). This is a new approach, which enables local reconfiguration of process plans when needed, where needed and as needed, while minimizing the extent of change/reconfiguration and its associated cost.

Initially, the new product/part is compared with the original existing one to identify the new and missing features/operations. The missing features are simply subtracted from the original sequence. For new features/operation, if the sequence is thought of as a genetic sequence, the added new features/operations would represent mutation of that sequence by optimally inserting new genes as shown in Fig. 10.1. This metaphor is consistent with the concept of evolving part families (ElMaraghy, 2006 and 2007). An innovative mathematical formulation, using 0–1 integer programming, was developed by Azab and ElMaraghy (2007a) and algo-

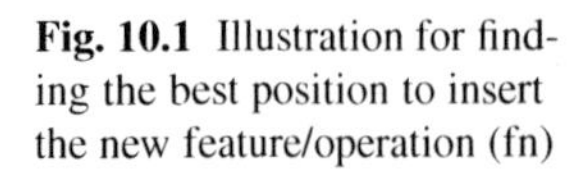

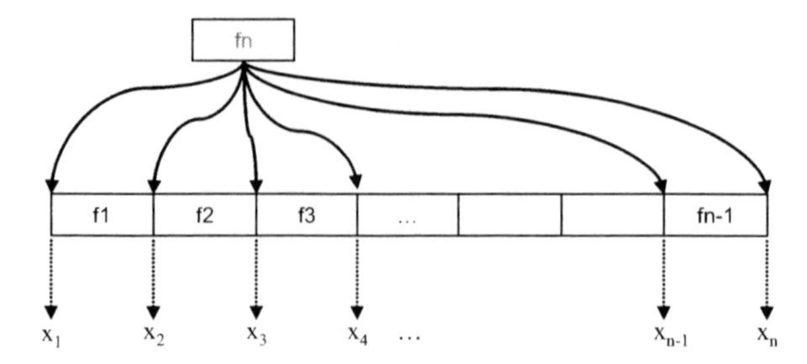

Fig. 10.1 Illustration for finding the best position to insert the new feature/operation (fn)

rithms for its automation and solution were presented. In this model, the new set of features/operations of the new part/product is inserted iteratively into the original sequence of the existing part/product. The problem is subject to a number of precedence constraints and the objective of minimizing the added handling time spent mainly in re-fixturing and tool changes. These handling time changes, to be minimized, imply the possible changes of setup and tooling, and the consequential other implications on the shop floor such as machine tool reconfiguration or re-assignment, the need to retrain personnel on new plans, possible resulting quality errors, downtime, opportunity cost, and so forth. A work piece is defined in this context as the stock in-process in the case of metal cutting or the assembly in-process in case of assembly.

10.4 Mathematical Modeling and Programming

The typical input to the process planning problem is the new part/product's CAD feature-model or blue prints. For the Reconfigurable Process Planning (RPP) problem, features/operations data for the old and new variants, such as setups and tools used, tool access directions and so forth, are input. It is required to reconfigure the process plan of the existing part/product to arrive at the new one minimizing the differential change between the two plans. The large number of interactions that exists between the different form features/operations constituting the part/product complicates the problem. These interactions generate precedence constraints for related features and operations and are modeled using Feature Precedence Graphs (FPG) and Operation Precedence Graphs (OPG). These are tree-like structures of nodes and arcs (directed graphs) where arcs between nodes represent features/operations precedence. FPG/OPGs for the existing part/product are retrieved and edited by adding and removing nodes and arcs. Only logical changes on the part/product level are taken into account; any feature design change that do not result into modifications in the logical precedence relationships are not considered changes from process planning point of view.

Different types of precedence constraints (or the so called anteriorities by Halevi and Weill (1995)), for which certain sequences cannot be reversed, do exist. Features' Precedence Graphs (FPGs) are manipulated, by adding and removing nodes, in order to accommodate additional as well as missing features in the new parts. They are translated into an Operation Precedence Graph (OPG), where each feature corresponds to one or more operation. A machining feature is defined in this work as a geometrical feature that requires processing by one or more operation. For metal-cutting, planning is carried out on the features level taking the following into considerations: Within each feature, logical sequence of operations is used to order the feature's sub-operations. Some features are represented by more than one node in exceptional cases due to interdependence of precedence relationships with other features. The ratio of the time required to re-fixture the work piece to that used for tool change is taken to be 2:1 based on practical experience. Planning for assembly is performed on the operation level.

The following notations are used: n denotes the problem size; it could also be interpreted as the total number of machining features/operations including the new machining feature/operation to-be-inserted; $C = [c_{i,j}]$ is the precedence penalty matrix; $S = [s_{i,j}]$ is the work piece repositioning on given fixtures (setups) time matrix; $Os = \{Os_i\}$ is the work piece repositioning on fixtures (setups) time matrix for original features/operations (i.e. not to include the new feature/operation) after subtracting the missing features/operation. $Tr = \{Tr_i\}$ is the right tool change time vector (i.e. the tool change between the new to-be-inserted feature/operation and every feature/operation in the old sequence from the right side). $Tl = \{Tl_i\}$ is the left tool change time vector (i.e. the tool change between the new to-be-inserted feature/operation and every feature/operation in the old sequence from the left side). $Ot = \{Ot_i\}$ is tool change time vector for original features/operations- not including the new feature/operation to-be-inserted.

The decision variables are: x_i is a 0–1 integer variable, where i runs from 1 to n; 1 if new feature/operation is inserted at position i; 0 otherwise. The position i takes the value 1 when the new feature/operation is inserted right before the first feature/operation of the original array of operations and takes the value n when it is positioned right after the last feature/operation of the original array, i.e. feature or operation $fn-1$.

Two criteria are considered (Azab and ElMaraghy, 2007a): 1) time for refixturing, and 2) time for tool change. The objective is to minimize the handling time. The time spent for rapid tool traverse from one feature/operation to the other is ignored due to its relatively minor contribution. Also the time required for transportation of the work piece between different workstations as well as that spent on adjusting process parameters are also ignored since these detailed parameters are not determined at the considered macro planning level. The objective function is:

$$\min \sum_{i=1}^{n} \sum_{j=1}^{n-1} C_{i,j}.x_i + \sum_{i=1}^{n} \left(\sum_{k=1}^{n-1} S_{i,k} \right) .x_i - \sum_{i=1}^{n} Os_i.x_i + \sum_{i=1}^{n} (Tr_i + Tl_i).x_i - \sum_{i=1}^{n} Ot_i.x_i \,. \tag{10.1}$$

The first term represents the penalty for violating precedence constraints, where the precedence relation between every feature/operation in the original sequence and the inserted new feature/operation is checked. The second term represents the re-fixturing time (i.e. setup change time as commonly referred to in the literature). The first summation of $S_{i,k}$ over k represents the work piece repositioning time on the different given fixtures/setups associated with a new sequence, i.e. between every pair of preceding features/operations in each new permutation. The terms Tr_i and Tl_i with their summation over i from 1 to n represent the tool change time. They account for the new tool change time due to the insertion of the new feature/operation between two existing features/operations in the original sequence- one to the right (Tr_i) and one to the left (Tl_i). Finally, the Os_i and Ot_i terms represent the time incurred due to changing the original precedence between the two features/operations that are separated by inserting the new feature/operation, and hence, the old re-fixturing and tool change time are subtracted.

$$\sum_{i=1}^{n+1} x_i = 1\,. \tag{10.2}$$

The constraints system of the RPP model is advantageously simple and is represented by Eq. 10.2. This constraint prevents a feature/operation from being inserted more than once at any position. The constraint equations of the RPP model are far less complicated than other classical models reported in literature.

10.5 A New Criterion in Process Planning

A new performance index has been formulated to evaluate the extent of reconfiguration of the process plan. The Plan Reconfiguration Index is used to evaluate the quality of the generated process plans. It is a measure of the extent of reconfiguration and changes that occur due to the reconfigured plans. This represents a new direction in process planning and a novel criterion aimed at minimizing the resulting disruption in downstream activities on the shop floor (Azab and ElMaraghy, 2007a).

The Plan Reconfiguration Index (RI_{Plan}), as expressed by Eq. 10.3, consists of two components: added handling tasks of re-fixturing/setups and tool changes. Weights are used to normalize the index and reflect the relative importance of its respective terms.

$$RI = \left(\begin{array}{l} \alpha \dfrac{\text{Number of new/missing acts of re-fixturing in new plan}}{\text{Total number of work piece repositioning of master plan}} \\ +(1-\alpha)\dfrac{\text{Number of added/missing tool changes in new plan}}{\text{Total number of tool changes of master plan}} \end{array} \right) \cdot 100\,. \tag{10.3}$$

The higher the value of RI_{Plan}, the more extensive is the process plan reconfiguration and its associated cost. The value of α is proportional to the average amount of time to reposition the work piece on a new fixture relative to the time required to change a tool. For example, if the ratio of time taken to re-fixture a work piece to that required to change a tool is presumed to be 3:1, then α would take a value of $3/(4+1) = 0.75$.

10.6 Computational Time Complexity

The developed method is far superior computationally compared with available methods in literature and in practice. The comparison of the computational time complexity of the proposed model with the classical re-planning using a TSP model raises some interesting observations. The Traveling Salesperson Problem (TSP) model is a network of nodes representing different features/operations of a part/product connected by arcs that represent the routes between them. The decision variable is the value associated with the arc; if the value is 1 then the route represented by this arc is in use; it is 0 otherwise. An arc in use means that the two features/operations are sequenced consecutively. This model is characterized by its exponentially growing solution space and the complexity of the sub-tour elimination constraints. The picture is completely different for the developed RPP model. The time complexity for inserting m feature/operation into an original sequence of size n grows polynomially (Azab, 2008); for one iteration, the solution space is n; the total solution space is of the form $n+(n+1)+\ldots+(n+m)$. The RPP optimization problem is by far more tractable since it offers a computational time complexity of $O(n)$ compared with the NP-complete exponentially growing classical TSP counterpart. Hence, typical industrial problems can be easily solved for optimality using the RPP model.

10.7 Application and Verification

10.7.1 Reconfigurable Assembly Planning of a Family of Household Products

The assembly process planning for a family of household products is the subject of this section. Two different variants of a small kitchen appliance (a kettle) were considered. The original and new product design as well their OPGs are shown in Fig. 10.2. Design changes were made as a result of performing Design For Assembly (DFA). The part count in the new Electric Kettle variant was reduced from 24 to 22. The components of the Power Base Lower sub-assembly were all combined into part

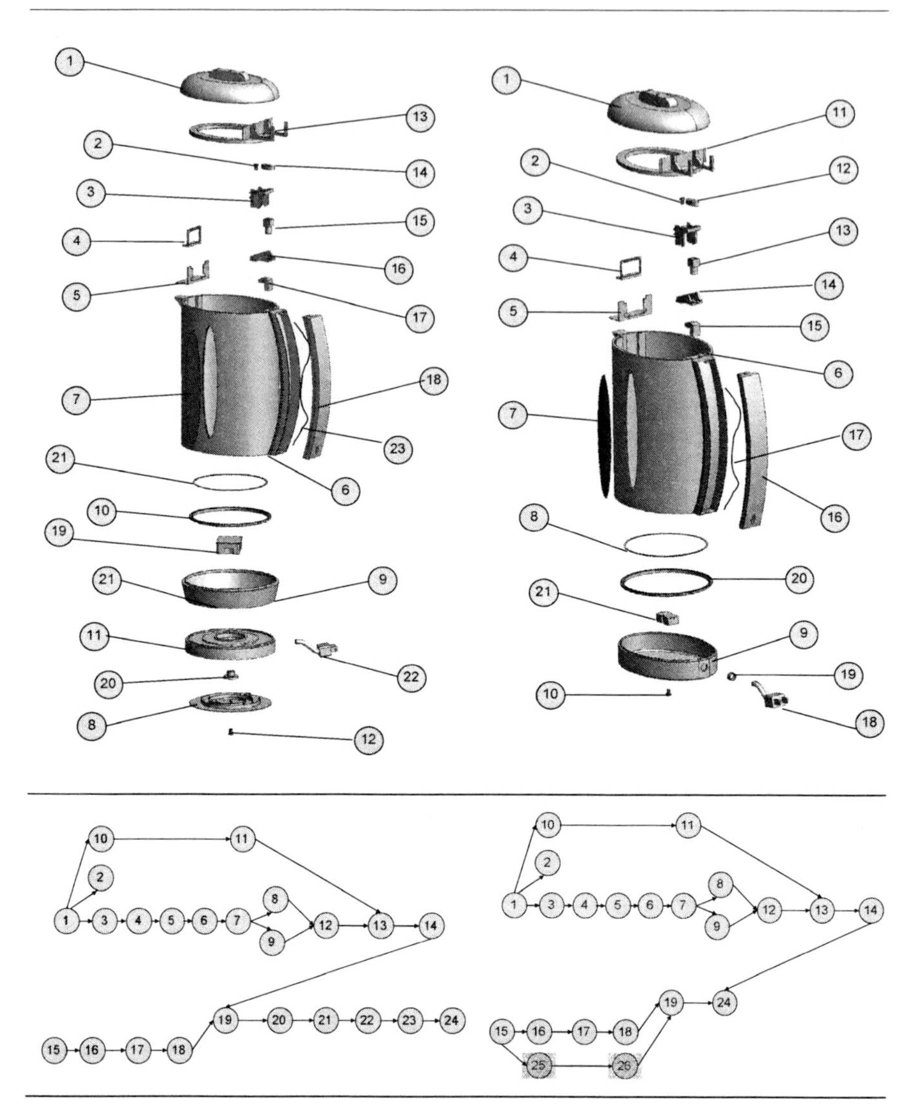

Fig. 10.2 Old/existing (*left*) versus new (*right*) electric kettle exploded diagrams and corresponding operations precedence graphs (OPGs)

#9 in the new Kettle variant. For more detailed account and full data of this case study, refer to Azab (2008). Sample details related to the assembly operation are given in Table 10.1. Three different setups are needed for the new design compared to two for the original design. Two setups are used for the main body in two opposite assembly directions (one position where the kettle would be upright and another

Table 10.1 Sample assembly operations details of the original electric kettle

Assembly Operation ID	Operation Description	Setup Used
1	Put main body (Part #6)	Vertical, Upright
2	Fix water indicator (Part #7)	Vertical, Upright
6	Insert steam tube (Part #15)	Vertical, Upright
7	Insert steam separator (Part #3)	Vertical, Upright
8	Fix screw (Part #2)	Vertical, Upright
16	Insert controller (Part #19)	Body lower Setup
17	Insert heating plate (Part #10)	Body lower Setup
18	Insert heating O'Ring (Part #21)	Body lower Setup

where it would be upside down), and a third setup to assemble the Body Lower sub-assembly.

Two iterations were carried out to insert the two new operations as highlighted in the OPG of the new product variant given in Fig. 10.2. In Tables 10.2 and 10.3, setup and tool change formulation matrices and vectors for the second iteration are given respectively. The precedence cost matrix is a sparse matrix, where elements $c(1,4)$, $c(2,4)$, $c(3,4)$, $c(4,4)$, $c(21,20)$ and $c(22,20)$ are assigned a relatively large penalty of a 1000 time units. Manual assembly is performed and hence, the tool change component in the handling time objective function is absent in this case, hence, all tool change vectors are zero vectors.

The given plan for the original variant of Kettle is {15, 16, 17, 18, 1, 10, 11, 2, 3, 4, 5, 6, 7, 9, 8, 12, 13, 14, 19, 20, 21, 22, 23, 24}. Missing assembly operations in the new kettle (variant 2) were subtracted resulting in the following sequence {15, 16, 17, 18, 1, 10, 11, 3, 4, 5, 6, 7, 9, 8, 12, 13, 14, 2, 19, 24}. Results of each iterative step of the RPP solution method are given in Table 10.4, where the new inserted operations are highlighted in bold face. The value of the objective function is 2 time units corresponding to 2 acts of re-fixturing. It should be noted that each act of re-fixturing is assumed to take an arbitrary time period of one unit time in this case study, since only re-fixturing of the work piece is considered.

This case study demonstrated the strength of the proposed approach. Only those design changes that cause logical/precedence changes on the product level, make a difference on a process planning level. For example, although the design of the Body Lower Sub-assembly is changed in the new product variant, its assembly is considered the same from planning perspective, since logically, the precedences are the same; both operations attach the lower sub-assembly into the main body regardless of the DFA enhancements in that lower assembly. On the other hand, some other operations in the original assembly and its corresponding ones in the modified one

Table 10.2 Old work piece repositioning time vector, S, for the second iteration

0	0	0	0	0	1	0	0	0	0	0	0	0	0	0	0	0	0	0	1	0	0

Table 10.3 Work piece repositioning time matrix C for the second iteration

0	0	0	0	0	1	0	0	0	0	0	0	0	0	0	0	0	0	0	1	0
0	0	0	0	0	1	0	0	0	0	0	0	0	0	0	0	0	0	0	1	0
0	0	0	0	0	1	0	0	0	0	0	0	0	0	0	0	0	0	0	1	0
0	0	0	0	0	1	0	0	0	0	0	0	0	0	0	0	0	0	0	1	0
0	0	0	0	0	1	0	0	0	0	0	0	0	0	0	0	0	0	0	1	0
0	0	0	0	0	1	0	0	0	0	0	0	0	0	0	0	0	0	0	1	0
0	0	0	0	1	1	1	0	0	0	0	0	0	0	0	0	0	0	0	1	0
0	0	0	0	1	0	1	1	0	0	0	0	0	0	0	0	0	0	0	1	0
0	0	0	0	1	0	0	1	1	0	0	0	0	0	0	0	0	0	0	1	0
0	0	0	0	1	0	0	0	1	1	0	0	0	0	0	0	0	0	0	1	0
0	0	0	0	1	0	0	0	0	1	1	0	0	0	0	0	0	0	0	1	0
0	0	0	0	1	0	0	0	0	0	1	1	0	0	0	0	0	0	0	1	0
0	0	0	0	1	0	0	0	0	0	0	1	1	0	0	0	0	0	0	1	0
0	0	0	0	1	0	0	0	0	0	0	0	1	1	0	0	0	0	0	1	0
0	0	0	0	1	0	0	0	0	0	0	0	0	1	1	0	0	0	0	1	0
0	0	0	0	1	0	0	0	0	0	0	0	0	0	1	1	0	0	0	1	0
0	0	0	0	1	0	0	0	0	0	0	0	0	0	0	1	1	0	0	1	0
0	0	0	0	1	0	0	0	0	0	0	0	0	0	0	0	1	1	0	1	0
0	0	0	0	1	0	0	0	0	0	0	0	0	0	0	0	0	1	1	1	0
0	0	0	0	1	0	0	0	0	0	0	0	0	0	0	0	0	0	1	1	0
0	0	0	0	1	0	0	0	0	0	0	0	0	0	0	0	0	0	1	1	1
0	0	0	0	1	0	0	0	0	0	0	0	0	0	0	0	0	0	1	0	1

Table 10.4 Solution iterations

Before Iteration 1	{15, 16, 17, 18, 1, 10, 11, 3, 4, 5, 6, 7, 9, 8, 12, 13, 14, 2, 19, 24}
Before Iteration 2	{15, 16, 17, **25**, 18, 1, 10, 11, 3, 4, 5, 6, 7, 9, 8, 12, 13, 14, 2, 19, 24}
Final Sequence	{15, 16, 17, 25, 18, **26**, 1, 10, 11, 3, 4, 5, 6, 7, 9, 8, 12, 13, 14, 2, 19, 24}

produced different logical precedence relationships; hence, in spite of them being technically identical they are considered different entities at the operations macro-planning level. The value of RI_{plan} is zero indicating the absence of any additional handling tasks (work piece re-fixturing) for the new product variant, hence, zero added assembly setups or workstations.

10.7.2 Reconfigurable Process Planning for Machining of a Front Engine Cover Part Family

An engine front cover family of parts is used in this example. The cover belongs to an aluminum single-cylinder, air-cooled engine with overhead valves. The aluminum front cover, shown in Fig. 10.3, is die cast to the near net shape; finish machining is required for precision features and the tapped holes. Two variants of the

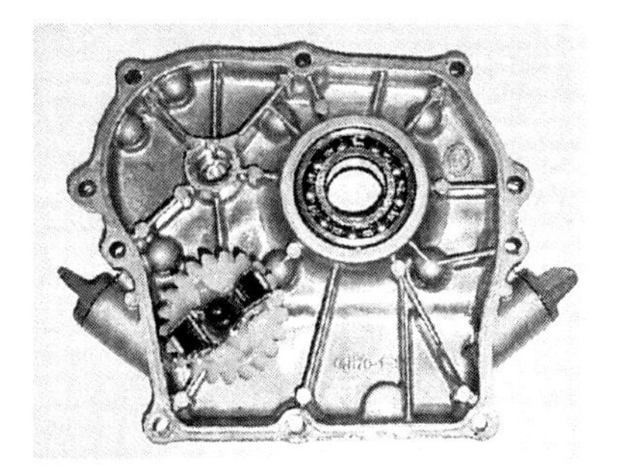

Fig. 10.3 Single cylinder engine front cover part

front cover are given: an original existing one, which is currently being machined on the shop floor and a new instance with new and missing features. For detailed description of the case study and its data, see (Azab and ElMaraghy, 2007c).

A three axis horizontal Reconfigurable Machine Tool is used to machine the original part, hence three setups are required to produce the part in order to access the features on the front and back faces and the side face. The ratio of the time required to re-fixture the work piece (composed mainly of unloading the work piece, cleaning the setup, and loading the work piece) to the tool change time is assumed 2:1 based on practical experience. The original three-axis horizontal configuration of the RMT was found not to be sufficient for producing the new part; an extra dedicated setup or two special angle head tools would be required to machine the tapped hole for the second oil plug, for example. Hence, it was suggested that the machine tool be reconfigured by adding an appropriate rotational axis of motion to the spindle or table, i.e. the RMT transforms into a 4-axis horizontal machining center; alternatively one more setup would have to be added.

The macro-process plan for the old engine cover, solved using a Genetic Algorithms toolbox originally developed by Azab (2003), is retrieved. The common features are extracted from this solution by subtracting those that are not found in the new part. The RPP method is applied to find the optimum insertion positions for the new features. Seven iterations were performed to optimally insert seven new features. The obtained solution and its corresponding objective function value for the obtained reconfigured part are shown in Table 10.5.

The Plan Reconfiguration Index (RIplan) was found to be 90%, indicating that the master plan of the current old part was significantly reconfigured, i.e. the obtained best plan sequence for the new part was significantly different as a result of the planning reconfiguration.

Table 10.5 Final RPP solution and corresponding objective function value

Solution	**Obj. fn. value**
{f1, f5, f20, f19, f6, f23, f3r, f17, f8, f4, f2, f7r, f18, f7f, f21,f22, f3f}	21 time units

10.7.3 Concluding Remarks

Two industrial case studies in assembly and metal cutting were used for application and verification of the newly developed concepts, model and method. Planning for a household product family was carried out to demonstrate the applicability and soundness of the developed new RPP concepts in assembly planning. The difference between RI_{Plan} (0 % for kettle assembly and 90 % for cover machining) illustrates the differences between these two extreme examples with very minor vs. significant differences respectively among the variants and consequently the process plans.

Finally, it should be stressed that the developed novel reconfigurable process planning concept is very suitable for industries, where progressive product design changes are day-to-day reality and for low- to mid-volume production, which is almost the case now for most companies. However, it is important to note that for mass production and products with less frequent design changes, where completely new lines of production are custom-designed and built for the new parts/products or when totally new facilities are first put up, then process planning from scratch would be used to arrive at a totally refined, highly optimized process plans.

10.8 Summary

This chapter addresses a new problem that arises due to the increased changes in products and manufacturing systems and the need to manage these changes cost effectively and with the least disruption of downstream production activities and their associated high cost. It presented novel solutions that were developed to satisfy the need to frequently plan and re-plan manufacturing processes. One of the main contributions is the development of a new mathematical model for solving the classical problem of process planning through reconfiguration. Conceptually, planning was changed from an act of sequencing to one of subtraction and insertion. It is applicable to industries with progressive design changes and for certain low- to mid-volume production. The developed methods enrich the science of manufacturing systems on both a theoretical and practical levels by providing an important logical enabler to cope with continuously evolving products, and production technologies and manufacturing systems paradigms. The proposed enabler is essential for the realization of Changeable/Reconfigurable manufacturing yet not limited to this particular paradigm since frequent product changes are experienced in almost all types of manufacturing nowadays.

The newly developed reconfiguration performance index has been presented. It measures the extent of change on the process planning level. RI_{Plan} evaluates the impact of the process plans changes on all downstream shop floor activities, and helps choose among alternate sequences with substantially similar total cost by opting for the one that causes the least changes on the shop floor, which saves other indirect cost components such as those related to errors, quality issues due to changes, training of labor, and lost opportunity. Reconfigurable Process Planning (RPP) offers localized optimal plans, which minimize the distance, in the process planning domain, between the original/existing or master plans and the new ones, and hence minimizes the reconfiguration effort.

One of the main benefits of the proposed methods is to reduce the time and cost required to generate a process plan. The overall proposed methodology is more advantageous than existing methods, such as pre-planning scenarios, where alternate process plans are developed and provided ahead of time in anticipation of future changes. In addition to the obvious cost and computational burden that is avoided by the developed approach, future changes in products and technology cannot be fully predicted; hence, the usefulness of pre-planned alternatives is diminished. Furthermore, pre-planned processes would likely become obsolete as manufacturing resources and technologies are changed.

The presented process planning methods can improve the efficiency of process planning activities and can help "manage changes" on the shop floor by introducing an important changeability enabler in the field of process planning.

Acknowledgements This research was conducted at the Intelligent Manufacturing Systems (IMS) Center. The support from The Canada Research Chairs (CRC) Program and the Natural Science and Engineering Research Council (NSERC) of Canada is greatly appreciated.

References

Azab A., 2003, Optimal Sequencing of Machining Operations, Department of Mechanical Engineering. Cairo, Cairo University. M. Sc

Azab A., 2008, Reconfiguring Process Plans: A Mathematical Programming Approach, Dept. of Industrial & Manufacturing Systems Engineering. Windsor, University of Windsor. Ph.D.: 150

Azab A., ElMaraghy H., 2007a, Mathematical Modeling for Reconfigurable Process Planning. CIRP Annals 56/1:467–472, 10.1016/j.cirp.2007.05.112

Azab A., ElMaraghy H., 2007b, Process Plans Reconfiguration Using QAP Mathematical Programming, CIRP International Digital Enterprise Technology (DET) conference, Bath, UK

Azab A., ElMaraghy H., 2007c, Sequential Process Planning: A Hybrid Optimal Macro-Level Approach, Journal of Manufacturing Systems (JMS) Special Issue on Design, Planning, and Control for Reconfigurable Manufacturing Systems 26/3:147–160

Azab A., Perusi G., ElMaraghy H. and Urbanic J., 2006, Semi-Generative Macro-Process Planning for Reconfigurable Manufacturing, CIRP International Digital Enterprise Technology (DET) conference, Setubal, Portugal. Also, appeared in Digital Enterprise Technology Perspectives & Future Challenges, P. F. Cunha and P. G. Maropoulos, Springer Science pp 251–258

Bley H., Zenner C., 2005, Feature-Based Planning of Reconfigurable Manufacturing Systems by a Variant Management Approach, 2005 CIRP 3rd International Conference on Reconfigurable Manufacturing

Chen C.L.P., 1990, And/or Precedence Constraint Traveling Salesman Problem and Its Application to Assembly Schedule Generation, Los Angeles, CA, USA, IEEE

Del Valle C., Toro M., Camacho E.F., Gasca R.M., 2003, A Scheduling Approach to Assembly Sequence Planning. IEEE, Besancon, France

ElMaraghy H., AlGeddawy, T., Azab, A., 2008, Modelling Evolution in Manufacturing: A Biological Analogy, CIRP Annals 57/1:467–472, doi:10.1016/j.cirp.2008.03.136

ElMaraghy H.A., 1993, Evolution and Future Perspecives of Capp. CIRP Annals 1/42(2):739–751

ElMaraghy H.A., 2006, Reconfigurable Process Plans for Reconfigurable Manufacturing, Proceedings of DET2006, CIRP international conference on digital enterprise technology

ElMaraghy H.A., 2007, Reconfigurable Process Plans for Responsive Manufacturing Systems, Digital Enterprise Technology: Perspectives & Future Challenges, Editors: P.F. Cunha and P.G. Maropoulos, Springer Sc., ISBN: 978-0-387-49863-8, pp 35–44

Galantucci L.M., Percoco G., Spina R., 2004, Assembly and Disassembly Planning by Using Fuzzy Logic &Amp; Genetic Algorithms. International Journal of Advanced Robotic Systems 1/2:67–74

Guan Q., Liu J.H., Zhong Y.F., 2002, A Concurrent Hierarchical Evolution Approach to Assembly Process Planning. International Journal of Production Research 40/14:3357–3374

Halevi G., Weill R.D., 1995, Principles of Process Planning: A Logical Approach, Chapman & Hall

Hetem V., 2003, Variant Process Planning, a Basis for Reconfigurable Manufacturing Systems, CIRP 2nd international conference on reconfigurable manufacturing, Ann Arbor, MI, USA

Huang K.I., Wu T.-H., 1995, Computer-Aided Process Planning for Robotic Assembly, Phoenix, AZ, USA

Jin Z., Song Z.-H., Yang J.-X., 2007, Process Route and Layout Design Method for Reconfigurable Manufacturing Systems. Computer Integrated Manufacturing Systems 13/1:7–12

Laperriere L., ElMaraghy H.A., 1994, Assembly Sequences Planning for Simultaneous Engineering Applications. International Journal of Advanced Manufacturing Technology 9/4:231–244

Park J.H., Chung M.J., 1991, Automatic Mechanical Assembly Planning for a Flexible Assembly System, Charlottesville, VA, USA, IEEE

Schuh G., Eversheim W., 2004, Release-Engineering – an Approach to Control Rising System-Complexity. CIRP Annals – Manufacturing Technology 53/1:167–170

Shabaka A.I., ElMaraghy H.A., 2005, Mapping Products Machining Requirements and Machine Tools Structure Characteristics in RMS, International Conference on Changeable, Agile, Reconfigurable and Virtual Production

Shabaka A.I., ElMaraghy H.A., 2006, A Ga-Based Constraint Satisfaction Model for Generating Optimal Process Plans, Proceedings of DET2006, CIRP international conference on digital enterprise technology

Shabaka A.I., ElMaraghy H.A., 2007, Generation of Machine Configurations Based on Product Features. International Journal of Computer Integrated Manufacturing 20/4:355–369

Song S., Li A. and Xu L., 2007, Study of Capp System Suited for Reconfigurable Manufacturing System, Shanghai, China, Institute of Electrical and Electronics Engineers Computer Society, Piscataway, NJ 08855-1331, United States

Tseng Y.J., Jhang J.F., Huang F.Y., 2007, Multi-Plant Assembly Planning Models for a Collaborative Manufacturing Environment. International Journal of Production Research 45/15:3333–3349

Warnecke H.J., 1993, The Fractal Company – a Revolution in Corporate Culture, Springer-Verlag

Xu H., Tang R.-Z., Cheng Y.-D., 2004, Study of Process Planning Techniques for Reconfigurable Machine Tool Design. Journal of Zhejiang University 38:1496–1501

Zhao J., Masood S., 1999, An Intelligent Computer-Aided Assembly Process Planning System. International Journal of Advanced Manufacturing Technology 15/5:332–337

Production and Capacity Planning and Control

Chapter 11
Adaptive Production Planning and Control – Elements and Enablers of Changeability

H-H. Wiendahl[1]

Abstract Rapid changes are characterizing the market and supply situation of manufacturing companies. Changeability ensures their ability to act successfully in this environment. Logistics is one topic of changeability and in particular Production Planning and Control (PPC). This chapter presents a framework to design a changeable PPC system and discusses the relevant enablers for changeability as well as required change processes. Examples illustrate how to achieve PPC changeability.

Keywords Production Planning and Control (PPC), changeability, turbulence

11.1 Introduction

Today's manufacturing companies are caught in a situation of rapidly changing market and supplier conditions. Figure 11.1 shows a case study of an equipment manufacturer for the semi-conductor industry. The example demonstrates how today's product demand varies in turbulent markets: Within five years the annual product demand for a specific machine type changed from 100 machines to 5 (in year 2 and 3) and back again from 20 to 120 machines (Figure 11.1a). According to the traditional understanding of logistics design, the company would have had to make fundamental changes to its plant layout as well as to the PPC logic three times in five years (Fig. 11.1b). This is not an unusual example and it illustrates the necessity of changeability.

Such changes to the physical systems can be achieved using the available hardware/technological solutions. But the logical systems, especially software tools for

[1] Institute of Industrial Manufacturing an Management (IFF), University Stuttgart, Stuttgart, Germany

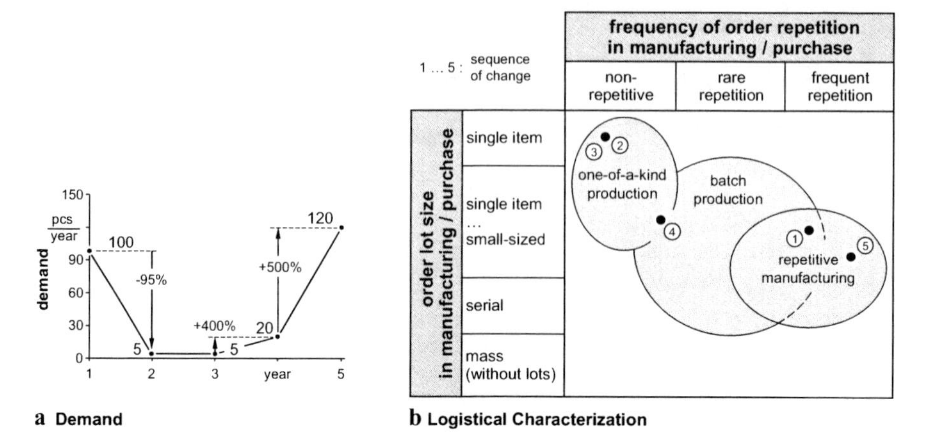

Fig. 11.1 a,b Case study on logistical requirements in turbulent markets (Schönsleben, 2007)

Production Planning and Control (PPC), cannot cope with this speed of change. A survey of 'user satisfaction with business software' seems to confirm this, where the number one problem was found to be limited software flexibility or adaptability (Sontow and Treutlein, 2004). Two reasons for the dissatisfaction relate directly to organizational issues and not to the software adaptability:

- *Division of labor*: This requires a person to provide information to the person next in the value chain without receiving a direct benefit. Often, information providers complain about the associated increase in the work effort.
- *Release readiness of software*: IT managers are especially worried about loss of capability by upgrading to the next release version of standard software. As a result, they limit the changes made to the used standard software.

In both situations the users are not satisfied and complain about inadequate software or their flexibility. To achieve an adequate adaptability in PPC software, the relevant influencing factors and boundary conditions have to be defined. First clues are discovered upon a closer examination of today's criteria for PPC software implementation projects (Sontow and Treutlein, 2004, Sontow et al., 2006):

- More than 60 % of the projects are initiated by IT related reasons, e.g. outdated software, software vendor gone out of business, etc.
- More than 20 % of the projects are initiated by internal organizational changes, i.e. reorganization of structures and processes or strategic business adjustment.
- About 10 % of the projects are initiated by 'quantity changes', i.e. growth, acquisition, downsizing of a business.

Therefore, the objective of this chapter is to establish a framework for a changeable PPC design and present indicators to review the PPC IT tools.

11.2 The PPC Framework

The Production Planning and Control (PPC) system is the central logistic control mechanism that matches the company's output and logistic performance with customer demands. Its general task is to allocate orders and resources over time, i.e. to plan and control the manufacturing of products. Typical decisions are capacity adjustments (e.g. adding shifts, over times, subcontracting) or to trigger purchase orders. In addition, PPC has to monitor and, in case of unforeseen deviations, adjust the order progress or the production plans (Wiendahl HP, 1994).

The traditional PPC understanding is a technical one, i.e. the primary point is to master the technical or logistical process flows. Therefore, PPC research was concentrated on the development of new functions and algorithms (Plossl, 1985, Vollmann et al., 1997), often based on operations research methods, see for example (Stadtler, 2002, Simchi-Levi, 2005, Balla and Layer, 2006), neglecting the analysis of the required pre-conditions such as organizational framework for PPC. These methods take no account of the great influence of the people involved in the supply chain whose ideas about best practice, their experiences and personal targets in planning and control are often disregarded. To overcome these deficiencies, a PPC design must therefore be based on a socio-technical approach.

According to this understanding, the term 'PPC system' encompasses here the entirety of tasks, tools and people necessary to plan and control the logistic processes in a manufacturing company. The scope of application includes the three basic processes: 'Source', 'Make' and 'Deliver' (SCOR, 2008). Like production itself, the input and output stores of a company are included in the PPC system (Wiendahl HP, 1994) as shown in Fig. 11.2. In this context, the PPC software is only one part of the PPC system. The software tool, however, is used to plan and control the logistic process chain as well as to store the production master data and feedback data.

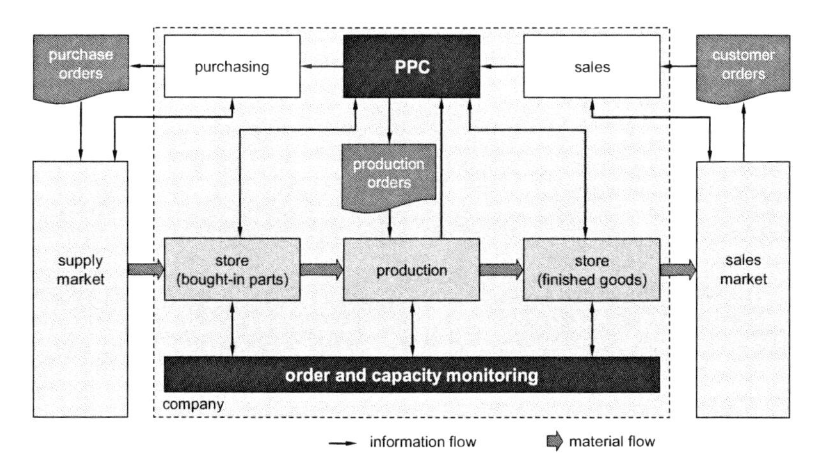

Fig. 11.2 Integration of PPC into the material and information flows (H.-P. Wiendahl, 2007)

11.2.1 Design Aspects of a Socio-Technical PPC System

Complex systems cannot be captured as a whole, which makes it necessary to model them aspect-by-aspect. Such a procedure examines the system from a certain point of view, thus reducing the complex reality by bringing one aspect to the foreground and moving other aspects to the background (Daenzer and Huber, 1997: 13ff).

In socio-technical systems, activities are the central criteria describing the system. Therefore, the Russian psychologist Leont'ev proposed an activity-based analysis (Leont'ev, 1977: 37f). Specker defined, based on these fundamentals, five perspectives to analyze and design IT systems (Specker, 2005: 33ff, 138). Their application leads to six design aspects of a PPC system (Wiendahl HH, 2006) as shown in Fig. 11.3:

- The *PPC targets* require decisions on logistical positioning. If necessary, different targets need to be defined for different departments or sub-processes. The classic example is the customer decoupling point: Upstream, the target priority is on high utilization and low work-in-process, downstream the priority is on short lead times and high schedule reliability (Hoekstra and Romme, 1992: 6ff, Nyhuis and Wiendahl HP, 2008:4).
- The *PPC functions* structure the decision activities that are required to plan, control and check the success of the manufacturing execution processes. PPC methods carry out functions based on defined algorithms and data.
- The *PPC objects* focus on the planning subjects of PPC. Most important are items (finished products, components or raw materials), resources (machinery, personnel, etc.), manufacturing processes and orders (customer orders, spare parts orders, etc.). Data models structure these objects and their relationships.
- The *PPC processes* structure the decision and execution activities as a logical or temporal sequence. Thus they define the work-flow of order processing along the logistic process chain, i.e. the business process. The process steps related to the material flow follow the same logic but are not a direct subject matter of the PPC system: they are modeled as PPC objects.

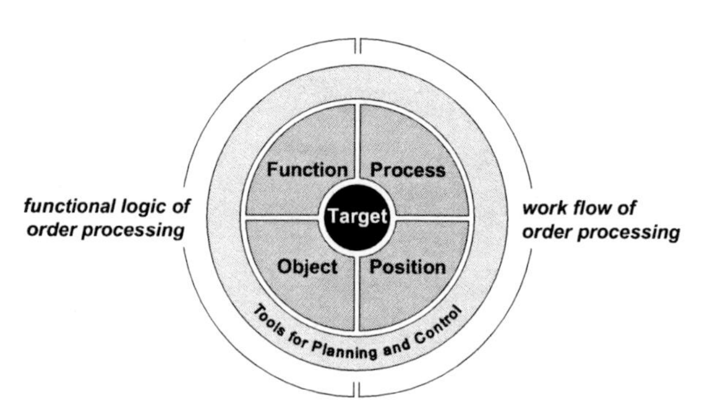

Fig. 11.3 Design aspects in PPC

- The *PPC positions* focus on the responsibility of staff members. The traditional PPC understanding assumes centralized decisions, ignoring this actor's aspect in PPC. A position defines performance requirements, which the owner must fulfill regarding competences and qualification.
- The *planning and control tools* support the operational order processing by semi-automated PPC activities. This creates daily business standards to increase efficiency. Therefore, the staff has more time available for the necessary planning and control decisions. This should increase the effectiveness of PPC.

The five design aspects of target, function, object, process and position described above constitute the logical core of a PPC system. The tools have to map the PPC design to the software effectively and efficiently.

In addition, the framework identifies the two main views on PPC (Fig. 11.3, outside): On the one hand, it indicates the functional logic of order processing, corresponding with the traditional technical view on PPC. On the other hand, the work-flow of order processing, corresponding with the required second view of structural and process organization as well as actors.

11.2.2 PPC Design Matrix

On one side, changes of a PPC design relate to the described design aspects in PPC. On the other, the scale of the change should be differentiated. Schwaninger distinguishes three levels: operative efficiency (doing the things right), strategic effectiveness (doing the right things) and normative legitimacy (task fulfillment from an overall perspective) (Schwaninger, 1994:51). The design matrix follows this idea and structures possible changes from a top down perspective. Figure 11.4 shows the design matrix with examples. Note that the matrix focuses on the logical core of PPC. It ignores the sixth design aspect 'PPC tools' because from user perspective a tool should only map the required design and changeability to software. Three change levels are relevant (Wiendahl HH, 2005):

1. *Parametrization:* The lowest changeability level in PPC is the adaptation of parameters. Examples are changed target values (design aspect target), changed process sequences (design aspect process often also position) or parameter adaptation such as 'planned lead times' (design aspect function shown in Fig. 11.4, bottom line). The central PPC planning parameters include the planned values for the offset and replenishment times, order throughput times and operation throughput times. With the help of these time-based parameters purchase orders are placed and production orders are scheduled. Hence the parameters represent the logical foundation of the entire due-date structure in a company. This is the most common level of change, but it is hardly ever performed systematically.
2. *Basic Configuration:* The second level represents the functional logic and work flow aspects of order processing. The functional logic is changed with respect to

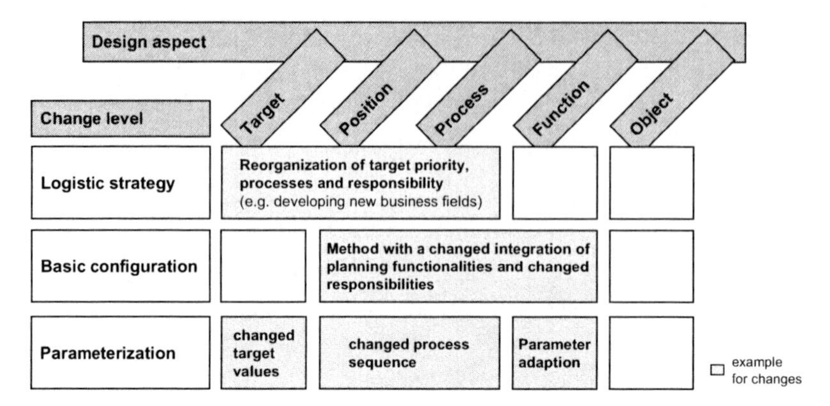

Fig. 11.4 PPC design matrix

the applied methods. An example is the change of the lot size rule from 'demand lot size' to 'Economic Order Quantity Model' (Harris, 1913). Most of today's PPC software includes this option. As a consequence, the PPC parameters on level 1 must also be changed. The introduction of a new method can also require new processes and responsibilities (Fig. 11.4, middle line).

3. *Logistic strategy:* The highest level of changes to PPC is the switch to another logistic strategy, usually along with a change of targets and their values. An example is the change from 'make-to-order' to 'make-to-stock' strategy. Consequently, new logistic targets are required, such as a change from 'schedule reliability' to 'avoidance of stock shortages'. Therefore, PPC methods (level 2) and the parameters (level 1) must be changed. In this case, at least the order generation rule must be changed from 'deterministic' to 'stochastic'. Fig. 11.4, upper line shows an additional example. In this case, development of new business fields requires bigger re-organization.

The levels of change are linked as described. Changes at higher levels usually trigger changes at lower levels.

11.3 Changeability of PPC Tools

The PPC design matrix focuses on the core logic of PPC and neglects the sixth design aspect, 'PPC tools', the relevant elements of which also have to be adaptable. As described in Chapter 2, certain features of an element describe its ability to change. These are called changeability enablers. Consequently also for the elements of a PPC tool first these 'enablers of changeability' need to be defined, followed by the PPC elements that may change. Relating enablers to elements reveals the 'building blocks of changeability' in PPC.

11.3.1 Change Elements of PPC

Apart from concepts and methods of IT technology, industrial experience showed eight change elements, which can be arranged into two categories:

- The *functional logic of order processing* is the logic of planning and control decisions. It is comprised of the PPC methods and their functional model, the data model as well as the data interfaces.
- The *work-flow of order processing* refers to the structural and process organization. It is comprised of the desired process steps and their sequences, the authorization concept, as well as the user interface.

11.3.2 Enablers of PPC Changeability

Existing definitions of changeability (Westkämper et al., 2000, Wiendahl HP, 2002, Hernández, 2002, Wiendahl HP et al., 2007) form the basis for defining specific enablers for PPC (Fig. 11.5). Each enabler focuses on a characteristic feature, which supports changeability, i.e. it represents a characteristic view on PPC's changeability. Therefore, some enablers may be linked in real cases:

- The ability to *adjust* to requirements concerning the functional logic of order processing, i.e. the weight of PPC targets and the importance of PPC functions.
- The ability to be *neutral* concerning different requirements of the work-flow of order processing, i.e. the defined process status and responsibility does not depend on the structural and process organization or the enterprise size.
- The ability to be *scalable*, i.e. the applicability is independent of the product, process, customer and supplier relationship complexity.
- The ability to be *modular*, i.e. the internal structure of the PPC design units follows the black box principle to achieve 'plug & produce' modules.

Enabler	Application Focus	Application Example
adjustability	functional logic of order processing	• generic PPC methods which change the order release algorithm only by parameter adaption (see chapter 18.4.2)
neutrality	work flow of order processing	• role-based authorization concepts which change the responsibility of process steps by adapting the assignment between roles and people
scalability	production complexity or quantity structure	• functional models or PPC methods which are applicable for different levels of detail, e.g. rough or fine planning and scheduling
modularity	internal structure of PPC design units	• generic functional models which change one logistical function independent from another (see chapter 18.4.1)
compatibility	external structure of PPC design units	• data models or interfaces which allow to add new production units or planning features (see chapter 18.4.3, and 18.4.4)

Fig. 11.5 Enablers of PPC changeability

• The ability to be *compatible*, i.e. the external structure of the PPC design units supports network-ability with each other.

11.3.3 Building Blocks of PPC Changeability

The combination of change elements with enablers of changeability allows two things: First, to rate the importance PPC enablers and change elements. Second, to find building blocks as a combination of an enabler and a change element (Fig. 11.6) and describe attributes that support PPC changeability. This will be essential for analyzing and designing changeable PPC systems.

The design scope 'target' is not mentioned separately. Practical experience shows that changing targets usually goes with an essential change in the PPC system. This underlines the great influence of targets and their priority in general. From a technical-logistical perspective, changes in targets are nothing but changes to the data model or the functional model. Therefore, one can neglect the design scope 'target', concentrating first on the necessary condition of changeability: These building blocks form the theoretical basis to develop adaptive PPC solutions and enable changeability. A defined change process is a sufficient condition for changeability. Both topics are exemplified in the following sections.

11.4 Adaptive PPC Solutions

The following section provides examples of changeable PPC solutions concerning the change elements functional models, planning and control methods as well as

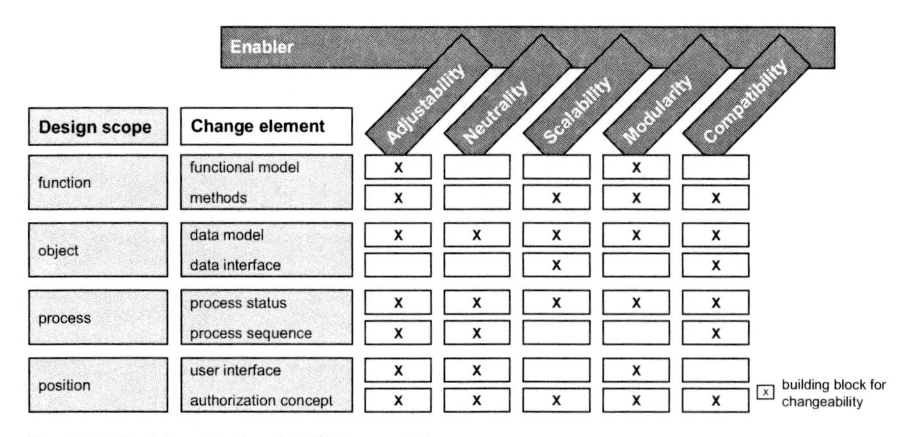

Design scope	Change element	Adjustability	Neutrality	Scalability	Modularity	Compatibility
function	functional model	x			x	
	methods	x		x	x	x
object	data model	x	x	x	x	x
	data interface			x		x
process	process status	x	x	x	x	x
	process sequence	x	x			x
position	user interface	x	x		x	
	authorization concept	x	x	x	x	x

Fig. 11.6 Building blocks of PPC changeability

the data models and their interfaces. It should be noted that sometimes changeability and systematic PPC design are confused: the latter analyzes, if the six aspects described in Fig. 11.6 are consistent. Consequently, stumbling blocks in PPC are attributed to internal mistakes in the configuration of these design aspects (Wiendahl HH, 2006, Wiendahl HH et al., 2005). Therefore, their removal can be understood as a pre-condition of adaptability and hence is not a changeability concern.

11.4.1 Functional Models

From a logical point of view, generic functional models support changeability. Lödding's 'influence model of manufacturing control' is one example: Its basic idea is to connect PPC functions with targets, as illustrated in Fig. 11.7 (Lödding, 2005:7ff). Four basic functions are relevant and they are coupled with the targets through manipulated and observed variables (Fig. 11.7a):

- *Order generation* determines the planned input, the planned output, as well as the planned order sequence.
- *Order release* decides when orders are passed to the shop floor (actual input).
- *Capacity control* determines the available capacity in terms of working time and staff assigned to work systems, and thus affects the actual output.
- *Sequencing* determines the actual sequence of order processing for a specific work system, and thus affects the actual sequence.

These functions affect the three manipulated variables *input, output* and *order sequence*. The discrepancies between two manipulated variables lead to the observed variables of manufacturing control:

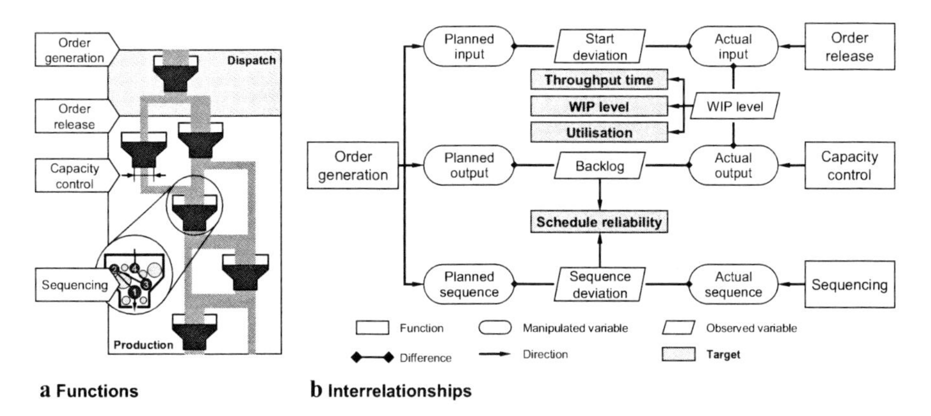

Fig. 11.7 a,b Influence model of manufacturing control (adapted from Lödding, 2005)

- The *start deviation* results from the difference between planned input and actual input,
- the *WIP level* results from the difference between actual input and actual output,
- the *backlog* results from the difference between planned output and actual output, and
- the *sequence deviation* results from the discrepancy between actual and planned sequence.

The observed variables affect the targets of PPC described above, i.e. *throughput time, WIP level, utilization and schedule reliability*.

Figure 11.7b shows the interrelationships between the elements: The functions define the manipulated variables, the observed variables result from the discrepancies between two manipulated variables, and the logistic targets are determined by the observed variables. The model easily and consistently supports a basic PPC design of the aspects 'function' and 'target' and supports a change of methods which provide the same function, e.g. order release.

11.4.2 Planning and Control Methods

In general experts emphasize the influence of PPC methods on PPC's adaptability (Kádár et al., 2005). To speed up implementation or to lower the threshold for changing PPC methods, two different approaches can be used:

- *Flexible methods:* One approach is to develop methods capable of dealing with various PPC requirements. Agent-based systems can be considered an example of this method. Their common properties (e.g. autonomy, intelligence, ability to interact) enable them to act in changing environments and their decentralized decisions enable the desired adaptability in accordance with the actual situation (Monostori et al., 2006, Hülsmann et al., 2007). Three PPC application domains for multi-agent systems (MAS) have been suggested (Monostori et al., 2006): production planning and resource allocation, production scheduling and control as well as an integrated scheduling and process planning.
- *Generic methods:* Another approach is to develop methods able to build different methods only by adapting parameters. The load-oriented order release (BOA) is an example. BOA fulfills the function of order release in Fig. 11.7.

Figure 11.8a shows the process logic of BOA, which is extended to a generic method by Lödding (Wiendahl HP, 1994: 206ff, Lödding, 2005: 374ff): The central PPC software generates a list of urgent orders via backward scheduling. An order will be released, if the WIP level does not exceed the load limit at any work system that will process the order. BOA consists of four elements:

- The *list of urgent orders* includes all known orders, which can be released for production. Their planned start date falls between today and the anticipation horizon.

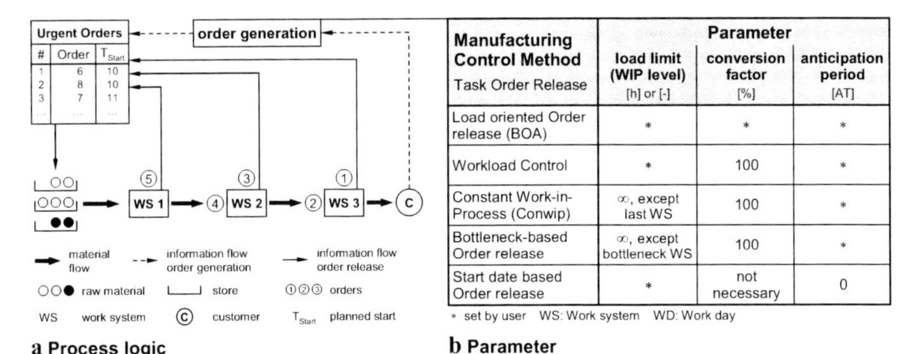

Manufacturing Control Method Task Order Release	Parameter		
	load limit (WIP level) [h] or [-]	conversion factor [%]	anticipation period [AT]
Load oriented Order release (BOA)	*	*	*
Workload Control	*	100	*
Constant Work-in-Process (Conwip)	∞, except last WS	100	*
Bottleneck-based Order release	∞, except bottleneck WS	100	*
Start date based Order release	*	not necessary	0

* set by user WS: Work system WD: Work day

b Parameter

Fig. 11.8a,b Generic PPC method BOA (Lödding, 2005)

- Each work system has a *loading account* to control its WIP level. Hence, the direct WIP (order currently processed at this work system) and the indirect WIP (order currently processed at preceding work systems) are distinguished. Orders in the direct WIP are considered with the full order time, orders in the indirect WIP are considered with a reduced – i.e. converted – order time.
- Each work system has a *load limit*. If exceeded, the release of the orders to be processed at this work system is stopped.
- Each work system has a *conversion factor*, which determines the reduction of the order time for following work systems.

Figure 11.8b shows which methods are included in BOA and how to implement these only by switching the three parameters 'load limit' (maximum permissive load), 'conversion factor' (how probable it is that the load arrives at the work station in this period) and 'anticipation period' (period for load balancing).

It should be noted that the criteria changeability and adequate functionality of a method are sometimes mixed up: *Changeability* describes the ability to adapt a method to different requirements and refers to the design phase in PPC. In contrast, the issue of *adequate functionality* refers to the operation phase and describes how an algorithm fits specific logistical requirements. Experience and research work shows that superior methods follow the closed-loop principle and are typically more sophisticated, thus guaranteeing a better logistical performance. Therefore, the chapter does not take a closer look on control feedback methods and their impact on improving the logistical performance. However, there is always a trade-off between changeable methods and superior functionality.

11.4.3 Data Models

Data models also influence changeability in PPC. Two alternatives are:

- *Hierarchical models:* This approach pre-determines structure and supports consistent, complete data. As a result, the user should not face problems with differ-

ent results based on different data requests because hierarchical models have, by definition, rigid boundaries for control and information sharing.
- *Non-hierarchical models:* This approach creates options to enable more flexibility. As a result the user should be able to adapt data models and the implemented information quickly but possibly jeopardizing data consistency and completeness. For example, it is possible to model different lead times of the same production part for each site or production segment.

The comparison of current Enterprise Resource Planning (ERP) software illustrates the differences between these approaches: Some suppliers provide an integrated solution with a hierarchical data structure. Its main advantage, consistent data, leads to a loss of flexibility under changing circumstances. Other suppliers provide data models for each site (non-hierarchical) with more flexibility concerning changing requirements (e.g. integration of new business units). However, data redundancy creates problems with data integration and consistency.

The analysis identifies the trade off between data consistency of the global enterprise data model (and its content) and the flexibility for local adaptations according to specific requirements. Non-hierarchical data models support changeability. However, they need higher effort and a better qualification of the employees to guarantee the required data consistency.

11.4.4 Data Interfaces

A complementary approach for achieving changeability is to standardize data interfaces. It assumes local data models with little need to change information and follows closely the black box idea. Practice shows two different solution ideas:

- *Complete approaches* of supporting data interchange by describing each case in great detail, resulting in a high implementation effort. One example is EDIFACT, the most common standard in various industrial sectors containing 550 elements within 100 segments.
- *Selective approaches* of supporting data interchange by focusing on relevant cases. This enables "lean" solutions with low implementation effort. E.g, the OpenFactory standard is limited to 20 messages, 185 elements (Meyer, 2005).

The underlying question – and the resulting design dilemma – is whether it is possible to perform 80 % of the transactions with only 20 % effort (Pareto Principle), while avoiding the numerous special cases in practice.

This dilemma describes an additional trade-off: Simple solutions support changeability by using the Pareto Principle, neglecting the exceptions. But each user wants the best fit to his business to achieve efficiency resulting in complex descriptions that obstruct changeability.

11.5 Change Process in PPC

A sufficient condition for changeability is the existence of a defined change process. A turbulent manufacturing environment requires a close link between the operational decision and execution activities in PPC and the change process, i.e. the design tasks in PPC (Wiendahl HH, 2005, Wiendahl HP, et al. 2007). The proposed process logic follows this closed loop principle. It combines the required activities in the four sub-phases: Plan, Do, Check and Act, as defined by the Deming cycle (Deming, 1992: 88f) and is adapted to PPC (Westkämper, Schmidt and Wiendahl HH, 2000: 849ff and Wiendahl HH, 2006). The three levels are Act 1: parametrization, Act 2: basic configuration, and Act 3: logistic strategy. They correspond to the change levels of the design matrix (Fig. 11.4). Figure 11.9a visualizes the control loop of operation and design tasks.

In accordance with the control loop idea, changes of the internal and external conditions require a periodic verification as follows (Fig. 11.9b):

1. *Detect necessity for change:* The ongoing review of relevant factors identifies abrupt as well as creeping changes.
2. *Implement only if change is deemed necessary:* This phase starts with an assessment of necessity and scope of change.
3. *Evaluate success of change:* A change process is successful if the targeted objectives are fulfilled.

For a successful change of a PPC system the described change process has to be accompanied by an adequate change management, which takes into account both the specific organizational and human aspects. The change processes in PPC are

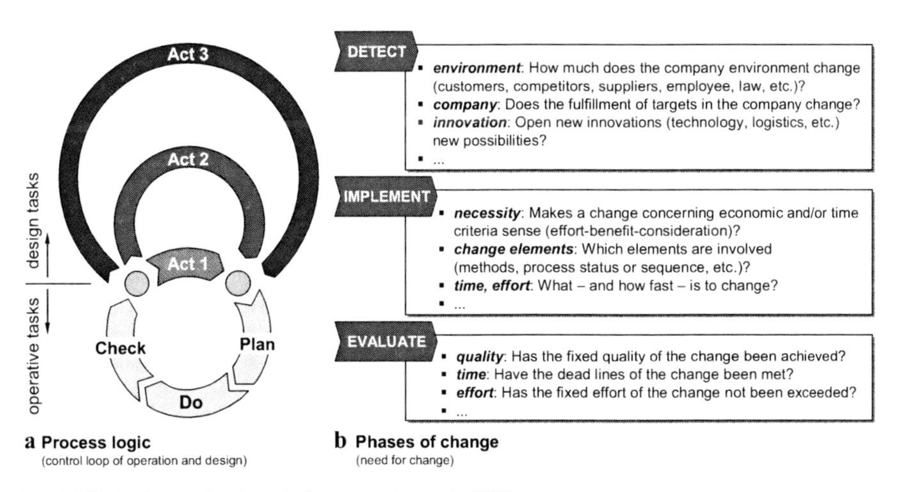

Fig. 11.9a,b Process logic and phases of change in PPC

in general similar to other organizational changes. Hence, available procedures and methods can be used as a template (Kotter, 1996, Kotter and Cohen, 2002), but have to be tailored for PPC (Kraemmerand et al. 2003, Wiendahl HH and Westkämper, 2004).

11.6 Summary and Further Research

This chapter described changeability from the perspective of Production Planning and Control: The presented PPC design matrix and change process build the framework, which structures PPC for a clear and consistent design. The building blocks of changeability form the basis to systematically derive software changeability and serve as a 'quick check' to assess changeability. Useful criteria were proposed to evaluate the implemented solution concepts, e.g. "how powerful is the configuration of the element 'authorization concept' by means of the implemented 'role-based authorization'?" (Fig. 11.5).

Today's ERP software users are not sufficient with the changeability in PPC. The analysis shows organizational as well as technical boundaries. Practitioners focus on technical aspects – typically neglecting organizational issues. Many concentrate on improving the PPC methods and underestimate the high impact of functional and data models as well as data interfaces on software adaptability.

According to this finding, further research should concentrate on four aspects: First, enlarge functional models to obtain a generic structure. In production control, initial results are available (Lödding 2005), but a generic model for planning and control functions is still missing. Second, develop a structure of basic PPC parameters, ideally arranged according to the functional models. Third, develop methods that are adaptable and scalable. These have to be robust with respect to demands concerning functional logic and complexity. These technical results would structure PPC more generically. This leads to the forth point: an investigation of organizational and human aspects that support a socio-technical understanding of PPC that would help improve changeability in practice.

Acknowledgements The results presented are part of the research project "Modellbasierte Gestaltung des Auftragsmanagements der industriellen Stückgüterproduktion" (Model-based order management design for industrial discrete manufacturing). The project is funded by the Deutsche Forschungsgemeinschaft DFG (WI 2670/1).

References

Balla J., Layer F., 2006, Produktionsplanung mit SAP APO-PP/DS (Production planning with APO-PP/DS). Galileo Press, Bonn

Deming W.E., 1992, Out of Crisis, 18th edn. Massachusets Institute of Technology, Cambridge/Mass (USA)

Daenzer W.F., Huber F., (eds)1997, Systems Engineering: Methodik und Praxis (Systems engineering: methodology and practice), 9th edn. Industrielle Organisation, Zürich

Harris F.W., 1913, How many parts to make at once? Factory 10/2:135–136, 152

Hernández R., 2002, Systematik der Wandlungsfähigkeit in der Fabrikplanung (Systematic changeability in factory planning). PhD University of Hannover. VDI, Düsseldorf

Hoekstra S.J., Romme J.H.J.M., (eds)1992, Integral logistic structures: developing customer oriented goods flow. McGraw-Hill, Berkshire (England)

Hülsmann M., Windt K., (eds)2007, Understanding Autonomous Cooperation and Control in Logistics : The Impact of Autonomy on Management, Information, Communication and Material Flow. Springer, Berlin

Kraemmerand P., Møller C., Boer H., 2003, ERP implementation: an integrated process of radical change and continuous learning. Production Planning & Control 14/4:338–348

Kádár B., Monostori L., Csáji B., 2005, Adaptive approaches to increase the performance of production control systems. CIRP Journal of Manufacturing Systems 34/1:33–43

Kotter J.P., 1996, Leading Change. Harvard Business School Press, Boston/Mass

Kotter J.P., Cohen D.S., 2002, The heart of change: real-life stories of how people change their organizations. Harvard Business School Press, Boston/Mass

Leont'ev A.N., 1977, Tätigkeit, Bewusstsein, Persönlichkeit (Activity, Consciousness, and Personality). Klett, Stuttgart, Dejatel'nost' – soznanie – licnost. Moskva: Polit. Literatury, 1975

Lödding H., 2005, Verfahren der Fertigungssteuerung (Methods for production control). Springer, Berlin

Meyer M., 2005, From Enterprise Resource Planning (ERP) to Open Resource Planning (ORP): The Open Factory Project. In: Taisch M, Thoben KD (eds.) Advanced Manufacturing – An ICT & Systems Perspective. Politecnico di Milano, Department of Economics, Management and Industrial Engineering, BIBA, University of Bremen

Monostori L., Váncza J., Kumara S.R.T., 2006, Agent-based systems for manufacturing. Annals of the CIRP 55/2:672–720

Nyhuis P., Wiendahl H.P., 2008, Fundamentals of Production Logistics. Theory, Tools and Applications. Springer, Berlin

Plossl G.W., 1991, Managing in the New World of Manufacturing. Prentice Hall, Englewood Cliffs

SCOR (2008) Supply Chain Operations Reference Model. http://www.supply-chain.org, 4.1.08

Schönsleben P., 2007, Integral Logistics Management. Auerbach, Boca Raton FL, USA

Schwaninger M., 1994, Managementsysteme (systems of management). Campus, Frankfurt

Simchi-Levi D., Chen X., Bramel J., 2005, The Logic of Logistics: Theory, Algorithms, and Applications for Logistics and Supply Chain Management, 2nd edn. Springer, Berlin

Specker A., 2005, Modellierung von Informationssystemen (Modelling of information systems), 2nd edn. vdf, Zürich

Sontow K., Treutlein P., 2004, Anwenderzufriedenheit ERP-/Business Software Deutschland (User satisfaction with ERP/business software Germany). Trovarit AG, Aachen

Stadtler H., 2002, Supply Chain Management – An Overview. In: Stadtler H., Kilger C. (eds) Supply Chain Management and Advanced Planning. Springer, Berlin, pp 7–27

Sontow K., Treutlein P., Scherer E., 2006, Anwender-Zufriedenheit ERP/Business Software Deutschland (User satisfaction with ERP/business software Germany). Trovarit AG, Aachen

Vollmann T.E., Berry W.L., Whybark D.C., 1997, Manufacturing planning and control systems, 4th edn. Dow Jones-Irwin, Homewood

Westkämper E., 2000, Ansätze zur Wandlungsfähigkeit von Produktionsunternehmen (Approaches towards changeable manufacturing). wt Werkstattstechnik 90/1-2:22–26

Westkämper E., Schmidt T., Wiendahl H.H., 2000, Production Planning and Control with Learning Technologies: Simulation and Optimization of Complex Production Processes. In: Leondes CT. (ed) Knowledge-Based Systems, Techniques and Applications. Vol 3. Academic Press, San Diego, pp 839–887

Wiendahl H.H., Westkämper E., 2004, PPC Design and Human Aspects. Annals of the German Acadamic Society for Production Engineering XI/1:129–132

Wiendahl H.H., Wiendahl H.P., von Cieminski G., 2005, Stumbling blocks of PPC: towards the holistic configuration of PPC systems. Production Planning & Control 16/7:634–651

Wiendahl H.H., 2005, Changeability in Production Planning and Control: A Framework for Designing a Changeable Software Tool. IFIP/TC5/WG5.7: Advances in Production Management Systems/CD-ROM: Modelling and Implementing the Integrated Enterprise, Rockville, USA, 18.–21.09.2005

Wiendahl H.H., 2006, Systematic Analysis of PPC System Deficiencies – Analytical Approach and Consequences for PPC Design. Annals of the CIRP 55/1:479–482

Wiendahl H.P., 1994, Load-oriented Manufacturing Control. Springer, Berlin

Wiendahl H.P., 2002, Wandlungsfähigkeit – Schlüsselbegriff der zukunftsfähigen Fabrik (Transformability: key factor of a future robust factory). wt werkstattstechnik online 92/4:122–127

Wiendahl H.P., ElMaraghy H.A., Nyhuis P., Zaeh M., Wiendahl H.H., Duffie N., Brieke M., 2007, Changeable Manufacturing: Classification, Design, Operation. CIRP Annals 56/2:783–809, Keynote Paper

Chapter 12
Component Oriented Design of Change-Ready MPC Systems

M.A. Ismail and H.A. ElMaraghy[1]

Abstract Agile manufacturing is defined as the capability of manufacturing systems to survive and prosper in a competitive environment of continuous and unpredictable change, by reacting quickly and effectively to changing markets driven by customer-designed products and services. A new agile Manufacturing Planning and Control (MPC) system design is needed to respond to the changeability of the underlying manufacturing system as well as to the uncertainty of the surrounding environment. It should be resilient to change and responsive to its environment. It should satisfy required performance measures and achieve the required competitive strategy. A new conceptual model and framework to handle MPC system problems from the system perspective is introduced. Component Based Software Engineering (CBSE) provides the tools and the power to design the proposed new system. The MPC system should be able to achieve the required balance between demands and supply, high service levels, low inventories and deal with volume-mix issues. The coordination and interactions between different system components achieve the required system resilience and peak system performance.

Keywords Changeable Manufacturing, Changeability, Manufacturing Planning, Aggregate planning, Agile Manufacturing, Component-Based MPC Systems, Object Oriented MPC Systems, Computer Integrated Manufacturing

12.1 Introduction

As the product life-cycle is shortened, high product quality becomes necessary for survival. Markets become highly diversified and globalized, and continuous and unexpected changeability become the key factors for success (Ramesh and Devadasan, 2007). The need for a method of rapidly and cost-effectively developing products,

[1] Intelligent Manufacturing Systems (IMS) Center, University of Windsor, Canada

production facilities and supporting software, including design, process planning, manufacturing planning and control systems has led to the concept of Agile Manufacturing (Gunasekaran, 1998). Academicians and practitioners define agility as the capability of operating profitability in an uncertain, unstable, continuously changing environment (Sahin, 2000). Information Technology enables the organization to move toward agility. It reduces cost and improves time-to-market customer responsiveness (Purohit et al., 2006).

Reconfigurable Manufacturing Systems (RMS) is a new manufacturing system paradigm that aims at achieving cost-effective and rapid system changes, as needed and when needed, by incorporating principles of modularity, integrability, flexibility, scalability, convertibility, and diagnosability (Koren, 2003; Mehrabi et al., 2000). RMS promises customized flexibility on demand in a short time, while Flexible Manufacturing Systems (FMSs) provides generalized flexibility designed for the anticipated variations and built-in a priori (ElMaraghy, 2005). The term "Changeability" was first introduced to define the boundary between flexibility and reconfigure-ability (Wiendahl, H.P. et al., 2007). The term "changeable manufacturing" is used as an umbrella term that addresses the whole manufacturing unit and its manufacturing support sub-systems. Changeable manufacturing tries to achieve agility at both the factory's logical and physical levels.

The changeability of both the reconfigurable manufacturing system as well as its uncertain environment calls for a new MPC system that not only adapts itself to the new changes but also is capable of evolving seamlessly. Defining Change-ready MPC system characteristics and adopting a new design approach for it is needed.

Several forces drive the change of MPC systems:

- **The Market:** Porter (Wheelen and Hunger, 2006) identified two dimensions to differentiate between the strategic scope and strategic strength. Strategic scope is a demand-side dimension that looks at the composition and size of the target market. Strategic strength is the other supply side dimension and considers the core competency of the manufacturing enterprise. Changeable MPC systems should be aware of these two dimensions and try to keep them in balance.
- **The Product:** product variety versus product volume, economics of scope versus economies of scale, placed certain constraints on the design of a manufacturing system and its MPC as well. Mass-customization is growing rapidly with fierce competition striving for lower prices. Companies now compete on being both responsive and efficient. A mix between agile and lean practices is essential to fit these new requirements.
- **The Manufacturing Systems:** manufacturing technology moves changeability boundaries and its limits forward, i.e. reconfigurable manufacturing systems (RMS) with its incremental change of functionality and capability versus Flexible Manufacturing Systems (FMS) with built-in abilities to change its functionality within a pre-defined scope. The evolution of RMS is uncertain, based on market and products requirements, and needs a co-evolving MPC system to effectively address its needs.

- **Information Technology:** advances in information technologies are the main drivers of the evolution MPC system design. Software engineering offers several solutions for changeability issues.
- **Organizational Structure:** the world now moves from taller hierarchies to flatter and matrix-like structures (Jones, 2006). The latter improves responsiveness and autonomy, which promotes the ability of the manufacturing enterprise to address the change. The MPC is a critical part of that enterprise and should embrace the same concept.

In this chapter, some of the changeable manufacturing planning and control system characteristics are introduced and the component based software engineering principles are used to design the proposed system. The motivation for breaking down the MPC system into multiple components is to loosen the coupling between these components and increase their capability to evolve independently. Since a component-based application consists of a collection of building blocks, any component can be added or removed as needed. When a component implementation is modified, the changes are confined to that component only. No existing client of the component requires re-compilation or re-deployment.

The component technology can be implemented in a full-fledged MPC system or simply in one of its sub-systems. A mini case study is introduced as an implementation, which represents a component based aggregate production system. The system is composed of several components, which interact together to achieve the required system performance and being change-enabled via its component-oriented architecture.

12.2 Related Review

Over the last two decades, the development of MPC frameworks and architectures has attracted the attention of many researchers as an active area of research. The work reported in the literature is significant and focuses on building integrated MPC systems that automate the manufacturing planning, scheduling, and control process; see for instance (Wu, 2000 and 2001, Monfared and Yang, 2007, Dah-Chuan and Yueh-Wen, 1997, Choi and Kim, 2002, and Devedzic and Radovic, 1999). Berry and Hill (1992, 1992), also updated in Vollman et al. (2005), developed a full-fledged MPC framework for choosing the appropriate type of planning and control system relative to a firm's market requirements. Most of these frameworks can be characterized as either Decision Support Systems or Automated MPC Systems.

The Object Oriented Analysis and Design (OOAD) principles are commonly used as an approach to master the complexity of building MPC frameworks. In OOAD, an object-oriented system is composed of objects, and the behavior of the system results from the collaboration of those objects. Embracing OOAD as an approach for developing MPC system frameworks can be found in, for example,

(Chang et al., 1990, Wache, 1998, Zhang et al., 1999, Shan et al., 2001, Tsu Ta and Boucher, 2002, and Tsai and Sato, 2004). Object-oriented MPC frameworks suffer from being only modularized at the logical level and hierarchical (via inheritance) which makes them inadequate for changeable MPC systems. Any update means the whole system should be replaced. The MPC system components should evolve independently to address the endlessly changing market needs and the changeability of the manufacturing system and manufacturing process.

With the advent of reconfigurable manufacturing systems, there is insufficient research that addresses the new challenges that face the MPC systems as a part of this new technology. Among several possibilities of research areas, the capacity scalability and production control received most attention. Examples include optimal capacity scheduling (Deif and ElMaraghy, 2007; Deif and ElMaraghy, 2006), capacity management with stochastic demand (Asl and Ulsoy, 2002) capacity scalability and line balancing (Sung-Yong et al., 2001), production control using control theory (Deif and ElMaraghy, 2006), to name but a few. The balance between the oscillating supply (scalable capacity) and uncertain demand is still not addressed. Excess demand means wasted opportunities while excess supply means wasted resources. Capacity management, MPC system stability and an agile MPC framework are still required.

The RMS research related areas gained a new leap with the introduction the concept of changeable manufacturing (Wiendahl, H.-P. et al., 2007). Different related areas of research are introduced such as Reconfigurable Process Planning (RPP) (ElMaraghy, 2006 and 2007), Adaptive Production Planning and Control (APP) and Transformable Factories (TRF) and their enablers (Wiendahl et al., 2007). The first part of this chapter elaborates on the subject of Change-Ready manufacturing planning and control where the component-based software engineering is introduced as a practical design solution for changeable MPC systems. Being change-ready through design only is not sufficient to address the changeable manufacturing environment, hence, a set of complementary characteristics are needed to achieve the required agility and resilience.

12.3 The New MPC System Characteristics

The new changeable manufacturing paradigm urges a better design of an agile manufacturing planning and control (MPC) system that is able to evolve and at the same time keep its stability in the face of changes in the manufacturing system and its environment. A set of system characteristics, which are perceived as the required qualifications of change-ready MPC systems, are summarized next.

Modularity: The System would be composed of loosely coupled sets of interacting components. Every component would have its own set of responsibilities and requirements. Components should encapsulate the core competencies as well as core values. Components can be added/removed to extend/change the system capabilities. They can also be updated or evolved to improve the system performance.

System-wide Performance Oriented: The system-wide performance should be the ultimate objective of the change-ready MPC system design. The behavior of the system as a whole depends on the components and their relationships and how the synergies between different components are utilized. Errors in forecasts can be corrected via cooperation with other system components, such as those concerned with sales or marketing.

Inter-activeness: The embedded algorithms should facilitate interaction with system users, top management and specialists, to improve its performance and guide the solution process. Realizing peak performance, through well-defined mathematical models and solution approaches, is not sufficient for the next generation of MPC systems. The interaction with system users, especially with senior management, is some times required to find solutions beyond the system capabilities. Corrective actions, continuous auditing of system performance, reviewing functional strategies and policies are needed for all activities of the proposed system.

Integrability: The MPC system should be integrated with the enterprise general system components. Therefore, the system reconfiguration is not only an operational decision but also a financial one, hence, financial personnel would also be involved. Furthermore, the MPC system would share the same database with other departments, such as procurement and sales via the Enterprise Resource Planning system (ERP). Integrability requires compatibility among these external components and even among the MPC components themselves. CBSE solves this problem by virtue of its intrinsic concept of interface-based design; this requirement cannot be achieved using OOAD.

Robustness: the system should be able to react appropriately to abnormal conditions. Robustness complements the system correctness. Correctness is concerned with conforming to specifications, while robustness goes beyond those specifications. Being robust means the system will not go into a catastrophic state when encountering any abnormal conditions. Any system inadequacies can be substantially overcome by consulting top management, experts, and involved parties.

Dependability: The system must be trustworthy. The output of the MPC system helps top management in making system wide decisions, which impact the whole business unit. The level of trust/confidence needs to be known to help system users take the right actions. This is particularly important for changeable or change ready MPC systems.

Resilience: the ability of the system to change and recover from bad experiences. For example, a demand forecasting causal model (Brockwell and Davis, 2003) can be used to develop planning forecasts for a certain product; a new substitute can be introduced by another competitor which necessitates a radical change of the underlying model, i.e. replacing the demand implementation model. Component-oriented design facilitates this changeability characteristic by updating, adding, or replacing already existing components.

Evolvability: The new system must be able to evolve to meet changing needs. Abnormal conditions, extending system capabilities, scalability, switching marketing strategies, coping with exceptional system failures, etc., all call for system evolution abilities facilitated by the proposed Component Oriented MPC system design.

Scalability: defines the system spectrum of activities as well as its capability limits needed to address these activities. Scalability limitations can be handled by the extendability feature, which is achieved via the component-oriented design or by setting specifications at the outset for the required system capabilities.

Genericity: The implemented algorithms, models, guiding policies etc. should be sufficiently generic to respond to different scenarios, and should be easily customized. Genericity is an object oriented programming concept, which represents a generalized set of classes with anonymous data types. It is used here to define generic algorithms, models, and guiding policies. Genetic algorithms, Tabu search, for example, are considered generic optimization algorithms and can handle several optimization problems encountered in the changeable MPC system.

Ease of use: People with different backgrounds and qualifications should be able to use the system. This will maximize system benefits and promote its evolution.

12.3.1 Component-Based Software Engineering (CBSE)

CBSE is the process of defining, implementing, and integrating or composing loosely coupled components system (Sommerville, 2004). CBSE emerged in the late 1990s as a reuse approach to software system development. The component technology enables implementing or maintaining several characteristics that are already defined in the previous section.

Component-Oriented Development (COD) enables systems to be constructed from pre-built components, which are reusable, self-contained blocks of code. These components have to follow certain pre-defined standards including interface, connections, versioning and deployment (Heineman and Councill, 2001).

The parallels between the characteristics of reconfigurable manufacturing systems and COD make it a natural choice for designing a Change-Ready MPC system.

There are three major goals of COP: conquering complexity, managing change and reuse (Wang and Qian, 2005):

- ***Conquering Complexity:*** COP provides an effective way to deal with the complexity of software; that is divide and conquer.
- ***Managing change:*** Software engineers have come to the consensus that the best way of dealing with constant changes is to build systems out of reusable components conforming to a component standard and plug-in architecture.
- ***Reuse:*** COP supports the highest level of software reuse including white-box reuse, gray-box reuse, and black-box reuse.

Component-enabling technologies such as COM (Box, 1998), J2EE (Johnson, 2002), CORBA (Pritchard, 1999; Slama et al., 1999), and .NET (Chappel, 2006; Grimes 2002) provide the "plumbing" or infrastructure needed to connect binary components in a seamless manner, and the main distinction between these technologies is the ease with which they allow connecting those components.

12.3.2 Component-Oriented Versus Object-Oriented Programming

The fundamental difference between the two methodologies is the way in which they view the final application. In the traditional object-oriented world, all the classes share the same physical deployment unit, process, address space, security privileges, and so on. On the other hand, a component-oriented application comprises a collection of interacting binary application modules that are bonded to each other via well-defined protocols or interfaces.

Component-oriented applications usually have a faster development time because they can be selected from a range of available components, either from in-house collections or from third-party component vendors, and thus avoiding repeatedly reinventing the wheel.

Component-oriented programming promotes black-box reuse, which allows using an existing component without being concerned about its internals, as long as the component complies with some pre-defined sets of interface requirements. Instead of investing in designing complex class hierarchies – White-box use, component-oriented developers spend most of their time factoring out the interfaces used as contracts between components and clients (Bruccoleri et al., 2003).

12.4 Mini-Case Study: Component-Based Aggregate Production Planning System Framework

12.4.1 System Architecture

The proposed system architecture has three tiers. The user interface is separated from the MPC framework, which in turn is separated from the data storage. The user interface can interact with the underlying framework, and the framework can interact with the data, but the user interface cannot directly interact with the data. A simple Three-Tier architecture is depicted in Fig. 12.1. Separating the system framework from the user interface bridges the gap between the ease of use of the MPC system and the intricacies of its built-in logical components, which can be updated seamlessly without bothering the system users with complexities of the underlying models and their solution algorithms.

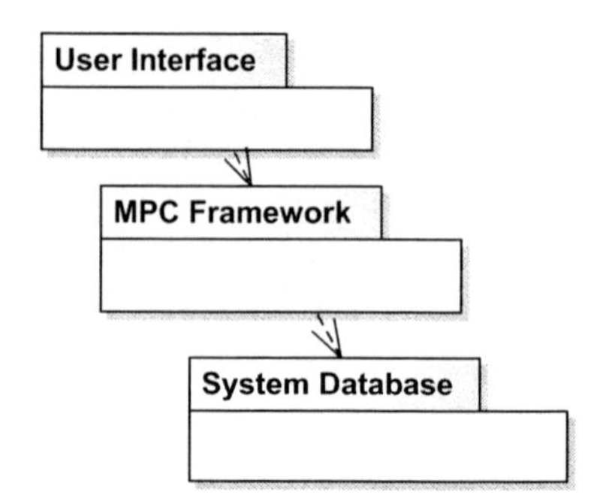

Fig. 12.1 The 3-tier MPC system architecture

12.4.2 Change-Ready MPC Framework

The proposed MPC framework will be broken down into several interacting components. These components include demand management, aggregate production planning, inventory management, operation management, in addition to some other services, such as the genetic algorithms and artificial neural networks libraries. The main MPC system components would utilize these libraries to help them achieve the required functionality, as it will be described later. Figure 12.2 depicts the components diagram of this framework.

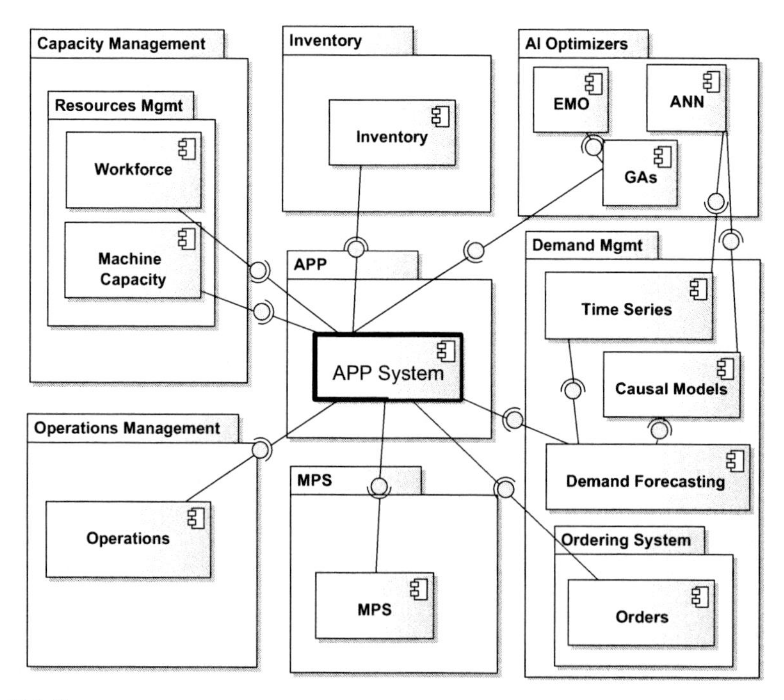

Fig. 12.2 System components

12.4.2.1 Multi-Layer Demand Forecasting Component

Accurate and timely demand plans are a vital component of any good MPC system and more so for a changeable one. Inaccurate demand forecasts would result in system imbalance between demand and supply as well as unsatisfied customers. For planning purposes, both long-term and short-term forecasts are needed. Inaccurate forecasts in the short-term means lost sales, lost customers, excess inventories and the like. Statistical models, such as time series forecasting, may be a good solution for short-term forecasts. Integration with other system components can even solve some of the short-term forecasting inadequacies, such as promotions.

Long-term forecasts are very important for capacity planning and strategic initiatives. These forecasts are more vulnerable to change and errors. Accuracy at aggregate levels is much more important. Determining product volumes is more important here than at the product mix. At this level, attention is paid to the production rate, inventory level and customer service level rather than which items are going to be produced and how many of them. The required production units for aggregate production and managing system resources (capacity scalability issues) should be expressed based on product family figures. Causal models can be used for this kind of forecasting, such as regression models.

Unlike statistical models, forecasting using artificial neural networks has the ability to capture demand non-linearity and does not assume a specific functional relation between the input data set (time or any causal set) and the resulting demand. Both the statistical and Artificial Neural Network (ANN) forecasting are embedded in the management component.

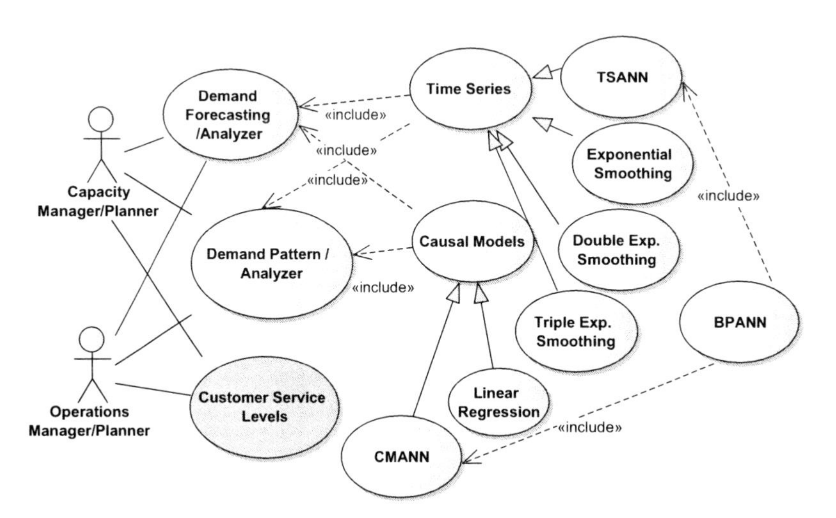

Fig. 12.3 Demand management use case model

The component technology and its main feature of interface based design can further extend this approach by defining the interfaces required for demand forecasting and leaving the implementation to a third-party consultant to evaluate the prospected forecasts. This might be useful for training the neural networks.

Figure 12.3 depicts the use case diagram of the demand management sub-system. This component is responsible for analyzing demand and recognizing its pattern via the demand analyzer. Both statistical and artificial neural networks are used to forecast the demand. The choice of the approach is left to the system user. Long-term forecasts are fed to both aggregate production planning and capacity planning components to manage the manufacturing system resources. The short-term forecast is fed to the master production schedule to determine the product mix. The ordering system (not implemented in this study for scope limitations but is still part of demand management component) is used to determine the customer service levels and customer back-orders.

12.4.2.2 Sales and Operations Planning Component

Most top-level decisions are taken through this unit. It is a very critical component through which the performance measures (service levels, capacity levels, inventory levels etc.) are set. Consequently, sales and operations plans are formulated here. These plans would be optimized to achieve the required system performance measures, mange the resources and respond better to customers. The Evolutionary optimization algorithms are used to introduce to the top management a set of trade-offs that they can choose from to achieve the required objectives.

Most of the system performance or system management would be conducted here. Identifying potential gaps and notifying decision makers about them would help greatly in achieving system resilience and the desired peak performance. Both the demand and the supply (oscillating capacity in the context of RMS) are uncertain and the system stability is an urgent requirement as well. Whenever the sales and operations figures deviate from either their goals or pre-allocated budgets, an immediate interaction should take place and the System should be considered out-of-control. These deviations, sometimes, go beyond the system specification or its capacities in the context of changeable manufacturing. Optimum plans or decisions do not exist in the context of changeable manufacturing and failing to achieve what is promised is not a bad symptom. A better approach is to keep operations and sales as realistic as possible. Sales figures can be re-adjusted and the operation plan can be re-initialized. The executives should take counteractive decisions in a timely fashion.

The aggregate production planning component facilitates achieving these requirements by being tied to several components, as depicted in Figs. 12.2 and 12.4. The system users can tweak several system inputs and the S&OP component can cooperate with other components to generate near optimal production plans. This component uses the system data, performance measures, objectives and constraints

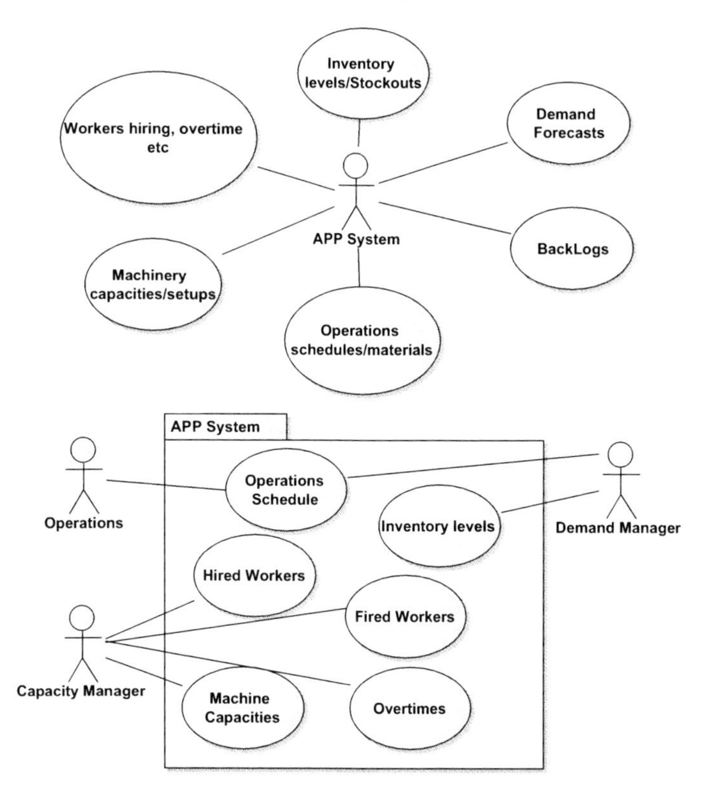

Fig. 12.4 The aggregate production planning use cases

to formulate the aggregate production planning model and solves it using genetic algorithms. The system can be updated at any time with new changes and a new efficient production plan is generated continuously.

12.4.2.3 Resources Management Component

Capacity will be used to track the already existing system resources, idle capacity, on-shelf machine modules, broken down or at work machines, work force levels, hiring and firing, costs and so on. Maintaining a balance between the oscillating capacity and the uncertain demand represents the most challenging problem to the MPC system and its senior management. System resources, on-shelf or in-service, are calling for new policies and better management approaches. The COD would be very useful regarding this issue through accumulated knowledge.

12.4.2.4 Other Components

Other components include MPC inventory (used to monitor inventory levels), Master production schedule (which takes the output of sales and operations and disaggregates it into weekly production plans) and operations management (where production related data are collected).

12.4.2.5 Case-Study Findings

This case study shows several new concepts and principles presented throughout this chapter. The Component based MPC system can be considered a reconfigurable version of MPC systems. The system can be updated, fixed, maintained, and evolved seamlessly. A change-ready MPC system is possible and can co-evolve with the underlying changes of a manufacturing system, market changes, and experience development. The MPC system alone is not sufficient to handle all challenges of the new environment unless there is a real collaboration with other enterprise-level system components and humans. The emergent behavior of all of these entities will define the system resilience and system stability required in facing any forthcoming change.

12.5 Discussion and Conclusions

In this chapter, the changeable manufacturing planning and control were discussed. Many contradictory and conflicting issues push MPC into quite disordered zones, form the market with harmonizing the strategic strength and strategic scope to manufacturing strategy of being lean or agile, and form the evolution of FMS with customized flexibility to vague boundaries of functionality and capability of RMS, and from limited product variety to the explosion of mass-customized products. Change is always constant and the question remains how to be change-ready surrounded with all these eminent chaos and uncertainties. As an initiative to answer this question from the MPC perspective, a new set of MPC characteristic was introduced. The CBSE was utilized as an enabling technology for the new MPC system. CBSE empowers the new system as well as system management with many characteristics that enable the design of a change-ready MPC system. A case study of component based aggregate production planning sub-system was introduced and some new challenges that face the main components were discussed.

Change-ready MPC systems, in spite of their versatile design and sophisticated algorithms, are not able to handle these new challenges alone. Human interaction and collaboration with other enterprise level sub-systems is critical to be able to face the new challenges. How to orchestrate all these entities together in order to keep or restore system balance is a real challenge and is a fertile area of research for years to come.

Acknowledgements This research was conducted at the Intelligent Manufacturing Systems (IMS) Center. The support from The Canada Research Chairs (CRC) Program and the Natural Science and Engineering Research Council (NSERC) of Canada is greatly appreciated.

References

Asl F.M. and Ulsoy A.G., 2002, Capacity management in reconfigurable manufacturing systems with stochastic market demand, American Society of Mechanical Engineers. 13:567–574

Berry W.L., Hill T., 1992, Linking systems to strategy. International Journal of Operations & Production Management 12/10:3–15

Berry W.L., Hill T., 1992, Linking systems to strategy. International Journal of Operations & Production Management 12/10:3–15

Box D., 1998, Essential COM, Addison-Wesley Professional

Brockwell P.J. and Davis R.A., 2003, Introduction to Time Series and Forecasting Springer; 2nd ed. 2002. Corr. 2nd printing edition

Bruccoleri M., Amico M., Perrone G., 2003, Distributed intelligent control of exceptions in reconfigurable manufacturing systems. International Journal of Production Research 41/7:1393–1412

Chang A.M., Kannan P.K., Wong B.O., 1990, Design of an object-oriented system for manufacturing planning and control. IEEE Comput. Soc. Press, Troy, NY, USA, pp 2–8

Chappel D., 2006, Understanding .NET, Addison-Wesley Professional

Choi B.K., Kim B.H., 2002, MES (manufacturing execution system) architecture for FMS compatible to ERP (enterprise planning system). International Journal of Computer Integrated Manufacturing 15/3:274–284

Dah-Chuan G., Yueh-Wen H., 1997, Conceptual design of a shop floor control information system. International Journal of Computer Integrated Manufacturing 10/1-4:4–16

Deif A.M., ElMaraghy H.A., 2007, Assessing capacity scalability policies in RMS using system dynamics. International Journal of Flexible Manufacturing Systems 19/3:128–150

Deif A.M. and ElMaraghy W.H., 2006, Architecture for decision logic unit in agile manufacturing planning and control systems, Singapore, Singapore, Institute of Electrical and Electronics Engineers Computer Society, Piscataway, NJ 08855-1331, United States

Deif A.M., ElMaraghy W.H., 2006, A control approach to explore the dynamics of capacity scalability in reconfigurable manufacturing systems. Journal of Manufacturing Systems 25/1:12–24

Devedzic V., Radovic D., 1999, A framework for building intelligent manufacturing systems. IEEE Transactions on Systems, Man and Cybernetics, Part C (Applications and Reviews) 29/3:422–439

ElMaraghy H.A., 2005, Flexible and reconfigurable manufacturing systems paradigms. International Journal of Flexible Manufacturing Systems 17/4:261–276

ElMaraghy H.A., 2006, Reconfigurable Process Plans for Responsive Manufacturing Systems. Keynote Paper, Proceedings of the CIRP International Design Enterprise Technology (DET) Conference, Setubal, Portugal

ElMaraghy H.A., 2007, Reconfigurable Process Plans for Responsive Manufacturing Systems, Digital Enterprise Technology: Perspectives & Future Challenges, Editors: P.F. Cunha and Maropoulos P.G., Springer Science, ISBN: 978-0-387-49863-8, pp 35–44

Grimes F., 2002, Microsoft .NET for Programmers, Manning Publications

Gunasekaran A., 1998, Agile manufacturing: Enablers and an implementation framework. International Journal of Production Research 36/5:1223–1247

Heineman G.T. and Councill W.T., 2001, Component-Based Software Engineering: Putting the Pieces, Addison Wesley

Johnson R., 2002, Expert One-on-One J2EE Design and Development, Wrox

Jones G.R., 2006, Organizational Theory, Design and Change, Prentice Hall

Koren Y., 2003, Reconfigurable manufacturing systems. Journal of the Society of Instrument and Control Engineers 42/7:572–582

Koren Y., Heisel U., Jovane F., Moriwaki T., Pritschow G., Ulsoy G., Van Brussel H., 1999, Reconfigurable manufacturing systems. CIRP Annals – Manufacturing Technology 48/2:527–540

Koren Y., Jovane H. U. and Moriwaki T. F., 1999, Reconfigurable Manufacturing System. Annals of CIRP 48

Mehrabi M., Ulsoy G., Koren Y., 2000, Reconfigurable manufacturing systems: Key to future manufacturing. Journal of Intelligent Manufacturing 11:403–419

Monfared M.A.S., Yang J.B., 2007, Design of integrated manufacturing planning, scheduling and control systems: A new framework for automation. International Journal of Advanced Manufacturing Technology 33/5-6:545–559

Pritchard J., 1999, COM and CORBA Side by Side: Architectures, Strategies, and Implementations, Addison-Wesley Professional

Purohit A., Pant R., Deb A., 2006, Role of rapid technologies as enablers for agile manufacturing in the automotive industry. International Journal of Agile Manufacturing 9/2:91–97

Ramesh G., Devadasan S.R., 2007, Literature review on the agile manufacturing criteria. Journal of Manufacturing Technology Management 18/2:182–201

Sahin F., 2000, Manufacturing competitiveness: Different systems to achieve the same results. Production and Inventory Management Journal 41/1:56–65

Shan F., Li L.X., Ling C., 2001, An object-oriented intelligent design tool to aid the design of manufacturing systems. Knowledge-Based Systems 14/5-6:225–232

Sharifi H., Colquhoum G., Barclay I. and Dann Z., Agile Manufacturing: a management and operational framework" Proc Instn Mech Engre 215

Slama D., Garbis J. and Russell P., 1999, Enterprise CORBA Prentice Hall

Sommerville, 2004, Software Engineering, Addison Wesley

Sung-Yong S., Olsen T.L., Yip-Hoi D., 2001, An approach to scalability and line balancing for reconfigurable manufacturing systems. Integrated Manufacturing Systems 12/7:500–511

Tsai T., Sato R., 2004, A UML model of agile production planning and control system. Computers in Industry 53/2:133–152

Tsu Ta T., Boucher T.O., 2002, An architecture for scheduling and control in flexible manufacturing systems using distributed objects. IEEE Transactions on Robotics and Automation 18/4:452–462

Vollman, Berry, Whybark and Jacobs, 2005, Manufacturing Planning and Control for Supply Chaing Management McGraw-Hill

Wache I., 1998, Object-oriented modelling of material flow controls. Journal of Materials Processing Technology 76/1-3:227–232

Wang A.J.A. and Qian K., 2005, Component-Oriented Programming., John Wiely & Sons, Inc.

Wheelen T. and Hunger D., 2006, Strategic management and business policy, Prentice Hall

Wiendahl H.P., ElMaraghy H.A., Nyhuis P., Zah M.F., Wiendahl H.H., Duffie N., Brieke M., 2007, Changeable Manufacturing – Classification, Design and Operation. CIRP Annals – Manufacturing Technology 56/2:783–809

Wu B., 2000, Manufacturing and Supply Systems Management: A Unified Framework of Systems Design and Operation, Springer

Zhang J., Gu J., Li P., Duan Z., 1999, Object-oriented modeling of control system for agile manufacturing cells. International Journal of Production Economics 62/1-2:145–153

Chapter 13
Dynamic Capacity Planning and Modeling Its Complexity

A. Deif[1] **and H. ElMaraghy**[2]

Abstract Uncertainty associated with managing the dynamic capacity in changeable manufacturing is the main source of its complexity. A system dynamics approach to model and analyze the operational complexity of dynamic capacity in multi-stage production is presented. The unique feature of this approach is that it captures the stochastic nature of three main sources of complexity associated with dynamic capacity. The model was demonstrated using an industrial case study of a multi-stage engine block production line. The analysis of simulation experiments results showed that ignoring complexity sources can lead to wrong decisions concerning both capacity scaling levels and backlog management scenarios. In addition, a general trade-off between controllability and complexity of the dynamic capacity was illustrated. A comparative analysis of the impact of each of these sources on the complexity level revealed that internal delays have the highest impact. Guidelines and recommendations for better capacity management and reduction of its complexity, in changeable manufacturing environment, are presented.

Keywords Complexity, Capacity Planning, Uncertainty

13.1 Introduction

13.1.1 The Dynamic Capacity Problem

Capacity planning has been the subject of great interest over the last 40 years due to the ability of capacity changeability to hedge against demand uncertainty. The typical problem in capacity planning is to decide on the timing and amount of in-

[1] University of Regina, Canada

[2] IMS Center, University of Windsor, Canada

vestment as well as the selection of the resources (equipment, facilities, systems, and people) to use in a manufacturing site at any time. However, there are many exogenous and endogenous parameters that make capacity planning complicated. Among these parameters are the dynamic and stochastic nature of demand, the availability of the required capacity and time to react to sudden changes either in demand (external) or in the manufacturing system itself (internal). The traditional trade-off between responsiveness and cost effectiveness adds another layer of complexity to the capacity management problem and various approaches were introduced to optimize both objectives. Extensive research has been conducted to study optimal capacity planning under different conditions (see Manne 1967, Luss 1982 and Mieghem 2003).

As manufacturing systems evolved through different paradigms from dedicated manufacturing all the way to changeable manufacturing, so did the capacity planning challenge in these systems. Examples of that evolution include not only considering the economy of scale but also the economy of scope in the capacity expansion/reduction decisions and reducing the reaction time to scale the capacity from years and months to weeks and even days. The modern infrastructure, based mainly on the modular and open architecture control design of machines and systems, in today's advanced manufacturing systems was one of the main enabler for such evolution.

The inherent complexity within the capacity planning process is one of the parameters that has a significant influence on the capacity management decisions and yet received little attention to date. Discussing the complexity of capacity planning requires positioning the capacity planning problem within the proper framework and determining the type(s) and sources of complexity present in this domain.

In today's competitive market, manufacturing enterprises face the challenge of being responsive to changeable market demands while keeping a cost effective level of production. Facing such a challenge would have been very difficult without the new manufacturing paradigms and the technological enablers to allow changing their functionality as well as their capacity (Wiendahl, et al. 2007). Such dynamic market environment, with the continuous advancement of technology, makes the management of the capacity change and reconfiguration very dynamic (Deif and ElMaraghy, 2006). Thus the capacity planning is inherently a dynamic problem.

13.1.2 Complexity vs. Uncertainty

Complexity covers a broad scope and is associated with systems that are difficult to understand, predict or control. It is usually measured with the quantity and quality of information required to describe or control the system (Suh, 2005). Thus if the information is easily captured, understood and manipulated the system is less complex than other systems with information that are difficult to capture or analyze. From that assessment angle, uncertainty is directly proportional to system complexity. In other words, the more uncertain the information required to describe and control the system, the more complex manging this system would be. This perspective of complexity is what the authors adopt in this research work.

Complexity can be generally classified into structural and operational complexity. Structural complexity refers to the static design dimension of the system (Deshmukh et al. 1998), and the different system's components and their relationship (ElMaraghy H., 2006 and 2005). Operational complexity on the other hand, is defined as the uncertainty associated with the dynamic manufacturing system (Frizelle, 1998).

13.1.3 Complexity in Dynamic Capacity Planning

Since capacity planning is dynamic in nature with various uncertainties associated with demand, capacity scaling time and manufacturing lead time, it can be associated with operational complexity. However, since there are various definitions of operational complexity, it is important to clarify how it is used in this research. Operational complexity is defined as the effort, expressed in terms of the magnitude and frequency of dynamic capacity planning, to determine when and by how much the capacity should be scaled in response to demand due to internal and external sources of uncertainty.

In today's market, the ability to frequently and effectively change capacity level is becoming a fundamental feature of any successful changeable manufacturing system; however, the operational complexity of this dynamic process is an obstacle in implementing such a strategy. This paper presents an approach to model and understand this operational complexity in an attempt to better manage dynamic capacity planning in changeable systems.

13.2 Literature Review

In this section, the two major dynamic methodologies that were used to handle the dynamic capacity planning are reviewed. The first methodology is the control-theoretic approach (mainly feedback control). The second methodology is system dynamics (SD) introduced by Forester (1961) and aims at understanding how the physical process, the information and the managerial policies of capacity scalability interact together.

A dynamic model developed by Duffie and Falu (2002) for closed loop production planning and control (PPC) was proposed to control work in process (WIP) and capacity using control theoretic approaches. They investigated the effect of choosing different capacity scaling controller gains as well as the WIP controller gains on system performance and how this can be used to achieve required system responses. Kim and Duffie (2004) extended this work to study the effect of capacity disturbances and capacity delays on system performance in single work stations and further applied it to multiple workstations in Kim and Duffie (2005). Their results indicated that if capacity can be adjusted more often with less delay, the system's performance would be highly improved in changing demand environments.

Another dynamic model that manipulates feedback control with the help of logistics operating curves, developed by Nyhuis (1994) to control work in process (WIP) and capacity of manufacturing systems, was presented in Wiendahl and Breithaupt (1999) and (2000). In this approach, the required capacity was found using flexibility curves. The capacity controller chooses the best capacity scaling decision based on balancing the backlog value and acceptable delay.

Asl and Ulsoy (2002) presented a dynamic approach to capacity scalability modeling in reconfigurable manufacturing systems (RMS) based on the feedback control principles and obtained sub-optimal solutions, partially minimizing the cost of capacity scalability, that were also robust against demand variations.

Deif and ElMaraghy (2006) developed a dynamic model for capacity scaling in RMS and analyzed the model based on control theoretic approaches to indicate the best design for the capacity-scaling controller. The results highlighted the importance of accounting for different physical and logical delays together with the trade-off decisions between responsiveness and cost. They further introduced an optimization unit to the capacity scalability model to optimally decide on the exact value of the scaling controller gain in Deif and ElMaraghy (2007).

Wikner et al. (2007) modified the automatic pipeline inventory and order-based production control system (APIOBPCS) used for "make-to-stock" to deal with "make-to-order" systems using the dynamic surplus capacity. These systems were shown to maintain agility and decrease the backlog levels by introducing a controller to account for the backlog resulting form the capacity scaling delays while responding to changing demands.

Examples of manipulating System Dynamics (SD) models for capacity planning include an attempt by Eavns and Naim (1994) to develop a SD model for supply chains with capacity constraints and study the effect of capacity constraints on the system's performance and overall cost.

Helo (2000) suggested a capacity-based supply chain model that includes a mechanism for handling the trade-off between lead-time and capacity utilization. It was shown that this capacity analysis (including surge effects) in supply chains would improve their responsiveness.

Goncalves et al. (2005) highlighted the issue of capacity variation in their push-pull manufacturing SD model through the effect of capacity utilization on the production start rate. They also showed how the sales and production effects interact to destabilize the system and degrade its performance.

Anderson et al. (2005) considered logical capacity scalability in supply chains for service and custom manufacturing. They showed the effect of reducing lead-time and sharing the demand information on improving the system performance. In addition they proposed polices to reduce backlog in these systems.

Vlachos et al. (2007) proposed a model to study the long-term behavior of reverse supply chains applied it to re-manufacturing. For that purpose, they examined efficient re-manufacturing and collection capacity expansion policies that maintain profit while considering direct and indirect factors.

Deif and ElMaraghy (2007) proposed a SD model for capacity scalability in make-to-order manufacturing. Various performance measures were used to examine the best scaling policy under different demand scenarios. They showed that the best scalability policy would be based on both the marketing strategy as well as the operational production objectives.

The previous dynamic approaches to model and analyze the dynamic capacity planning problem focused on either controlling the capacity scalability process or exploring different policies to hedge against various internal and external disturbances. Although they offered good solutions for both problems, no work has been reported to study the associated operational complexity. Thus the work presented in this paper is motivated by the need to understand sources of operational complexity and their degree of influence in dynamic capacity planning. It is believed that such understanding would result in reducing the complexity of the dynamic capacity planning and management.

13.3 System Dynamic Model for Multi-Stage Production

A stochastic dynamic model for capacity planning and associated different sources of complexity has been formulated. Figure 13.1 contains a dynamic model of three-stage instantiation for n-stage ($n > 1$) serial production system. It is important to note that a continuous-time model is used because it provides an acceptable approximation of the continuous dynamic capacity scaling process at that level of abstraction. Both the operations management and system dynamics literature support the use of continuous models for capacity planning (e.g.: Anderson, et al. 2005, Sethi and Thompson, 2000, Holt, et al. 1960). Finally, similar dynamic characteristics can be obtained using discrete-time models (John, et al. 1994).

13.3.1 Multi Stage Production System

A manufacturing system in which several production activities have been functionally aggregated into different production stages is considered. There are many reasons for wanting to aggregate production activities into stages. First, in most manufacturing systems, production activities are typically grouped into identifiable stages. Second, when dealing with multi-product systems, changing of setups to switch from one product to another are often performed on major sub-systems of machines rather than on individual machines. Finally, having fewer points to control

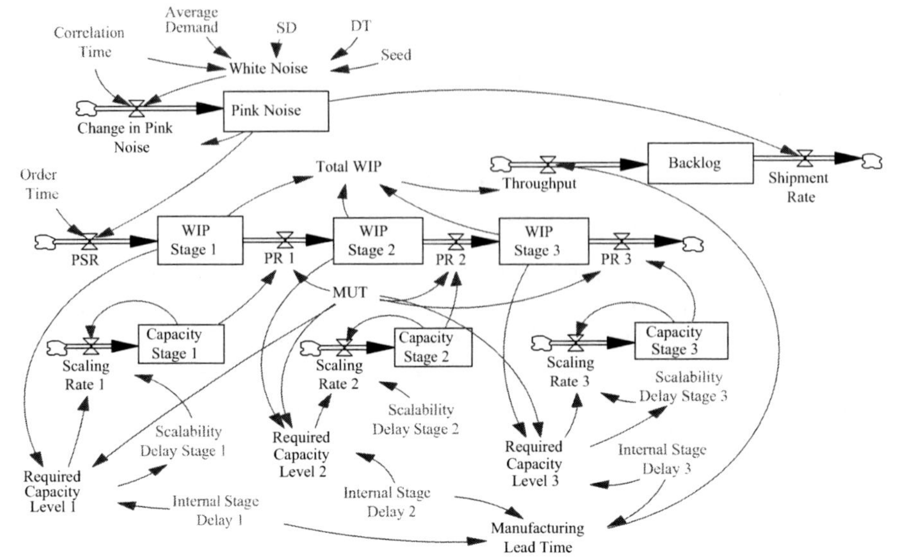

Fig. 13.1 Dynamic capacity model in multi-stage production

makes the dynamic capacity problem simpler and the implementation of a capacity management policy easier.

A WIP-based control multi-stage production system is considered where the WIP level is observed and controlled by varying the production rate through utilizing the dynamic capacity property of these systems. In addition, the backlog of the system is monitored as a performance measure for the responsiveness of the system. Backlog is calculated based on overall throughput of the system and its manufacturing lead-time.

13.3.2 Model Nomenclature

$Ci(t)$ = capacity level at time t at stage i
$B(t)$ = backlog level at time t
$WIPi(t)$ = WIP level at time t at stage i
$PRi(t)$ = production rate at time t at stage i
$PSR(t)$ = production start rate at time t
$AD(t)$ = average demand at time t
CT = correlation time. This constant captures the degree of inertia (dependence) in noise process
SD = standard deviation for the normal demand distribution
DT = time step
$Seed$ = the seed for randomly generated variates of the stochastic demand data

$Th(t)$ = system throughput at time t
$ShR(t)$ = shipment rate at time t. It is the rate of physical product leaving the system
$TWIP(t)$ = total WIP of the system time t
MLT = manufacturing lead time. It is the time required to process products
$RCi(t)$ = required capacity at time t at stage i
$SDTi$ = scalability delay time. Time require to scale the capacity at stage i
$SRi(t)$ = scalability rate at time t at stage i
$ISDi$ = internal stage i delay
MUT = manufacturing unit time

13.3.3 Mathematical Model

13.3.3.1 Stochastic Market Demand

The market demand is modeled as a stochastic parameter with dependent distribution or pink noise. The noise is an expression used to reflect the random variation in the data due to the stochastic nature of the process that follows a certain distribution. While convenient statistically, the independent distribution assumption of demand, or white noise (as in the case of most of previous dynamic capacity analysis), does not hold for real world cases (Sterman, 2000). To have a better assessment of the impact of demand uncertainty on dynamic capacity complexity, it is necessary to model demand forecast as a process with memory in which the next value of demand does not depend on the last demand but rather on the history of previous forecasts.

The demand in this model is assumed to have a continuous cumulative Normal Distribution Function. Huh et al. (2006) state that demand should have a continuous distribution because demand is inherently continuous; the variance in demand is often high, and finally because continuous demand distribution may generate a more robust capacity plan than finite number of discrete scenarios. Equation 13.1 formulates the demand as white noise with a normal distribution.

$$\text{White Noise}(t) = AD(t) + \left[SD^2 * \frac{(2-(DT/CT))}{(DT/CT)}\right]^{0.5} * \text{Normal (0, 1, Seed)} \tag{13.1}$$

Equations 13.2 and 13.3 display the values for the demand pink noise and the change in demand pink noise respectively

$$\text{Pink Noise (t)} = \text{Change in Pink Noise} - \text{Pink Noise}_0 \tag{13.2}$$

$$\text{Change in Pink Noise} = \frac{\text{Pink Noise}(t) - \text{White Noise}(t)}{\text{CT}}. \tag{13.3}$$

13.3.3.2 Dynamic Capacity Planning and Control

Capacity scaling decisions at each production stage (i) are controlled through scaling rate in Eq. 13.4.

$$\dot{C_i}(t) = SR_i(t) - C_{i-1}(t) \tag{13.4}$$

The scaling rate at each stage is determined by the required capacity together with the scalability delay time (Eq. 13.5).

$$SR_i(t) = \frac{C_i(t) - RC_i(t)}{SDT_i} \tag{13.5}$$

The scalability delay time in this model is a varying parameter, which is function of the type of capacity resource(s) to be scaled (Eq. 13.6). This is an important assumption meant to capture the real world scenarios since the time for example to add a spindle to a machine is indeed less than that required to add a machine to an existing line. Classical capacity planning work used either a simple assumption of instantaneous or fixed time for scaling the system's capacity.

$$SDT_i = F\,\{X_i\}\,\mathrm{i}\,, \tag{13.6}$$

where X_i is the type of capacity to be scaled at stage i.

The required capacity (Eq. 13.7) is calculated based on the WIP level since this is a WIP-based controlled system as explained earlier.

$$RC_i(t) = \left(\frac{WIP_i(t)}{ISD_i}\right) * MUT \tag{13.7}$$

The internal stage delay, sometimes referred to as production lead-time, is in general difficult to calculate (Hoyte, 1980) because of the different sources of variability within production systems (Schmitz et al. 2002). Thus the typical assumption of a deterministic value for such a parameter is highly questionable. In this model, a stochastic variable function is used to calculate the Internal Stage Delay (Eq. 13.8). The ISD function depends on the different processes and activities in each of the system production stages.

$$ISD_i = \text{Random } f(\min, \max, \mu, \sigma, s)\,, \tag{13.8}$$

where "min" and "max" are the minimum and maximum values that the probabilistic function will return, μ is the mean of the random distribution, σ is the standard deviation of the distribution, and s is the seed for the randomly generated numbers of the probability distribution.

13.3.3.3 Production Control

The WIP level at each stage is determined by the difference between the production rate of the current and the next stage (Eq. 13.9).

$$W\dot{IP}_i(t) = PR_i(t) - PR_{i+1}(t) \tag{13.9}$$

The production start rate is set to be equal to the demand or the pink noise (Eq. 13.10). The production rate is controlled by the capacity scaling level since this is the typical case in systems with dynamic capacity (Eq. 13.11).

$$PSR(t) = \text{PinkNoise}(t) \tag{13.10}$$

$$5PR_i(t) = \frac{C_i(t)}{MUT} \tag{13.11}$$

13.3.3.4 Backlog Calculation

The backlog level is generally used as indicator for the responsiveness level of the manufacturing system. In this model, it is defined as the difference between the shipment rate (which is assumed to be exactly equal to the demand as in Eq. 13.1 and the system throughput (Eq. 13.12).

$$\dot{B}(t) = ShR(t) - Th(t) \tag{13.12}$$

The throughput of the system is calculated based on Little's law as the function of the total WIP and the manufacturing lead-time (Eq. 13.13). The total WIP is calculated using the maximum WIP level accumulated in the production stages (Eq. 13.14). The manufacturing lead-time is also calculated based on the maximum internal stage delay in the system (Eq. 13.15).

$$Th(t) = \frac{TWIP(t)}{MLT} \tag{13.13}$$

$$TWIP(t) = MAX(WIP_i(t)) \tag{13.14}$$

$$MLT = MAX(ISD_i) \tag{13.15}$$

In summary, the dynamic capacity scaling model is composed of three main units, the first captures the demand as a stochastic process; the second handles the dynamic capacity decisions and incorporates the uncertainty of both internal stages delay and capacity scaling delay time, and finally the third models the multi-stage production line and calculates the different production control parameters.

13.4 Numerical Simulation of Industrial Case Study

In this section, the application of the developed model is demonstrated by applying it to a multi-stage engine block production/assembly plant shown in Fig. 13.2. Numerical simulation of the case study are conducted to determine and analyze the dynamic capacity planning complexity. Check the text in the following boxes, the graphics in the middle is not visible, improve or modify or remove.

13.4.1 Overview of the Multi-Stage Engine Block Production Line

The first stage in this production system is the machining stage and contains three lines, one for the manufacturing of the cylinder head (with 84 CNC machines), the second is for the manufacturing of the cylinder block (with 8 metal cutting machines) and the third is responsible for manufacturing of the crankshaft (using casting moulds).

The second stage is the engine main assembly where the three lines of the previous stage feed into the assembly line. This stage is composed of around sixty assembly stations and is considered the bottleneck of the whole plant. It contains both manual and automated wok station.

The assembled engines go to the final stage for overall inspection and different tests. The processes in this stage involve pressure test, functional test and overall sampling and inspection. This stage is composed of 4 testing lines.

13.4.2 Input Data

Two types of data are required to demonstrate the developed model, the demand data and the production system data. The demand data is shown in Table 13.1, while the production system data is shown in Table 13.2.

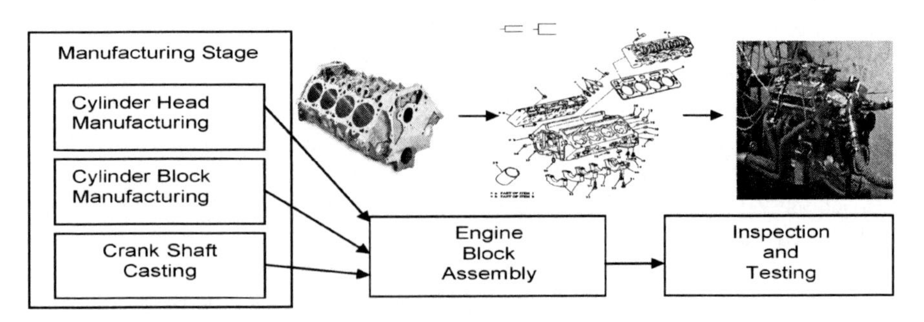

Fig. 13.2 Schematic model of multi-stage production of engine blocks

Table 13.1 Data for demand

Data	Value	Comments
Average Demand (AD)	30 000 engines/month	This is the value of the average batch per customer order per engine model.
Standard Deviation (SD)	±5000 engines/month	This reflects a high degree of market demand fluctuation
Correlation Time (CT)	4 months	This means that each demand forecast depends on the actual data of the proceeding 4 months
Time Step (DT)	0.125 month	Time step will be every 3.5 days
Seed	9	Used to generate random variates for the normally distributed demand data

Table 13.2 Data for the three production stages of the engine block production system

	Internal Stage Delay ISD (weeks)	Scalability Delay Time SDT (weeks)
Stage 1:	RANDOM UNIFORM (0.8, 1, 0)	IF THEN ELSE (Required Capacity Level 1 > 5000, 0.7, 0.25)
Comments	The internal stage delay varies randomly between 0.8 and 1 month with a uniform distribution. This stage has the shortest delay due to its automatic nature. The variation sources are the variability in the processing times of the different machines in this stage.	If the required capacity to be scaled is below 5000 engines/month, then the CNC machines are reconfigured by adding more spindles or axes. This requires 0.25 month for installation, reprogramming and ramp up. If the required capacity to be scaled is above 5000 engines/month, then extra CNC machines are added to the specific line. This requires 0.7 month for installation, calibration and ramp up.
Stage 2:	RANDOM UNIFORM (1.2, 1.5, 0)	IF THEN ELSE (Required Capacity Level 2 > 12000, 0.4, 0.6)
Comments	The variation is larger than previous stage due to the many workers involved with different learning curves. This is also the reason for this stage having the longest delay	The capacity scaling in this stage is achieved by hiring more workers or adding extra shifts. The delay is due to different administrative and training procedures involved.
Stage 3:	RANDOM UNIFORM (0.9, 1.2, 0)	(IF THEN ELSE (Required Capacity Level 3>15000, 0.7, 0.5))
Comments	The variations in this stage are due to the variability of both the testing stations machines as well as the manual labor involvement in this stage.	The scalability options here are either hiring more workers (delay is 0.5 month) if the required capacity change is less than 15K engine/month or increasing the test stations (delay is 1 month) if the required capacity is more than 15K engines/month.

It is important to note that although the delay due to adding capacity may be different form removing capacity, for simplicity, it is assumed that both delay times are equal.

13.4.3 Numerical Simulation Results

In this section, the results of various simulation experiments conducted to investigate the impact of the sources of operational complexity on the dynamic capacity planning are reported. The scaling rate is used in this analysis as the main performance measure that can give insight into the complexity of the dynamic capacity planning problem in terms of effort and cost. In addition, backlog and throughput are also used as performance measures in some of the conducted analysis to evaluate the responsiveness and efficiency of the developed capacity planning system.

13.4.3.1 Comparing Stochastic and Deterministic Analysis

The first analysis compares the cases where the three main sources of complexity in the developed dynamic capacity system are modeled with bot stochastic and deterministic data. The main objective is to highlight the impact of these factors on the complexity of the dynamic capacity planning. Figure 13.3a–c compares the scaling rate, as a performance measure of capacity scalability, in the stochastic case (left side) with that in the deterministic case (right side) at each stage in the engine block production system. Analysis of Fig. 13.2 reveals the following observations (DC refers to Dynamic Capacity):

- The levels of the scaling rate in each stage for the two scenarios illustrate the effect of the various sources of complexity. The magnitude of the scaling rates in the stochastic case has much higher values than those of the deterministic case. The stochastic case experiences more oscillations compared with the deterministic case (except for the 3rd stage), and the later even reaches stability at the value of zero after some weeks. Thus incorporating the sources of complexity into capacity planning increases the operational complexity of the scaling decisions by increasing their number and frequency.
- The desirable dynamic behavior of the deterministic case compared with the stochastic case points to a fundamental trade-off decision in dynamic capacity planning. The planner has to balance the need for accurate representation of the scaling process against the desire to keep an acceptable level of controllability of that process.
- The results also show the occurrence of the "bullwhip" effect, which is the variance in the processing rate and, hence, the next stage's demand becomes greater than that of input tasks (Frank, et al. 2000). This adds another dimension to the complexity of the decision regarding the level of aggregation when designing dynamic capacity planning systems.

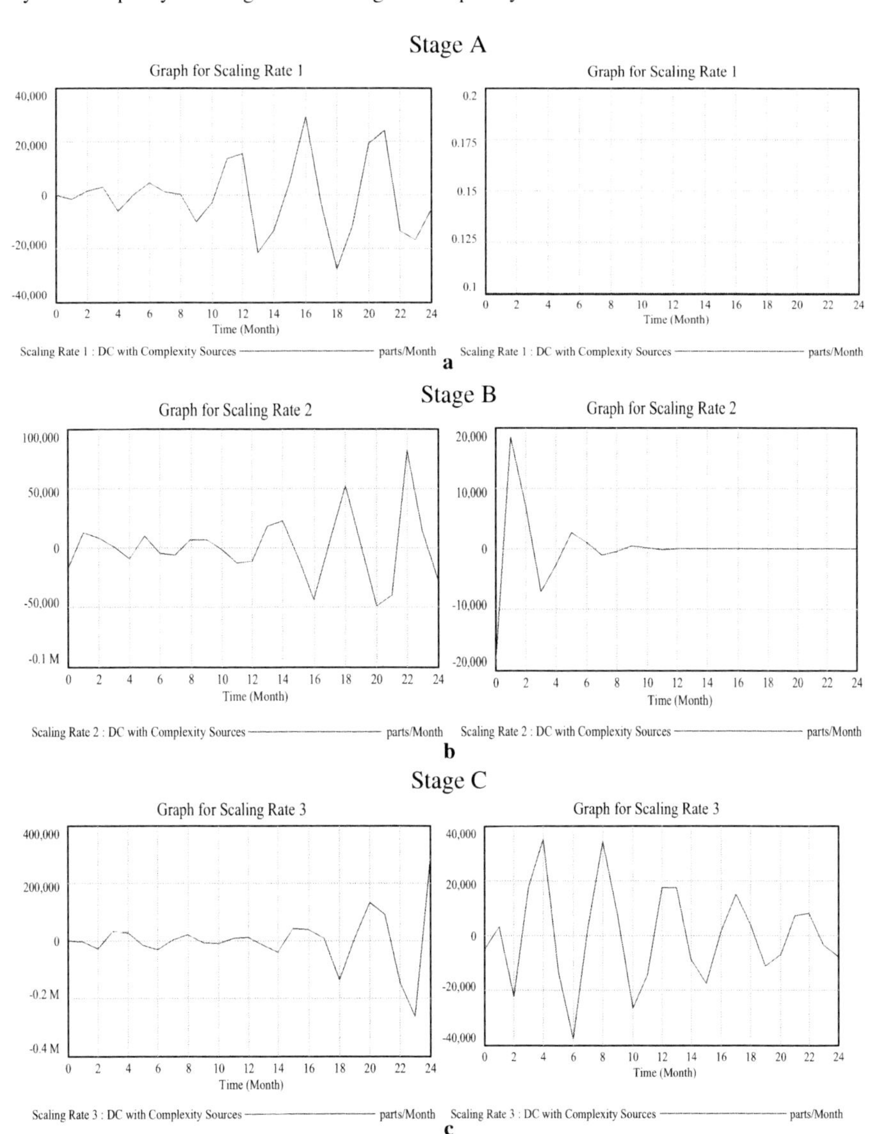

Fig. 13.3a–c Scalability rate in cases of stochastic analysis (*left side*) and deterministic analysis (*right side*) **a** first stage, **b** second stage, **c** third stage

13.4.3.2 Impact of Operational Complexity Sources on Production Systems' Performance and Responsiveness

Throughput is a fundamental performance measure of a production system (Hopp and Spearman, 2000). Figure 13.4 compares the evolution of the throughput of the developed multi-stage production system in the stochastic case (a) and deterministic case (b).

The result shows that the variations in the sources of complexity negatively affect the performance of the system in comparison with the case where these variations are eliminated. The uncertainty associated with these sources led to higher than the required level of throughput in addition to dynamic oscillations that will affect the stability of the system. This leaves the capacity planner with another trade-off decision between efficient production in terms of cost (less inventory and oscillation) and using a realistic abstraction and representation of the sources of operational complexity. In other words, if uncertainty sources are ignored, the production can be better efficiently planned, however, the ability of such a plan to hedge against internal and external disturbances will be questionable.

Backlog is also a crucial indicator for the degree of responsiveness especially in systems employing dynamic capacities to maintain a short market lead-time. Figure 13.5 shows the backlog level in both cases of stochastic and deterministic analysis of the developed dynamic capacity model.

Results in Fig. 13.5 highlight that, in general, the backlog level in the deterministic case is much lower than in the stochastic case. This indicates that ignoring the uncertain nature of the complexity sources can lead to false assessment of the level of responsiveness of the production system. In other words, having 50 K engines over the 2 years only as an average backlog indicates a level of responsiveness that

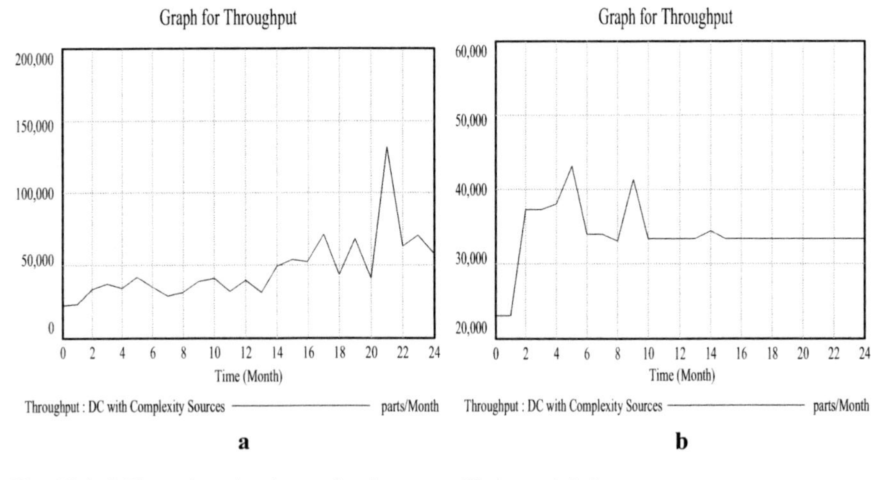

Fig. 13.4a,b Throughput level **a** stochastic case and **b** deterministic case

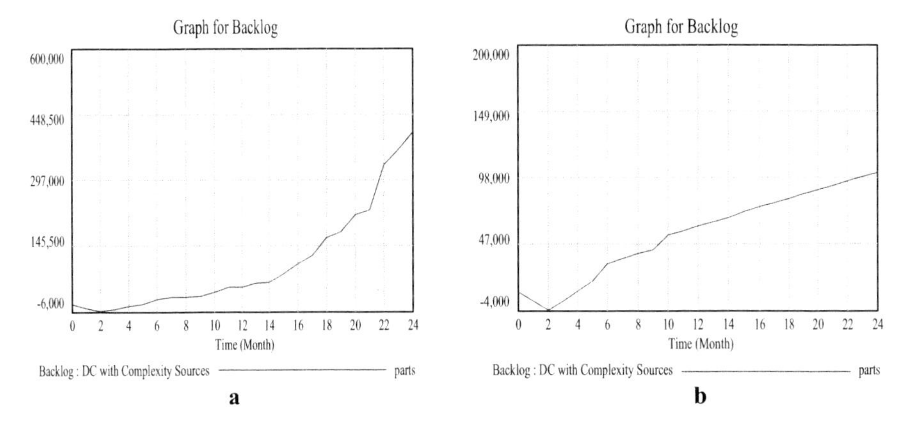

Fig. 13.5a,b Backlog level **a** stochastic case and **b** deterministic case

is much higher than the real level with the complexity sources considered, where the average backlog should be 150 K engines (almost three times the deterministic case).

13.4.3.3 Comparative Assessment for Sources of Operational Complexity in Dynamic Capacity Planning Systems

In this section, the impact of each of the three sources of operational complexity considered in this study on the dynamic capacity planning is discussed. The used performance measure is the scaling rate as an indicator of the required capacity planning effort and cost. Two of the three sources of uncertainty are kept constant, where a deterministic value of their average is used, while observing the impact of the third source over the scaling rate.

Figure 13.4.3.3a–c shows the impact of each of the three considered sources of operational complexity over the scaling rate at each of the three stages of production. The following three main observations can be deduced based on the obtained results:

- The internal stage delay is the main source of operational complexity in the capacity scaling process. This is demonstrated through having the greatest number of oscillations for the scaling rate across the three production stages. In addition, the scaling rate has the highest value at each stage with the internal stage delay. It is important to note that the scaling rate experiences many oscillations due to the stochastic demand, which makes it the second source for the operational complexity in the capacity scaling process.
- An interesting observation is that the scalability delay time, based on the magnitude and number of oscillation, has a minimal contribution to the operational complexity, as the scaling rate tends to reach zero after a period of time. This is

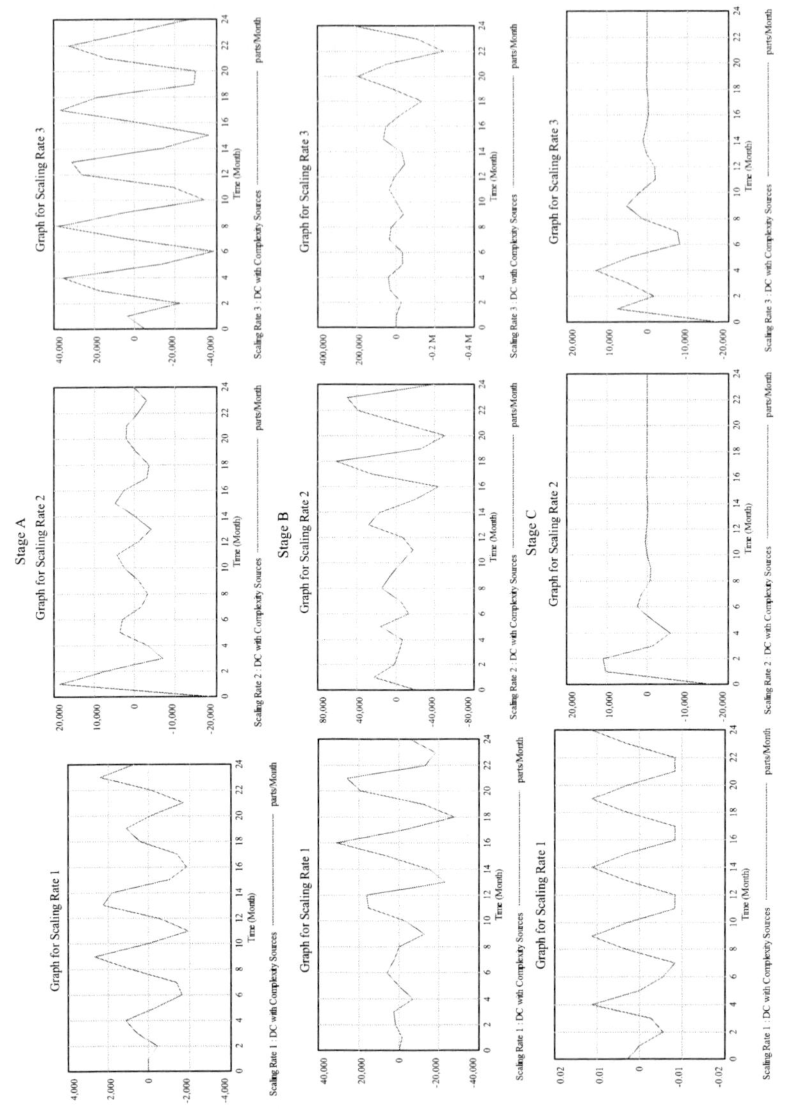

Fig. 13.6a–c Impact of each of the three considered sources of operational complexity over the scaling rate at each of the three stages of production **a** stochastic demand, **b** internal stage delay and **c** scalability delay time

because after a period of time, and since demand is assumed to be stable, the production (after some capacity changes) will be able to exactly match the demand and thus no further capacity scaling is required. This leads to the fact that the share of the scalability delay time in the operational complexity of the scaling process is proportional to the stability of the market demand.
- The "bullwhip" effect is clear in the impact of the three operational complexity sources over the scaling rate across the production stages. This suggests bullwhip as another source of operational complexity in capacity planning and highlights the importance of studying the conditions under which this phenomenon occurs to better manage capacity planning and its complexity.

13.5 Conclusions

The presented study of the dynamic capacity scalability in multi-stage production systems is focused, for the first time, on the intuitive understanding of the operational complexity associated with the capacity scaling process in these changeable systems. An approach based on system dynamics was presented to model the dynamic nature of capacity scaling in changeable manufacturing environments.

The unique feature of this modeling approach is that it identified and quantified the three main sources of the operational complexity relevant to this problem. These sources are stochastic demand, internal stage delay and capacity scaling delay time. The developed approach was illustrated by a case study for a typical industrial multi-stage engine block production system. Several results were demonstrated using simulation, which lead to conclusions applicable to dynamic capacity planning in changeable manufacturing as follows:

- The uncertainties associated with the considered sources of complexity were quantitatively proved to increase the level of operational complexity of dynamic capacity planning.
- A trade-off between the complexity and controllability of capacity scaling must be exercised by the capacity planner. A desirable high level of controllability requires capturing the stochastic characteristics of the sources of uncertainty in-capacity scaling, which would lead to increasing the operational complexity of the planning decisions.
- The performance, in terms of throughput and responsiveness, was negatively affected by the considered sources of operational complexity.

The reported results together with the previous conclusions lead to the following recommendations to better manage the operational complexity in this changeable environment:

- Reducing randomness and uncertainty through better information management and/or tighter control information sources is essential to decrease the degree of

uncertainty associated with demand forecasting, manufacturing lead-time and capacity scalability delay time.
- More effort should be devoted to the stabilizing and/or accurate calculation of the internal stage delay, which was shown to have the highest impact on the operational complexity level.
- The conditions under which the bullwhip effect occurs should be determined and used as constraints for capacity scaling decisions. It was demonstrated that the bullwhip effect contributes to the operational complexity of dynamic capacity planning in multi-stage production system.

The obtained results are not limited to the investigated case study but are also applicable to other multi-stage production systems that share similar structure. The relaxation of some of the assumptions considered in the proposed model concerning the stochastic and other time parameters should be investigated further. In addition, the system dynamics approach was shown to be highly capable in capturing the dynamic behavior of operational complexity, but it should be noted that the approach as presented does not offer a numerical evaluation of the operational complexity. Such evaluation would be a natural extension of the proposed research.

Further work is also required to investigate the effect of other sources of uncertainty on operational complexity. In addition, studying the relationship between structural and operational complexity can lead to the development of a general framework for optimal capacity management in changeable manufacturing environments.

Acknowledgements This research was conducted at the Intelligent Manufacturing Systems (IMS) Center where the first author held a Post Doctoral Fellowship. The support from The Canada Research Chairs (CRC) Program and the Natural Science and Engineering Research Council (NSERC) of Canada is greatly appreciated.

References

Anderson E., Morrice D., Lundeen G., 2005, The "physics" of capacity and backlog management in service and custom manufacturing supply chains. System Dynamics Review 22/3:217–247

Asl R. and Ulsoy A., 2002, Capacity management via feedback control in reconfigurable manufacturing systems, Proceeding of Japan-USA symposium on Flexible Manufacturing Automation, Hiroshima, Japan

Deif A., ElMaraghy H.A., 2007, Assessing capacity scalability policies in RMS using system dynamics, International Journal of Flexible Manufacturing Systems (IJFMS) Special Issue on Capacity Planning in Flexible and Dynamic Manufacturing 19/3:128–150, doi: 10.1007/s10696-008-9031-2

Deif A., ElMaraghy W., 2006, A control approach to explore capacity scalability scheduling in reconfigurable manufacturing systems. Journal of Manufacturing Systems 25/1:12–24

Deif A.M., ElMaraghy W.H., 2007, Integrating static and dynamic analysis in studying capacity scalability in RMS, International Journal of Manufacturing Research (IJMR) 2/4:414–427

Deshmukh A., Talavage J., Barash M., 1998, Complexity in manufacturing systems, part 1: Analysis of static complexity. IIE Transactions 30:645–655

Duffie N., Falu I., 2002, Control-theoretic analysis of a closed loop PPC system. Annals of CIRP 52/1:379–382

ElMaraghy H.A., 2006, A complexity code for manufacturing systems, 2006 ASME Int. Conference on Manufacturing Science & Engineering (MSEC), Symposium on Advances in Process & System Planning, Ypsilani, MI, USA

ElMaraghy H.A., Kuzgunkaya O. and Urbanic J., 2005, Comparison of manufacturing system configurations – A complexity approach, 55th CIRP Annals, 54/1:445–450

Evans G. and Naim M., 1994, The dynamics of capacity constrained supply chains. Proceedings of International System Dynamics Conference, Stirling: 28–35

Frank C., Drezner Z., Ryan JK., Simchi-Levi D., 2000, Quantifying the bullwhip effect: the impact of forecasting, lead-time and information. Management Science 46/3:436–443

Frizelle G., 1998, The Management of Complexity in Manufacturing. Business Intelligence, London

Goncalves P., Hines J., Sterman J., 2005, The impact of endogenous demand on push-pull production systems. System Dynamics Review 22/3:217–247

Helo P., 2000, Dynamic modeling of surge effect and capacity limitation in supply chains. International Journal of Production Research 38/17:4521–4533

Holt C.C., Modigliani F., Muth J.F., Simon H.A., 1960, Planning production, inventories, and work force. Prentice-Hall, Englewood Cliffs, NJ

Hopp and Spearman, 2002, Factory Physics, McGraw Hill

Hoyt J. 1980, Determining Lead Time for Manufactured Parts in a Job Shop, Computers in Manufacturing, (ed) J.J. Pennsanken. pp 1–12

Huh W.T., Roundy R.O., Cakanyildirim M., 2006, A general strategic capacity planning model under demand uncertainty. Naval Research Logistics 2:137–150

John S., Naim M., Towill D.R., 1994, Dynamic analysis of a WIP compensated support system. Intl Journal of Manufacturing System Design 1/4:283–297

Kim J.-H., Duffie N., 2004, Backlog control design for a closed loop PPC system. Annals of CIRP 54/1:456–459

Kim J.-H., Duffie N., 2005, Design and analysis for a closed loop capacity control of a multi workstation production system. Annals of CIRP 55/1:470–474

Luss H., 1982, Operation research and capacity expansion problems: A survey. Operation Research 3/5:907–947

Manne Alan S., 1967, Investments for capacity expansion, size, location, and time-phasing. The MIT Press, Cambridge, MA

van Mieghem J., 2003, Capacity management, investment and hedging: Review and recent developments. Manufacturing and Service Operation Management 5/4:269–302

Nyhuis, P., (1994), Logistic operating curves – A comprehensive method for rating logistic potentials, EURO XIII/OR36, University of Strathclyde Glasgow

Schmitz J.P.M., van Beek D., Rooda J., 2002, Chaos in Discrete Production Systems. Journal of Manufacturing Systems 21:23–35

Sethi S.P., Thompson G.L., 2000, Optimal control theory: Applications to management science and economics. Kluwer, Boston, MA

Sterman, J.D. (2000), Business Dynamic – Systems Thinking and Modeling for a Complex World. McGraw-Hill

Suh N.P., 2005, Complexity in Engineering. CIRP Annals 54:581–598

Wiendahl H.P., Breithaupt J., 1999, Modeling and controlling of dynamics of production system. Journal of Production Planning and Control 10/4:389–401

Wiendahl H.P., Breithaupt J., 2000, Automatic production control applying control theory. International Journal of Production Economics 63:33–46

Wiendahl, H.-P. ElMaraghy H.A., Nyhuis P., Zaeh M., Wiendahl H.-H., Duffie N. and Kolakowski M., 2007, Changeable manufacturing: classification, design, operation, Keynote Paper, CIRP Annals, 56/2

Wikner J., Naim M., Rudberg M., 2007, Exploiting the order book for mass customized manufacturing control systems with capacity constrains. IEEE Transaction on Engineering Management 54/1:145–155

Vlachos D., Georgiadis P., Iakovou E., 2007, A system dynamics model for dynamic capacity planning of remanufacturing in closed-loop supply chains. Journal of Computers and Operation Research 34:367–394

Part IV
Managing and Justifying Change in Manufacturing

Products and Systems Design, Planning and Management

Chapter 14
Design for Changeability

G. Schuh, M. Lenders, C. Nussbaum and D. Kupke[1]

Abstract Numerous markets are characterized by increasing individualization and high dynamics. A company's ability to quickly adjust its production system to future needs and conditions with minimum effort is a key competitive factor. Especially in high-wage countries, two conflicts increasingly complicate the design of production systems: the conflict between scale and scope on the one hand and the conflict between a high planning orientation and maximizing value-added activities on the other hand. For future production systems in high-wage countries, effective means are needed to minimize the gaps resulting from this poly-lemma. This contribution introduces a measurable target system to assess the degree of target achievement with regard to these criteria. Based on this target measurement system, a new approach that introduces object-oriented-design to production systems is presented. The central element of object-oriented design of production systems is the definition of objects, e.g. product functions, with homogeneous change drivers, which are consistently handled from product planning up to process design. Both product and process design are driven by interfaces between the defined objects and their inter-dependencies. The findings show that a consistent application of object-oriented design to production systems will significantly increase the flexibility in implementing product changes, minimize engineering change and process planning efforts and support process synchronization to achieve economies of scale more efficiently. Two case studies illustrate the implementation and impact of this approach.

Keywords Complexity, Production system, Production management, Object-oriented design

---

[1] WZL at RWTH Aachen University, Aachen, Germany

14.1 Production Trends in High-Wage Countries

The majority of design problems are driven by trade-offs between numerous conflicting effects. If an improvement is achieved in one field, a change for the worse in another field may arise. This is also true for product and production design problems. Most traditional design approaches follow an analytical, target oriented problem decomposition to structure and resolve these trade-offs. While analytic approaches are successful in stationary environments with good predictability, they increasingly fail when dynamics grow to become the determining factor.

From an economic perspective, globalized and heavily segmented markets increase dynamics for the production systems and lead to the requirement of a thoroughly differentiated product offering and changeable organization of production to assure a sustainable business development (Wiendahl et al. 2007).

Regarding product and production design, companies today generally face two dilemmas: the dilemma between scale and scope on the one hand and the dilemma between a high plan- and a high value-orientation on the other hand (Fig. 14.1) (Schuh et al. 2007). In order to stay competitive, companies are forced to optimize their production systems towards one position on the continuum of both dilemmas.

The dichotomy "scale vs. scope" is characterized by highly synchronized systems and low flexibility ("scale") on the one hand and by one-piece-flow and high flexibility ("scope") on the other hand. Low total unit cost can be achieved by designing the production system for economies of scale. Economies of scale are particularly achieved by the higher efficiency of strictly synchronized systems but implicate a limited changeability of the production system. Economies of scope are achieved when high adaptivity is implemented. This means that the systems are designed in order to enable several pre-defined degrees of freedom. However, additional investments or a higher number of manual tasks are required, leading to higher unit cost in comparison with scale optimized production. Having moved away from job shop production, numerous companies in high-wage countries maximize their economies

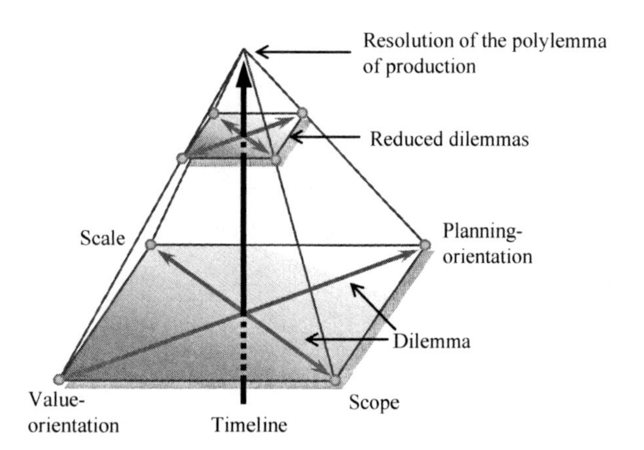

Fig. 14.1 Resolution of the poly-lemma of production (Schuh et al. 2007)

of scale – that is, utilize relatively expensive production means and resources to an optimum degree. These companies try to cope with increasingly individualized market and changing customer needs by way of customization and fast adaptations to market needs, often at the cost of optimum utilization of production means and resources. Thus, realizable economies of scale decrease. Resorting into sophisticated niche markets as a general strategy does not seem to be as promising anymore (Schuh et al. 2007).

The dichotomy of "value-orientation vs. planning-orientation" is characterized by less planning and standardized (work) methods ("value-orientation") on the one hand and by extensive planning, modeling and simulation ("planning-orientation") on the other hand. A planning-oriented production system can ensure optimum utilization of production means and resources (e.g. batch sizes or logistics planning), but at the cost of high planning efforts and most of all reduced flexibility. In comparison to this, value-oriented production systems demand less planning effort being based on a continuous process cycle and focused on the value adding activities. However, it is not guaranteed that optimum operating points will be identified.

Today's high relevance of scope and value-orientation for companies in high-wage countries is caused by an increasing introduction of dynamics to production systems. Whenever complex, individualized products undergo frequent changes, high economies of scope and low planning-efforts promote successful adaptation. Without a substantial influence of this kind of dynamic on a production system, scope and value-orientation would almost not have any relevance for a production system. Without this influences companies could straighten their production planning oriented to well known conditions.

To achieve a sustainable competitive advantage for production in high-wage countries, it is not sufficient to achieve a better position within one of the dichotomies "scale vs. scope" and "planning-orientation vs. value-orientation". The objective for future production systems has to be the resolution of both dichotomies, the poly-lemma of the production (Schuh et al. 2007). The vision of the future production system for high-wage countries is achieving an individualized and flexible production system at the cost of today's mass production.

14.2 Introduction of a Target System for Complex Production Systems

14.2.1 Holistic Definition of Production Systems

In order to master the resolution of the described poly-lemma of production systems, a suitable understanding of production systems is inevitable. According to the holistic definition underlying further research, the basic elements of a production system are the product program (the product program is the sum of all product fam-

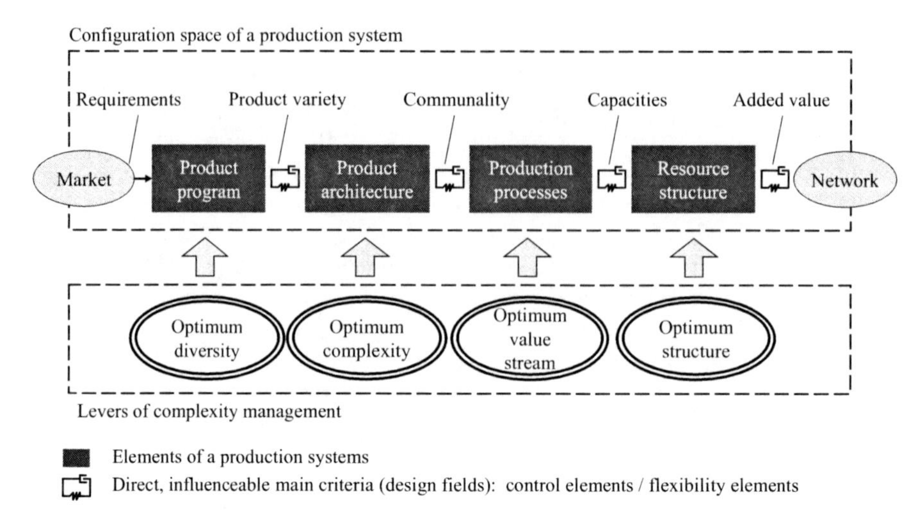

Fig. 14.2 Elements of a production system according to holistic research definition

ilies), the product architecture, the production processes and the resource structures (Schuh et al. 2007). They define the configuration space of a production system (Fig. 14.2).

Product type, variant, quality and quantity are defined within the product program, which will be offered (Bleicher et al. 1996). One of the main challenges is to define optimum product diversity within the product program. The product architecture is the sum of product structure and functional structure as well as the transformation relationships between the two. Every physical element of the product structure can be described with the attribute's function, technological concept and interface (Meier 2007). The goal is finding the optimum degree of complexity in the product architecture to meet the manifold requirements. The core of a production system is the production process itself because it constitutes the physical value creation and has to be optimized in terms of value stream. The resource structures, such as supply chain management and quality management, are further downstream elements of a production system included within this definition. The improvement of resource structures in terms of process optimizations is the main challenge in this field.

14.2.2 Target System for Complex Production Systems

It is the target of the described production research to minimize the poly-lemma explained in Fig. 14.1. In order to measure, manage and control the impact of changes to a production system, a collectively exhaustive set of key performance indicators is required. These indicators define the target system for complex production systems and are necessary to evaluate the degree of target attainment and to further illustrate the understanding of the pyramid's relevance for today's production systems.

Figure 14.3 provides an overview of the main aspects, which have to be considered in terms of the poly-lemma of production systems.

The dilemma between scale and scope is characterized by the ratio of costs per piece and the produced quantity (Fig. 14.3, upper left chart). The general objective is the overall reduction of costs per piece. However, especially for production systems in high-wage-countries and for individualized products, the disproportionate decrease for small quantities is crucial. The theoretical optimum would be achieved at the theoretical minimum cost per piece (horizontal dashed line) with costs being independent of the produced quantity. The dilemma between scale and scope can also be described based on life-cycle sales over time (Fig. 14.3, lower left chart). The aims are to reduce development expenses, to quickly achieve the break-even point and to ensure a steeper rise of sales right after market entry. The dichotomy between a planning-oriented and value-oriented production system is reflected in the robustness of production (Fig. 14.3, upper right chart). The main targets are to reach the theoretical maximum capacity faster and at the same time to increase process robustness. The dichotomy between value-oriented and planning-oriented production systems is furthermore represented by the ratio of output to production planning (Fig. 14.3, lower right chart). The ratio of output and production planning is introduced as integrativity of a production system. The aim is the maximization of its value.

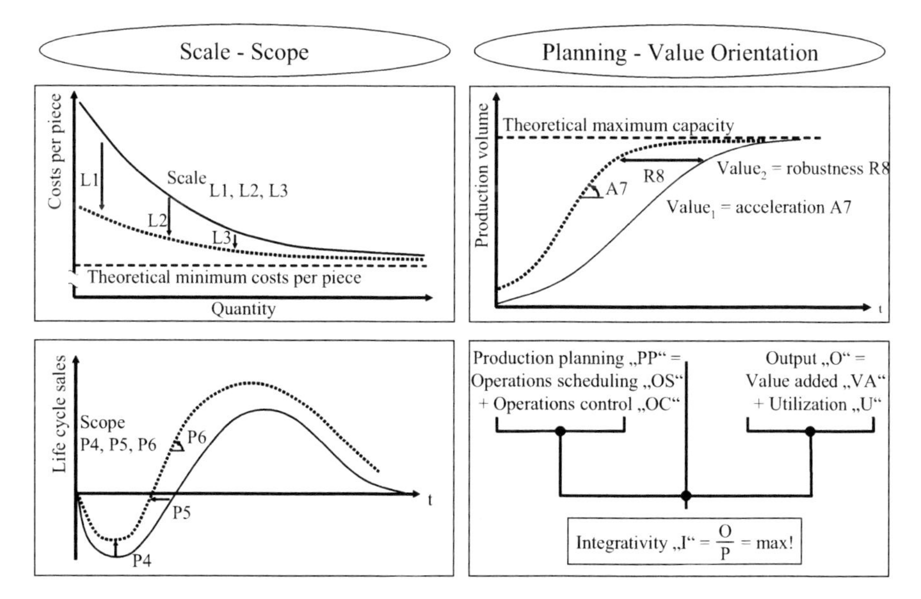

Fig. 14.3 Four basic goals for the minimization of the poly-lemma

It is the objective of any improvement measure of a production system to enhance at least one or ideally several of the four basic goals while not deteriorating any of the other goals at the same time.

14.2.3 Differentiation Between Complicated Systems and Complex Systems

Nowadays, production systems are particularly affected by the increasing dynamics of market requirements as already pointed out above. A new generic understanding and categorization of the environment of production systems is necessary to distinguish system requirements into a time-dependent and a variety-dependent part. This differentiation will allow a thorough differentiation between complex and (merely) complicated system elements (ElMaraghy et al. 2005).

Complexity is mainly characterized by two elementary system conditions: on the one hand by the impossibility to interrelate all elements of a system to each other, and on the other hand by the in-determination and unpredictability of a system's behavior (Schuh 2005b).

The composition of a system is also determined by the number and variety of the elements and their connections. System complexity depends on the changeability of system parameters over the course of time. Four basic types of systems can be distinguished (Fig. 14.4):

- **Simple systems**: few elements, inter-dependencies, and behavior possibilities
- **Complicated systems**: many elements and inter-dependencies; system behavior is deterministic
- **Complex systems**: few elements and inter-dependencies; high number of behavior possibilities; entire controllability is not possible
- **Complex and complicated systems**: many elements and inter-dependencies; high changeability of system elements over time.

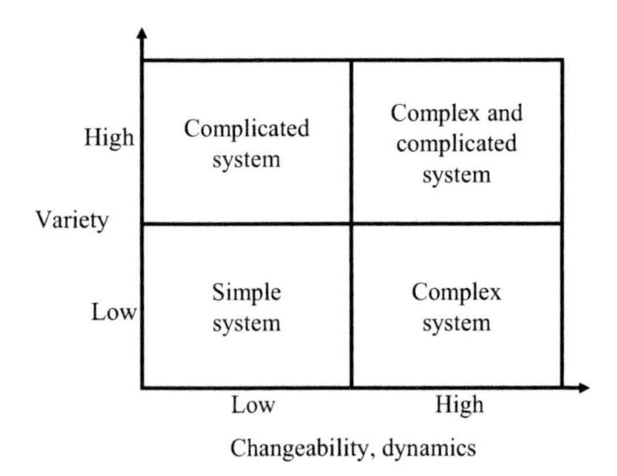

Fig. 14.4 Basic system types according to differentiation of variety and changeability (Grossmann 1992)

Hence, complexity is a result of *product and process* variety influenced by *external* dynamics. The complexity problem can be divided into static and dynamic parts, which is helpful for the understanding and resolution of the complexity problem (Reiss 1993).

The merely complicated part of a production system is characterized by a large number and variety of system elements, which have many inter-dependencies. However, the varieties and their inter-dependencies can be precisely described and are thus not complex. Solving complicated – but not complex – tasks can be achieved through an "explanatory" approach using models, methods, planning and simulation. Whereas the complicated part is characterized by predictability and determination, the complex part of the production system is characterized by its unpredictable and undeterminable nature. In short, complexity exists when "surprise" comes into play.

14.3 Approach to Mastering Complexity in Production Systems

One of the key issues of future production systems design will be to identify the optimum internal complexity corresponding to variety required externally. Every production system is designed to master a certain (today possibly very low) share of complexity – i.e., system elements without precisely predictable states or conditions – as opposed to deterministic (complicated) system conditions.

14.3.1 Object-Oriented Design

The central approach to mastering complexity in production systems will be the application of object-oriented design throughout the entire value chain from product program to resource structures. Object-oriented design is focused on an interface and interdependency driven design of systems.

An object-oriented method, especially for facility layout planning, has been developed at the Laboratory for Machine Tools and Production Engineering (Bergholz 2005). Using this approach, organizational units and processes shall be treated as encapsulated modules with defined interfaces so they can be configured in an object-oriented way (Gottschalk 2006). Based on a temporary cross-linking of these modules, changeability can be achieved to face the dynamic challenges in the field of production systems by a flexible adaptation of single modules simultaneously resulting in robust structures.

Based on certain parallels, the theory and development of object-oriented software engineering inspires facility layout planning (Bergholz 2005). The software industry is affected by very fast hardware development cycles in combination with rising software complexity. Hence, software industry is a very dynamic industry as well (Balzert 1998). Against the background of increasing customer requirements, particularly large software systems must be capable of being reconfigured with little time and effort. Software changes are to be minimized to keep development efforts as small as possible. Despite external dynamics, a high level of system stability has to be achieved. In software engineering, the principle of object orientation for the support of versatile software has been widely established (Oestereich 1998).

14.3.2 Object-Oriented Management of Production Systems

The described approach for object-oriented design of production systems consists of four steps (Fig. 14.6). Steps one to three describe how to identify, analyze and classify the complexity drivers and how to specify the production system. Step four explains how the complexity of production systems can be controlled by object-oriented design.

The four steps are explained as follows:

1. Identify and classify the change drivers
   In the first step, the reasons for dynamic changes are analyzed and the necessity for changeability is determined. The changeability requirements of a production system can be described by so called change drivers (Wiendahl et al. 2007). Change drivers are characteristic of a specific production system and can therefore not be generalized. At high level aggregation, it is possible to differentiate the following types of change drivers (Schuh et al. 2005a):

   - Product-related change drivers can be identified along the product structure, in most cases defined by the product assembly process (e.g. geometry changes of certain parts)
   - Volume-related change drivers can be decomposed into few basic mechanisms: Adding of resources, integration and separation of processes into resources, substitution (e.g. manually by automated) and optimization (e.g. slow by fast tooling).
   - Technology-related change drivers can be classified into product- and process-related change types (e.g. new joining technique).

1	Identify and classify the change drivers
2	Description of the production system: Detailing and evaluation of change profiles
3	Description of the production system: Define interdependencies and interfaces
4	Object-oriented design: Separation of merely complicated and really complex elements

Fig. 14.5 Four steps for object-oriented design of production systems

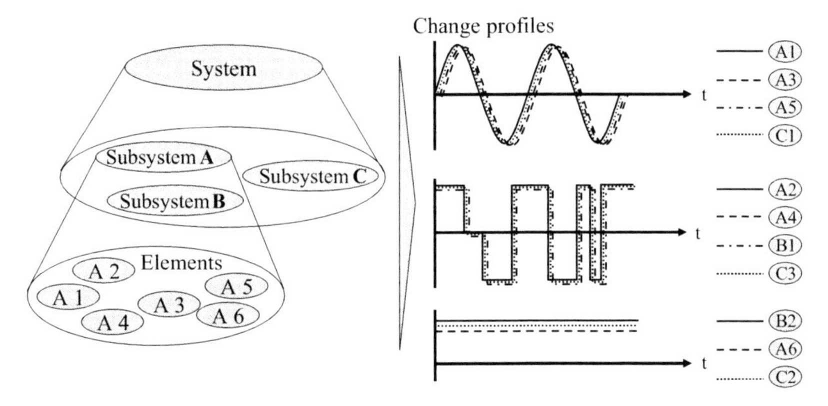

Fig. 14.6 Detailing of the production system and evaluation of change profiles (Schuh et al. 2005a)

For an object-oriented design (Step 4) it is important to identify these change drivers and to classify them with regard to their attributes (entry frequency, cause etc.). The analysis of change drivers reveals when, how often and why a system has to change. In addition, it must be shown how accurate the predictions of changes are.

2. Description of the production system: Detailing and evaluation of change profiles
   In the second step, the production system is analyzed. Systems can be detailed into multiple subsystems, whereas higher system levels always contain the lower ones. The smallest parts in such decompositions are called elements (left half of Fig. 14.7).
   With regard to production systems, e.g. the structure of a factory, they can be detailed in several production lines that again consist of several workstations (Schuh et al. 2003).
   The possible level of detail depends on the application case and planning status. The intention of detailing is the identification of system elements whose inter-dependencies and properties are focused on in the next steps.
   Based on the analyzed change drivers, the properties of the system elements have to be examined. To minimize the system changes caused by change drivers, it is important to figure out the dependencies of the change drivers and system elements. Change drivers cause different change profiles (amplitude or frequency of the changes, right half of Fig. 14.7). The elements can be classified by allocation of the system elements to different change profiles. This classification is important for object-oriented design (step four).
3. Description of the production system: Define inter-dependencies and interfaces
   The third step focuses on the inter-dependencies between the identified elements. The inter-dependencies between the individual elements will now be analyzed (Fig. 14.8).

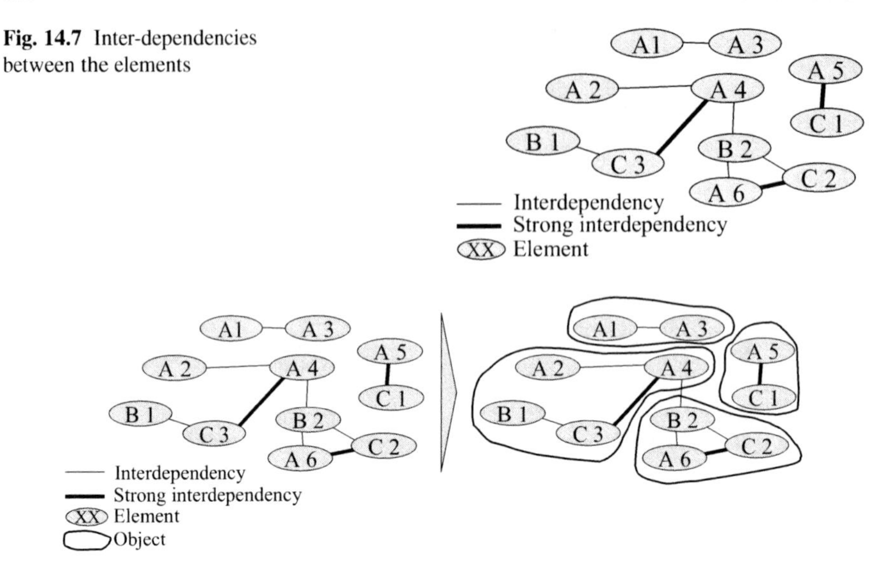

Fig. 14.7 Inter-dependencies between the elements

Fig. 14.8 Creation of objects

Typical relationships between production elements are material and information flow, energy flow, value streams, spatial proximity or other physical ties, work progresses etc. (Daenzer and Haberfellner 1999). In addition, it is important how often an element is influenced by another. Strong inter-dependencies (e.g. A4–C3) can be characterized by many and frequent interaction.

4. Object-oriented design: Separation of complex and complicated system elements.

In order to reduce system dynamics, system elements that are merely complicated need to be isolated from system elements that are subject to complex system behavior. The objective of this separation is the definition of objects that can be well planned for future applications, where possible, and that can be adapted flexibly to future requirements, where necessary. The approach of object-oriented design is based on the encapsulation of certain elements to objects. This encapsulation should allow – among other things – an easy interchangeability or transformation of objects.

With regard to an easy interchangeability or transformation of objects it is highly relevant to encapsulate most of the inter-dependencies within the objects. Moreover, the focus has to lie on intense inter-dependencies, which often play a central role in mastering complexity. Hence, only a few and at the same time weak inter-dependencies between the objects remain (e.g. A4–B2 in Fig. 14.9).

In this manner it can be ensured that the transformation of one object has only low or no influence on other objects.

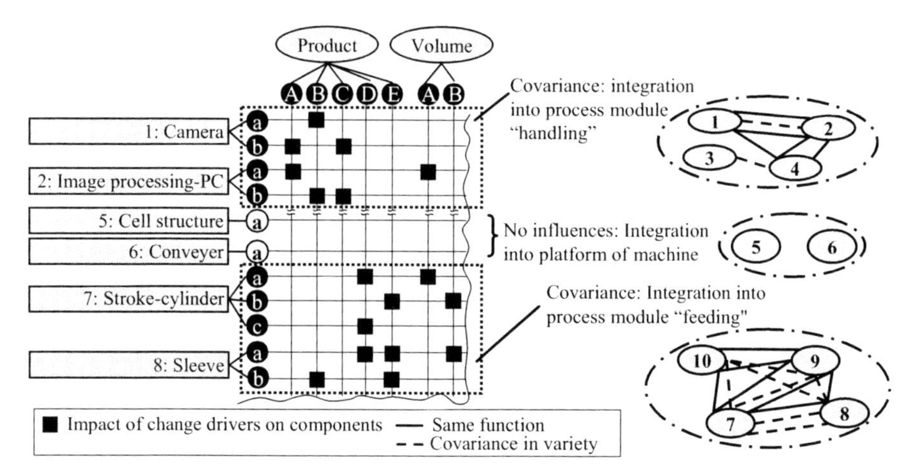

Fig. 14.9 Design of a flexible automated changeable feeding system (Schuh et al. 2005a)

In case of changing requirements, the system capacity can be scaled by the reduction or rise of the number of redundant objects. With changing processes, single modules can be substituted and processes can be reorganized.

Taking into account the change drivers, the change profiles of the elements must be considered. System elements that tend to change at the same time for the same reason are potentially being en-capsuled into one object. Thus, system elements that do change for different reasons (e.g. A2, A4, B1, C3 and A6, B2, C2) are being separated (Schuh et al. 2005a). Therefore, the influence of the change drivers can be limited to a very small system area.

After realization of these four steps, it is ensured that companies can adapt their production system fast and with very low effort to new requirements. Based on object-oriented design, system changes only affect single objects and not anymore parts of the whole system. In case of reconfiguration, the affected objects can be adapted (or replaced) to new processes easily.

14.4 Case Studies

The application of the presented approach for object-oriented design is shown illustrated based on two case studies. Case A describes the design of a flexible automated changeable feeding system for series assembly in the automotive industry. Case B shows the implementation of an object-oriented approach to mastering complexity in automotive product development. These case studies still focus on partial elements of a holistic production system (production layout in Case A, and product architecture in Case B). However, the consistent application of object-oriented design to an entire production system from product program to resource structures still needs to be carried out in future.

14.4.1 A: Object-Oriented Production Design

The aim of this case study was to design a flexible automated changeable feeding system for series assembly in the automotive industry. As automated feeding systems are usually designed specifically for the fed parts, frequent product and volume changes and the proliferation of product variants present a particular challenge for such systems (Schuh et al. 2005a).

Accordingly, part and process related properties (e.g. part size, weight, complexity, material, assembly direction etc.) as well as production volume/cycle time have been identified as relevant change drivers in the first step (Schuh et al. 2005a).

In the second step, the system was decomposed into components (elements) and the impact of change drivers and their specifications on functions and potential components of the feeding system were identified. Many change drivers were only affecting certain product components (Fig. 14.10) (Schuh et al. 2005a).

If two components were affected by the same change driver specification (coherence), the impact was modeled as a relationship between the components in the third step. The probability of occurrence for each specification was used as a measure for relationship intensity (Schuh et al. 2005a).

Based on the identified relationships, the components of the system (1: camera, 2: image processing-PC, 3: ground-plate, 4: robot, 5: cell structure, 6: conveyer, 7: stroke-cylinder, 8: sleeve, 9: rotary-plate, 10: bowl) were structured into objects in the fourth step. Thereby, the ordinary relationships (material and energy flow, fulfillment of the same function etc.) between the components also had to be considered (Schuh et al. 2005a).

Result of this process was a modular changeable system design consisting of three different objects: a process module 'feeding', responsible for part storage, sorting and orientating; a process module 'handling', responsible for separation, gripping and assembly; and a platform module, responsible for providing the basis and housing, electrical energy etc. for the process modules (Schuh et al. 2005a).

The new design features several advantages with regard to changeability compared to the usual integral designs: The feeding components are affected by product changes only. Therefore, the isolation of these components in separate modules permits an easy changeability of feeding components in case of product. The flexible handling module represents the bottleneck in terms of production volume. Adding an additional handling module allows easily scaling the system with changing volume changes. Finally, the platform is affected neither by product nor by volume changes. Thus, its life cycle can easily span several new product launches. Due to the object-oriented design the modules can be developed and changed independent of each other (Schuh et al. 2005a).

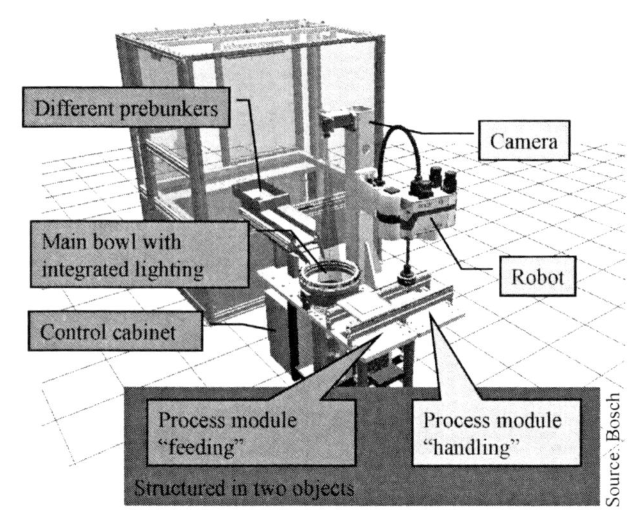

Fig. 14.10 Flexible automated changeable feeding system (Schuh et al. 2005a)

14.4.2 B: Release-Engineering in the Automotive Industry

Development processes in the automotive industry have many constraints resulting from many design changes of different, highly interdependent components over their life cycle. Insufficiently coordinated product changes are a substantial complexity driver.

The decoupling of product structure elements into objects is the solution to manage the dichotomy between rising development efforts for product or component changes and required economies of scale of the entire product.

The definition of object-oriented design within the product structure enables the establishment of a release-oriented engineering ("Release-Engineering"), which is based on significantly lower influences of inter-dependencies due to a bundling of product changes in releases (Schuh 2005b).

The realization of the full potential of Release-Engineering requires a new way of product modularization. Release units have to be optimized in terms of inter-dependencies and their planned innovation frequencies. The formation of release units can be divided into four stages:

- Segmentation and clustering of components
- Classification of inter-dependencies
- Optimization of inter-dependencies
- Definition of release units (objects within product structure) and release cycles.

In the first step, the product components have to be divided based on a modular product structure. The accurate identification and classification of change drivers is

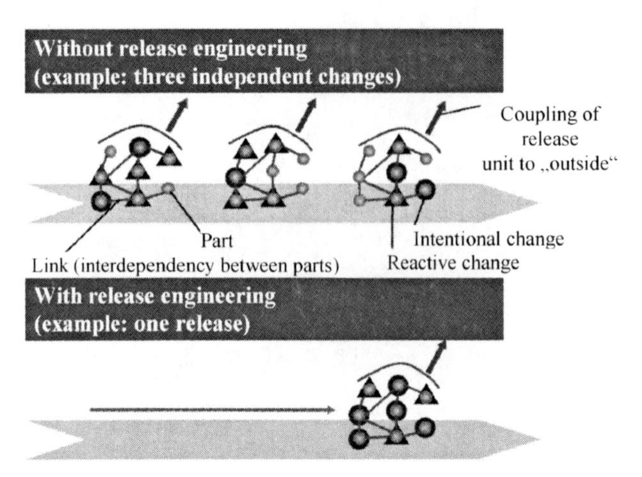

Fig. 14.11 Abstract model for release units and parts inter-dependencies

crucial at this stage. Components are classified in predefined clusters according to their innovation frequencies.

In a second step, the different inter-dependencies have to be classified to bundle components. Therefore change profiles have to be detailed and evaluated whereas the level of detail depends on the application case. The analysis of dependencies between change drivers and system elements is the core part in this step.

The third step consists of the optimization of inter-dependencies mentioned above. Thereby, the product architecture has to be designed according to criteria exceeding mere functional or spatial considerations by additionally analyzing inter-dependencies in terms of different innovation and change cycles. An abstract model can illustrate the bundling of parts to releases and the inter-dependencies between these parts. The release unit as such is symbolized by a composition of individual parts that are interlinked and illustrating interdependency (Fig. 14.11).

A differentiation has to be made between intended changes and reactive changes, i.e. those that are provoked by an intended change but do not represent any added value. In Fig. 14.11 the consolidation of three independent changes to one release is shown. As a result, the number of intended changes remains the same (five, highlighted by a dark background) at the same time the quantity of reactive changes (bright background) is reduced from ten to three. This example illustrates the potential of engineering in releases.

In a last step, objects and corresponding release cycles are defined according to the inter-dependencies and actual change cycles by a separation of complex and complicated system elements. The focus is placed on a prearrangement of change cycles such that not each modification or change will be allowed or implemented unless the time frame permits delays. As a result, changes appear bundled within each release.

The concept of release engineering was exemplified at a large first tier automotive supplier that produces steering columns. In this case, the steering column module consisted of 80 parts that were all subject to potential changes. The entire module was marked by an average change index of 3.5 during a 9-month time span between the release of means of production and SOP (start of production). Multiplied with its quantity of parts, it resulted in a total quantity of 280 part changes during the described period. The exemplified steering column module for a particular type of car is sold in 90 variants. For approximately 10 percent of all individual variants, changes have to be executed. The 280 part changes multiplied by 10 percent times 90 variants result in a total of 2,520 changes that were subject to our considerations. Roughly estimating a share of 60 percent for reactive changes (being caused by other, intended changes), this number divides into 1 512 reactive changes and 1 008 intended changes. Assuming five intended changes on average per change process, the company needs to carry out approximately 202 intended change steps in 39 weeks. The described consolidation of changes to release being performed every second week leads to a number of approximately 52 intended changes per step, assuming a fixed number of intended changes. The reactive changes per step summed up to approximately 30. Hence, the resulting quantity of executed reactive changes adds up to 585 (vs. 1 512 changes before) and the total number of changes now equals 1 170 (vs. 2 520 changes before). Looking at percentage changes, this means a 61 percent and 54 percent reduction in reactive and total changes respectively. Thus, Release-Engineering leads to a reduction of the addressed poly-lemma in modern production systems

In terms of an object-oriented method, release engineering increases development efficiency by adopting this development principle from software engineering and introducing it to the field of mechanical engineering. The synchronization of changes and innovations enables the bundling of changes. As a result, unnecessary change processes can be eliminated and large savings potentials regarding change efforts can be uncovered and utilized.

14.5 Summary

Markets are characterized by an increasing individualization and high dynamics. Consequently, companies have to be able to adjust their production system to actual and future conditions quickly and with low effort. Therefore, companies have to resolve the two dichotomies "scale vs. scope" and "value-orientation vs. planning-orientation" to achieve a sustainable competitive advantage. In order to measure, control and manage the impact of changes on a production system, a target system was explained. These target system can be used to evaluate the degree of target achievement further on and simplify the understanding of their inter-relationships and their relevance for today's production systems.

A central approach to mastering complexity in production systems is an object-oriented method based on analogies to object-oriented software engineering. The described approach for the object-oriented design consists of four steps. Step one to step three describe how to identify, analyze and classify the dynamic drivers and how to specify the production system. Step four explains how the complexity of production systems can be controlled by object-oriented design.

The application of the approach for the object-oriented design has been shown based on two examples.

References

Balzert H., 1998, Textbook of software technique – Software-management. Springer, Heidelberg

Bergholz M, 2005, Object-oriented factory design. Dissertation, RWTH Aachen

Bleicher K., Hahn D., Warnecke H.J., Schuh G., Hungenberg H., 1996, Strategic production management. In: Schuh G., Eversheim W. (eds) Betriebshütte – Production and management. Springer, Berlin, pp 5-1–5-52

Daenzer W., Haberfellner R., 1999, Systems Engineering. Verlag industrielle Organisation, Zürich

Dörner D., Buerschaper C., 1997, Thinking and acting in comlpex systems. In: Ahlemeyer H.W., Königswieser R. (eds) Management of complexity. Gabler, Wiesbaden

ElMaraghy H.A., Kuzgunkaya O., Urbanic J., 2005, Comparison of Manufacturing System Configurations – A Complexity Approach. CIRP Annals 54/1:445–450

Gottschalk S.F., 2006, Dedicated Flexibility. Shaker, Aachen

Grossmann C., 1992, Accomplishment of complexity in management. CGN, St. Gallen

Meier J., 2007, Types of product architecture of globalized companies. Dissertation, RWTH Aachen

Nicolis G., Prigogine I., 1987, Exploration of complexity. Piper, München

Oestereich B., 1998, Object- oriented software development. Oldenbourg, München

Reiss M., 1993, Complexity management I. In: WISU, Wirtschaftsstudium No. 1, pp 54–60

Schuh G., Van Brussel H., Boer C., Valckenaers P., Sacco M., Bergholz M., Harre J., 2003, A Model-Based Approach to Design Modular Plant Architectures. In: Proceedings of the 36th CIRP International Seminar on Manufacturing Systems, Saarbrücken, pp 369–373

Schuh G., Harre J., Gottschalk S., 2005a, Design for Changeability (DFC) in Product-oriented Production. In: CIRP Journal of Manufacturing Systems, 34/5

Schuh G., 2005b, Management of product complexity. Hanser, München

Schuh G., Orislki S., Kreysa J., 2007, Chances for manufacturing in high-wage countries. In: Schuh G. (ed) Excellence in Production, 1st edn. Apprimus-Verlag, Aachen

Schuh G., UAM J., Schöning S., Jung M., 10/2007, Individualized Production. In: ZWF Zeitschrift für wirtschaftlichen Fabrikbetrieb. Carl Hanser Verlag, München, translation, p 630–634

Wiendahl H.-P., ElMaraghy H.A., Nyhuis P., Zäh M.F., Wiendahl H.-H., Brieke M., 2007, Changeable Manufacturing. CIRP Annals 56/2:783–809

Wohland G., Wiemeyer M., 2006, Tools of thinking for dynamic markets. Verlagshaus Monstein und Vannerdat, Münster

Chapter 15
Changeability Effect on Manufacturing Systems Design

T. AlGeddawy and H. ElMaraghy[1]

Abstract The changeability of manufacturing systems enhances their adaptation to the increasingly dynamic market conditions and severe global competition. The effect of manufacturing systems changeability objective on their design process is discussed in this chapter at different levels, from the general frameworks, where the main objectives are stated, to the finest synthesis details, where product and production modules are designed. The conventional design frameworks are discussed and critiqued, and the tendency of most manufacturing systems design processes to be uni-directional is pointed out. Furthermore, a new manufacturing systems design framework is proposed to overcome the uni-directionality drawback of conventional design frameworks and help achieve a closer integration of both products and systems design and evolution.

Keywords Manufacturing, systems, Design, Synthesis

15.1 Introduction

Manufacturing systems design aims to find the optimum selection of manufacturing physical components such as machines; facility layout and structures, products design. . . etc., which enables the system to achieve the manufacturing requirements. In the context of changeability, these needs may include optimum product modular design to facilitate the system reconfiguration, best machines structure for reconfiguration, and the most adequate system design to make future changes possible and easily implemented. Several frameworks, policies, and guidelines were established for that purpose; providing a road map to lead producers and manufacturers through the complex task of designing manufacturing systems that are easy to configure and change.

[1] Intelligent Manufacturing Systems (IMS) Center, University of Windsor, Windsor, Canada

Literature in the area of manufacturing systems design started long before introducing changeability to these systems. Some design tasks were well established, such as facility layout design, selection of machines and determining their location, which results into major long-term commitments. Other system design activities are still being debated, such as the various systems general frameworks. However, the emerging paradigms within the changeability context prompted the system designers, industry and researchers to revisit and review those well-established design tasks to take the new paradigms into consideration.

Changeability is introduced to manufacturing systems to increase their adaptability to varying market conditions, competition, rapid product changes and increased customization and to make effective and rapid reaction possible and smooth. The surveyed literature in this chapter is concerned with the design of manufacturing systems in light of the emergence of the different changeability classes such as agility, flexibility and reconfigurability. The general understanding needed for the problem of manufacturing systems design at its various levels is established and their related activities are discussed. A new design framework for manufacturing systems is introduced to better aid in integrating changeability in those systems.

15.2 Synthesis of Manufacturing Systems

Design of manufacturing systems is mainly concerned with synthesizing their main parts and subsystems. Synthesis can be literally defined as aggregating individual parts or elements to form a whole, or the combination of separate elements of thought into a whole, as in combining simple into complex conceptions, or individual propositions into systems (Ueda, 2001). Synthesis of manufacturing systems in this chapter is further classified into four levels; system, factory, machine, and product levels. Those levels and their associated activities are shown in Fig. 15.1. Such activities are generally performed in any manufacturing system design process; yet they are also affected by the notion of changeability.

15.2.1 Enabling Changeability in Systems Frameworks

The synthesis at the system level includes the construction of the big picture that encompasses the rest of the manufacturing levels; therefore, it is placed at the top of the hierarchy of the system design levels. Synthesis at the system level generates the required system framework through which the physical system components will be integrated and controlled on the shop floor. Typically such frameworks include a preliminary step to decide the product and its market dimensions in the customer's domain, hence defining the strategic targets and production objectives of the manufacturing system. System frameworks should also recognize the elements

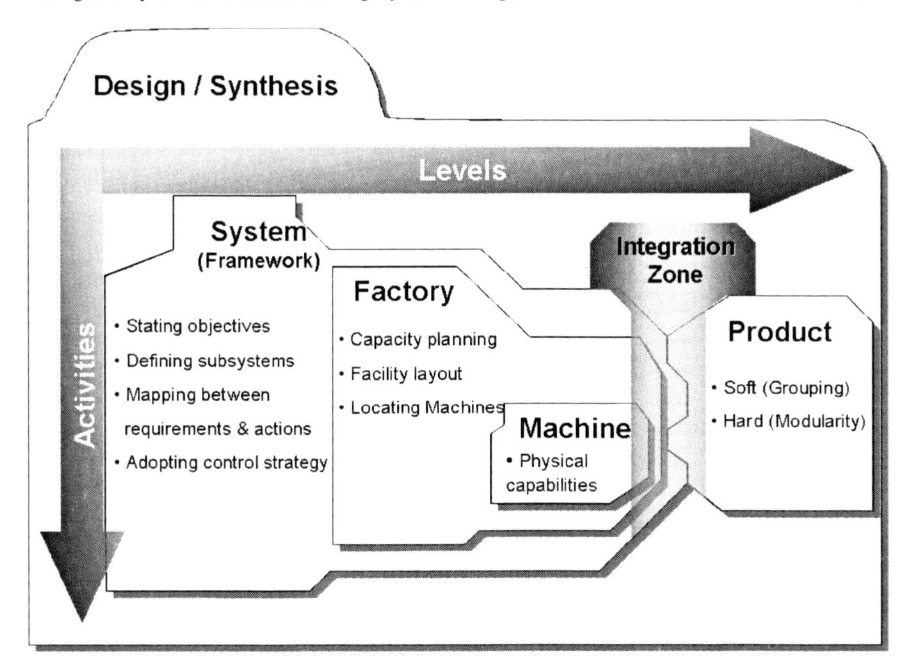

Fig. 15.1 Main levels and activities in manufacturing systems design process

(sub-systems) of the manufacturing environment, as they are well defined, orientated, and managerially organized through adopting an adequate control strategy. Any system framework should include constructing for the production facility in the functional (what is required) and physical (how it can be done) domains, and how it is managed in the process domain, with some optimization schemes employed throughout, which is called the mapping process.

Introducing changeability as a design objective added more considerations and requirements that had to be implemented in the deployed design frameworks via new design approaches, to take full advantage of the power of these new paradigms. Many approaches were used to account for Product Flexibility within the system structure, such as using expert systems (Mellichamp et al., 1990) for designing Flexible Manufacturing Systems (FMS), and following heuristics and knowledge rules to handle bottlenecks and other critical problems that appear in such systems. Requirements Driven Design (RDD) languages that have a behavior notation, such as executable IDEF0 by Alford and Skipper (1992), claimed to rapidly prototype products behavior and system processes in system frameworks to facilitate the implementation of changes. An integrated framework of the system structure and its control parameters can be found in the unified structural procedural approach (USPA) (Macedo, 2004) where four design steps are included; identification of the target market requirements, conception of the manufacturing system target structure (production resources attributes and their organizational relationships), design of the manufacturing system structural improvements and design of the manufacturing system procedural improvements (operating values of the system components). Many approaches were also used to achieve system Agility such as the holistic enterprise approach (Vaughn and Shields, 2002), which aims at considering all products or product lines for a high performance manufacturing systems in the aerospace industry that needs high responsiveness, and collaborative manufacturing platform by Sluga et al. (2005). System frameworks had to also change to account for Reconfigurability, as one of the important changeability classes, and allow for the resulting change in manufacturing paradigms and philosophy and changing machines capabilities. Abdi and Labib (2003) proposed a general framework for designing Reconfigurable Manufacturing System (RMS) using the Analytical Hierarchical Process (AHP) approach. Also a network-like framework was proposed by Cunha et al. (2003) where the specialized support is provided by an engineering center in the middle of the network. Tang and Qiu (2004), integrated supply chain management, enterprise resources planning, sales and service management, product and process engineering, manufacturing execution systems, and shop floor controls in a generic RMS model, based mainly on database programming languages. Deif and ElMaraghy (2004) used systems design architecture to prescribe the different design activities starting from capturing market demands to the system-level configuration and the component-level implementation, along with the control layer of each level. The new approach of emergent synthesis (Ueda, 2007) can handle the vagueness in design, where neither complete manufacturing environment description nor complete system specifications may exist. The reconfigurable systems are characterized by their un-predictable design path and evolution, which indicates that emergent synthesis approach would be useful for the design and control of such systems. To establish a system framework, Bi et al. (2008) defined the manufacturing requirements from manufacturing systems in a changeable environment, where four requirements were recognized. They proposed the strategies that can satisfy those requirements including: 1) Short lead-time; by reducing or eliminating indirect activities, i.e. transferring, buffering; reducing time for direct activities by increasing system capacity and reducing system ramp-up time; and operating the system concurrently, i.e. allowing overlapping among manufacturing activities, 2) More product variants; by optimizing a product platform; increasing variants or versatility of manufacturing resources; and increasing variants or versatility of assembly resources, 3) Low and fluctuating production volumes; by modularizing the product platform, i.e. basic parts would be interchangeable in the same product family so that the demands of the products of the same family can be maintained even if the volumes of some specific products are reduced; and changing manufacturing or assembly resources dynamically, and 4) Low product price; by reducing the cost caused by indirect activities; by reducing the cost caused by direct activities; and by reducing the cost by system integration. RMS was recognized to be the best manufacturing system that can meet these requirements and easily apply the strategies mentioned earlier. Bi et al. (2008) identified general RMS design directions including: 1) Architecture design, which determines system components and their interactions. System components are encapsulated modules. Interactions are the options when the modules are assembled. RMS architecture has to be designed to produce as many system variants as possible, so that the system can deal with changes and uncertainties cost-effectively, 2) Configuration design, which determines system configuration under given system architecture for a specific task. A configuration is an assembly of the selected modules; which fulfills the given task optimally, and 3) Control design, which determines appropriate process variables, so that a configuration can be operated to fulfill the task satisfactorily.

One of the main approaches used in realizing systems frameworks is Axiomatic Design (AD), for its power in identifying the Functional Requirements (FRs) in the functional domain, and Design Parameters (DPs) in the physical domain of a certain problem, and trying to reach a decoupled Design Matrix (DM) that captures their relationships. At the high level, the FRs were defined by Suh et al. (1998) to maximize the return on investment, while DPs were selected to provide products at minimum cost, and to provide high quality products that meet customer needs. A transformation to changeability context, from process orientation to cellular orientation, was introduced by Kulak et al. (2005), where a four-level decomposition was made to realize couplings between FRs and DPs, ranging from the most abstract FRs such as classifying products and machines into groups to the most detailed ones like eliminating inappropriate part assignment and their related DPs. Increasing the system intelligence in FMS was also attempted by implementing the AD framework (Babic, 1999 and Gu et al. 2001), while quality issues were addressed for example by Liu (2004) in the FRs to attain the best diagnosability.

15.2.2 Effect of Changeability Enablers on the Factory Level Design

More detailed sub-levels are needed after determining the general framework within which the manufacturing system would operate and its parts would communicate. The factory level design provides more details for establishing a physical facility, as well as bridging the gap between the theoretical ideas and guidelines established at the system level, and the practical reality on the shop floor. Capacity planning, facility layout, and machines locations are the details decided at this level.

Capacity Planning is concerned with evaluating the amount of manufacturing capacity needed to meet market demands over the entire production planning horizon, and indicating the time of its installation while minimizing cost. This problem was conventionally handled in both a deterministic and a stochastic manner. However **capacity flexibility** is a major objective of changeability that added some new challenges to capacity planning in FMS, where not only additional capacity is determined, but also the parts are assigned to particular machines since the same set of products can be produced using different machines (operation flexibility). When designing an FMS, one would ideally like to determine the optimal configuration (including both the physical and the control aspects), for example, determine the lowest cost configuration that achieves a desired production volume with a given level of flexibility. This problem is a very difficult one although some efforts have been made in this direction by Vinod and Solberg (1985), Dallery and Frein (1988), and Lee et al. (1989). Buitenhek et al. (2002) presented an iterative approach for determining the production capacity of an FMS with several part types, dedicated pallets, and fixed production ratios among the different part types. Also, Liberopoulos (2002) presented a deterministic formulation for the capacity planning problem, and used graph theory to solve it.

Scalability is one of the reconfigurability enablers that changed the approach to capacity planning drastically, as a Reconfigurable Manufacturing System (RMS) facilitates having exact manufacturing functionality and capacity where and when they are needed. It is difficult to achieve good scalability using conventional systems design methods because they aim at designing systems to be optimal for a fixed capacity. Towards this end, Asl and Ulsoy (2003) proposed that capacity management be performed by observing the current capacity and the probability distribution of the market demand at each time period, and making optimal decisions to change the capacity. They presented new stochastic approaches based on Markov decision process and feedback control. The feedback policy creates sub-optimal solutions, which are more robust to unexpected events, as they are less sensitive to changes in input parameters. Deif and ElMaraghy (2006) presented an approach to model the capacity scalability scheduling in RMS, using an objective function that includes both the cost of physical capacity unit and cost of reconfiguration associated with the system reconfiguration. Spicer and Carlo (2007) defined the set of system configurations that an RMS assumes as it changes over time along the system configuration path. They also investigated the minimization of investment and reconfiguration costs over a finite horizon with known demand by determining the optimal RMS configuration path. A cost model was presented to compute the reconfiguration cost between two RMS configurations that includes labor cost, lost capacity cost, and investment/salvage cost due to system reconfiguration and ramp up. Then an optimal solution model was proposed for RMS configuration using dynamic programming, and finally a combined integer programming/dynamic programming heuristic was utilized to allow the user to control the number of system configurations considered by the dynamic programming to reduce the solution time.

On the factory level, the **Facility Layout** is determined, which influences the long-term commitment regarding the physical arrangement of different facility departments. The problem usually is dependent on estimates and rough figures for market demand, inventories, and management policy. To achieve **transformability** as a main objective for **factory flexibility**, machines **mobility** emerged as one of its enablers. The layout of transformable factories allows changes to take place at certain points on the time scale, which are related to the intervals of each configuration. To allow such layout change and machines mobility logically, a flexible layout mathematical model was proposed by Yang and Peters (1998), to identify the best layout for the present time interval, and the easiest to be reconfigured for the next interval. Similarly, a dynamic facility layout methodology was introduced by Kochhar and Heragu (1999), where the material handling system reconfiguration cost and relocation of layout cost were taken into consideration as new criteria for the facility layout problem. A layout optimization method for manufacturing cells accompanied with location optimization method for transportation robots was proposed by Yamada et al. (2003), and a 3D graphics simulator demonstrated the procedure to meet the manufacturing task. A four-phase approach was introduced for the reconfigurable layout problem for multi-planning periods by Meng et al. (2004) in an attempt to connect the plan of product variety mix with the design of cellular manufacturing layout.

Further details are determined during the **Machines Location** determination, which handles the arrangement of machines in each single department or production cell. This is also affected by machines **mobility** as one of the changeability enablers in a **transformable factory**. In an RMS, each machine is assumed to be movable from one location to another. Such location is dependent on demands and product mix of the current interval. Reconfiguring the system by changing machines locations in each cell was discussed by Hu and Koren (2005), while an AHP model was introduced by Abdi (2005) to validate reconfiguring machines location. A reconfiguration smoothness measure was introduced by Youssef and ElMaraghy (2006), to be taken into consideration when changing machines layout from one period to another according to demands, also Youssef and ElMaraghy (2007) introduced an optimization model for RMS configurations with multiple-aspects to incorporate the effect of machine availability using the universal generating function (UGF), which is a polynomial function that represents all the possible states of the system by relating the probability of each system state to the expected performance of the system in that state.

15.2.3 Changeability Effect on Machine Level Design

Determining **Machines Capabilities** is the main concern at the machine level analysis, which means defining the structures, attributes, and designs needed to perform the required jobs by the machine. As **reconfigurability** is a main class of changeability, one of its enablers at the machine level is **modularity**, which calls for new machine tools design concepts to be achieved. A concept for a 3D reconfigurable machine structures was proposed by Murata (1998), where a suggested base building block was able to expand in both directions of X, Y, and Z axes, then similar building blocks can be stacked together in numerous ways to build up the required structure. This process was proposed to take effect autonomously and even remotely as needed. A modular design was presented by Moon and Kota (2002), for a Reconfigurable Machine Tool (RMT), where modules are selected from a module library. A general framework to introduce modularity in a RMT was proposed by Perez (2004), based on the knowledge of the requirements of the machine builders. A selection framework to optimize module selection for a RMT was presented by Chen et al. (2005). To build flexibility into machine tools, mechanical systems that perform work piece and tool changing, re-positioning, work piece handling, and tool handling should be further developed. Those systems run-time represents the secondary processing time of production, which needs to be minimized. Fleischer et al. (2006) presented an approach for the assessment of the technological effectiveness of work piece and tool handling systems for machine tools, giving an overview of the most recent developments concerning those systems. An architecture for modular RMS was devised by Abele et al. (2007), which can be structured hierarchically. At the highest system level, reconfigurable machine tools are linked into sequential or parallel manufacturing systems, where each RMT consists of modules that can be arranged by means of a common platform. Module functions include: machining operations, work-piece handling operations, and quality control tasks.

Introducing scalability to RMT was proposed by Spicer et al. (2005) where a validating architecture was proposed to achieve scalability in machines. Liu and Liang (2008) identified three main goals for an RMT design; maximizing the number of configurations of a RMT; minimizing the cost of all machine modules; and minimizing the tool position errors caused by the interfaces between machine modules, thereby improving process accuracy. Their design optimization process consisted of two stages; generation of alternative designs; and selection of the best design alternatives, to solving five single objective problems which are; the configurability sub-problem, the cost sub-problem, and the accuracy sub-problem in X, Y and Z directions.

Modular and Reconfigurable machine tools, transfer lines, and manufacturing systems are now offered by machine builders such as the Mori Seiki machine tool company (Mori Seiki Co., 2007) and others.

15.2.4 Product Design Directions

The product level complements all previous system design levels, as the design of a manufacturing system follows the product(s) concept, and also the subsequent sub-levels of factory and machine design are affected by the product(s) choice. Hence, an integration zone exists (Fig. 15.1) between the product design and other design levels, rather than a hierarchical relationship. Changeability did not only affect the system design domain, but also the product design domain, influencing the manner in which products are designed for better integration with the system domain. Two main product design directions support the changeability to the system; products **grouping** and **modularity**.

Grouping of products into families of products to best fit a certain manufacturing system can be considered a soft design task, as it involves reorganization of existing data, and rarely results in new product designs. Grouping of products is not a new field; it began with the emergence of Group Technology (GT), cellular manufacturing, and FMS. However, the need for ideas to support system **reconfigurability** is becoming imperative. A stochastic model was proposed by Xiaobo et al. (2000) for RMS, to choose a candidate family of products and best system configuration in a stochastic environment. The main idea of Abdi and Labib (2004) is to select a candidate family of products for each configuration stage, with no machines assigned a priory, using Group Technology, but based on their Operational Similarities. That step results in several families presented; then the AHP technique is used to select a single parts family that satisfies market and manufacturing criteria for the whole planning horizon. A suggestion made by Jose and Tollenaere (2005) is to increase **standardization** of components – which is one of changeability enablers - when establishing different families of products along with minimal architectural changes when introducing different products and for future development. An innovative notion of evolving product families was introduced by ElMaraghy (2007), where families of products lose and gain features as new parts/products evolve, which is analogous to the biological evolution and its definition. A pioneering application of this evolution concept of evolving families of products was the development of a Reconfigurable Process Planning (RPP) introduced by ElMaraghy (2006). Azab and ElMaraghy (2007) presented a new method where process re-planning and re-configuration is achieved by adding/removing features and operations akin to inserting/removing genes in a chromosome. This innovative approach optimally reconfigures process plans to account for new products that have changed and evolved beyond the boundary of the original product family, while minimizing the extent of the resulting changes in tooling, set-ups and other downstream activities on the shop floor to minimize the cost of change. Two reconfiguration indices are proposed to measure the extent of process changes and to select the best reconfiguration algorithm. Galan et al. (2007) implied that the effectiveness of an RMS depends on the formation of the best set of product families. RMS requirements such as modularity, commonality, compatibility, re-usability, and product demand should be taken into consideration when forming those families. The methodology starts by calculating a matrix that summarizes the similarity between pairs of products for each of the previous requirements. An Average Linkage Clustering (ALC) algorithm is then applied, where the selection of families is determined by the costs incurred when a product could not be manufactured within its cell, and the costs incurred when a machine within a cell could not manufacture a product of its associated family. Finally, a dendogram is formed, which is a hierarchical classification (inverted tree structure) that illustrates the different grouping (parts or cells) that can be formed depending on the similarity of parts within a family or machines within a cell.

Modularity of products is a pre-requisite for enabling system **reconfigurability**. This approach results in hard/physical design changes to achieve the desired modularity. Simpson (1998) emphasized the role of picking the best modular design methodology using the AHP approach, and applying the principles of design platforms. Yigit and Allahverdi (2003) presented a systematic methodology of manufacturing modular products in RMS environment. The procedure consists of constructing the relationships between design parameters in each product module and satisfying the customer requirements. The objective is to generate the different values of design parameters to optimize the objective (design) function and determine the performance measures. This results in a large range of values that represent the possible configurations of each module. A quality loss function is used to relate the chosen module configuration with the percentage failure to meet optimum performance. Next, the best combinations of product modules are chosen to minimize configuration cost and quality loss using mathematical optimization.

15.3 Changeability Integration into the Design Process

15.3.1 The System-Product Changeability Design Loop

It is informative to view research concerning changeability in manufacturing system design in terms of Axiomatic Design. As introduced earlier in this chapter, axiomatic design relates the functional requirements (FRs) with the design parameters (DPs) of an object or a system, through a design matrix. Careful abstraction process is needed to formulate the elements of the design matrix and decompose them into their most fundamental components. The objective of any designer is to arrive at a design configuration that minimizes the number of none-sparse design matrix elements to establish the maximum possible uncoupling of FRs and DPs relations.

Some manufacturing systems design methods used axiomatic design to identify their needs in terms of FRs, and their components in terms of DPs. However, all surveyed design approaches, viewed in the axiomatic design sense, exhibit a uni-directional logical flow of tasks starting from FRs, using abstraction to get design matrix and ending with DPs. The flexibility or reconfigurability features were only reflected in the design matrix in terms of attributes related to these manufacturing technologies and manufacturing systems design.

A fundamentally different view is required in future research, a view that integrates changeability and its enablers into the manufacturing systems design, not only in the relations between inputs and outputs, FRs, DPs, and design matrix elements, but also in the very essence of the design logic and flow, which should capture a closed loop cycle of symbiotic relationships between changes occurring in the product domain, and those that follow in the system domain, and vice versa. The difference between the common design process and the proposed one is illustrated in Figs. 15.2 and 15.3. The current design process flow is uni-directional, with no means to accommodate future changes, unless they were predicted a prior and taken into consideration in the functional domain, the mapping process, and physical domain. Even the zigzagging process (back and forth) through the mapping scheme is used only during the abstraction process to realize the relationships between parameters.

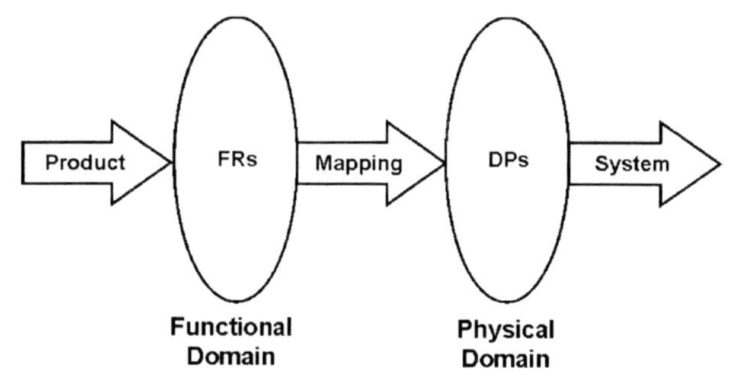

Fig. 15.2 The current uni-directional design framework

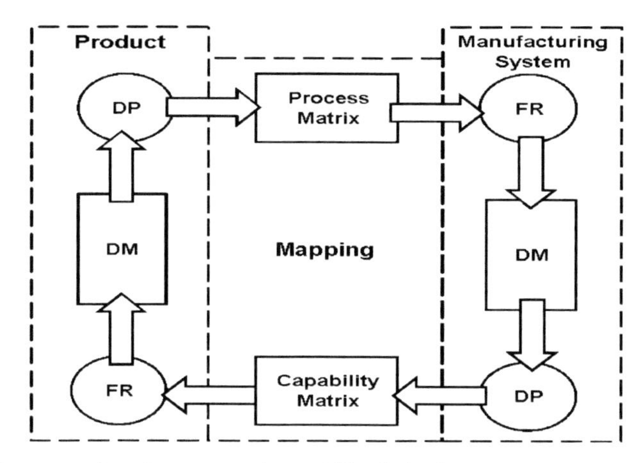

Fig. 15.3 The proposed product-system changeability design loop

The proposed view offers a new design framework, where the gains of past research are preserved through the use of the same abstraction process, but the design process flow becomes a loop that relates both product and system components designs. This loop is meant to capture the natural progression of products design, or technology breakthroughs by expressing their close interdependence and symbiotic relationship. The terms of Axiomatic Design are used here again for better illustration. A change in product design can be translated through a process matrix to the manufacturing domain, which would cause changes in the system design unless the current system capabilities are sufficient to accommodate the product changes. The modified system capabilities in turn would present new opportunities for processing additional features as the products evolve. These new manufacturing abilities are mapped from the system domain to the product domain through a capability matrix, which would be the inverse of the process matrix.

The proposed system design loop should be integrated in the production system structure. Therefore, an open system perspective is suggested in Fig. 15.4. The pro-

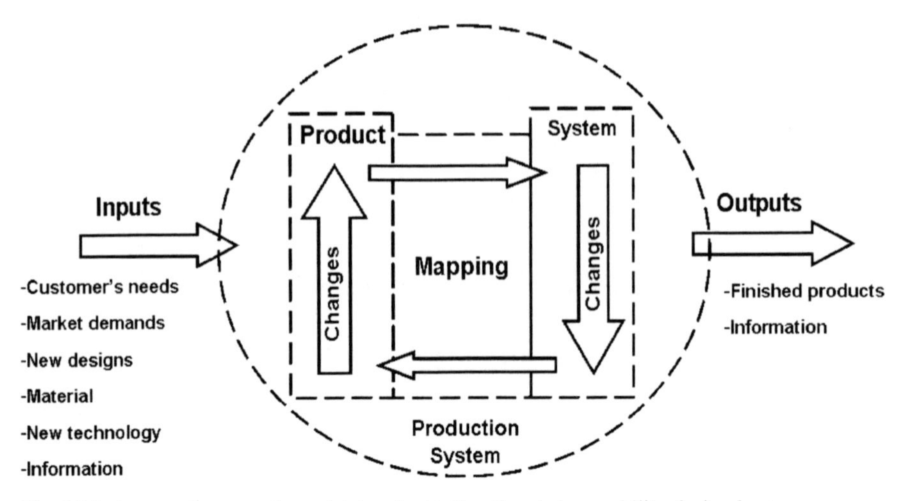

Fig. 15.4 An open framework containing the bi-directional changeability design loop

posed open system would exchange information, energy, material and products with the environment, while it continuously changes internally in an attempt to reach its steady state. The suggested closed changeability design loop working inside the open system would be the mechanism through which such a system can evolve and grow as if it were a living being maintaining its own life.

15.3.2 Biological Evolution/Co-Evolution Analogy

Design for changeability in both product and system domains, is directed basically towards accommodating changes in both design domains. A new changeability design loop was advised for this purpose. Changes take place in every aspect in our surrounding environment, there are examples that can be brought forward, especially from nature, that can help us visualize the intended design process for changeability.

The example used here is "Evolution", which is a biological terminology that can be defined as the process of change with time in the characteristics of organisms. Heritable characteristics of an organism are encoded in the genetic material of that organism. Evolution results from changes in this genetic material, and the subsequent spread of these changes within a population of a species, and inheriting these changes through the generations, resulting in new different species (Ridely, 2004). The famous "Charles Darwin" defined evolution as "Descent with Modification" (Darwin, 1859), and the word "Descent" refers to the way evolutionary modification takes place in series of populations that descended from one another. Evolution does not only indicate a local temporary change in attitude or even in morphology, but rather describes the wider inheritable changes transferring to a predecessor from its ancestor. Therefore, the main qualities of the evolution process are dominated by the occurrence of the change, and the ability to preserve and transfer that change over time. A more precise biological terminology would call the latter quality as "Isolation of Characters".

The wellness of adaptation can be the generator for a series of changes that an organism undergoes, however, some of these changes will allow the organism to better adapt to the situation. This same scenario can be seen in the manufacturing environment; where changes are always driven by a desire for adaptation. That is why new technologies, paradigms and products are continuously introduced to increase the ability of manufacturing entities to adapt to changes due to market circumstances and competition requirements.

Moreover, there exists a process called "co-evolution" in nature, where two or more species influence each others evolution and it is often invoked to explain co-adaptation between species. The common example from biology is that of flowers and butterflies co-evolution. Natural selection favors those flowers whose pollen is transported only to other flowers of the same species, because if the insect flies to another flower species, the pollen will be wasted. A flower may put its nectar reward in a place that can only be reached by insects with a specialized organ, such as a long tongue butterfly, which will be rewarded by the hidden nectar, and which has little competition. The parallel change process can continue, as the plant places its nectar deeper and deeper, and the butterfly evolves longer and longer tongue (Ridely, 2004).

Such evolution and co-evolution natural processes can inspire the construction of a design model, which integrates both products and manufacturing systems design in a design process cycle that allows changes to play a key role in shaping its internal structure.

15.4 Final Remarks

Manufacturing system design in this chapter refers to the synthesis of the physical structure of the manufacturing system components. Four design levels and their related activities were further detailed. On a *system level*, the big picture framework of the manufacturing system is realized, and its inputs, outputs, processes, relations, and control are determined. Product flexibility, agility, and reconfigurability were some of the classes introduced to the system framework that address changeability. Many approaches were used to implement these attributes, such as expert systems, Requirements Driven Design (RDD), emergent synthesis, Unified Structural Procedural Approach (USPA), Analytical Hierarchical Process (AHP), collaborative network platform, supply chain management, database management, and axiomatic design. The design synthesis at the *factory level* aims at defining the physical facility, by introducing production capacity plans, factory layout, and further detailing the location of machines in each department or cell. Changeability at this level is manifested by the expansion flexibility, capacity scalability, and mobility through machines re-location and re-configuration.

On the *machine level*, the machines capabilities are established. The need for achieving changeability lead to introducing new machine design concepts such as the reconfigurable machine tools (RMT). The *product synthesis* level interacts with all other levels. Changeability affected product design, because of the rapidly evolving products and their variants and families. Grouping products into families and introducing modularity to product design are the main design directions affected by changeability where the main objectives were to find the best families of products for a reconfigurable system, produce modular product designs to facilitate the system subsequent configurations, and introduce better product designs to accommodate system reconfiguration.

Finally, it was observed that the surveyed design frameworks are uni-directional in that they always start with a product design and progress through mapping procedures to a system design. Those mapping procedures embody a specific design methodology, and changeability objectives and enablers (Functional requirements and Design Parameters). However, a bi-directional design framework that better integrates changeability into the structure of the design process closes the design loop between product and system design and captures the two way interactions between the two domains. The applicability of this concept to the evolution of products in manufacturing has been illustrated by ElMaraghy, AlGeddawy and Azab (2008). Further research is needed to exploit these natural processes in the industrial domain and convert them into mathematical models.

Acknowledgements This research was conducted at the Intelligent Manufacturing Systems (IMS) Center. The support from The Canada Research Chairs (CRC) Program and the Natural Science and Engineering Research Council (NSERC) of Canada is greatly appreciated.

References

Abdi M.R., 2005, Selection of Layout Configuration for Reconfigurable Manufacturing Systems Using the AHP. ISAHP, Honolulu, Hawaii

Abdi M.R., Labib A.W., 2003, A Design Strategy for Reconfigurable Manufacturing Systems (RMSs) Using Analytical Hierarchical Process (AHP): A Case Study. International Journal of Production Research 41:2273–2299

Abdi M.R., Labib A.W., 2004, Grouping and selecting products: the design key of Reconfigurable Manufacturing Systems (RMSs). International Journal of Production Research 42:521–546

Abele E., Wörn A., Fleischer r. J., Wieser J., Martin P., Klöpper R., 2007, Mechanical module interfaces for reconfigurable machine tools. Production Engineering Research Development 1:421–428

Alford M. and Skipper J., 1992, Application of Requirements Driven Development to Manufacturing System Design. 2nd International workshop on Rapid System Prototyping, IEEE, 177–178

Asl F.M., Ulsoy G., 2003, Stochastic Optimal Capacity Management in Reconfigurable Manufacturing Systems. CIRP Annals Manufacturing Technology 52:371–374

Azab A., ElMaraghy H.A., 2007, Mathematical Modeling for Reconfigurable Process Planning. CIRP Annals 56/1:467–472

Babic B., 1999, Axiomatic Design of Flexible Manufacturing System. International Journal of Production Research 37:1159–1173

Bi Z.M., Lang S.Y.T., Shen W., Wang L., 2008, Reconfigurable manufacturing systems: the state of the art. International Journal of Production Research 46:967–992

Buitenhek R., Baynat B., Dallery Y., 2002, Production Capacity of Flexible Manufacturing Systems with Fixed Production Ratios. International Journal of Flexible Manufacturing Systems 14:203–225

Chen L., Xi F. and Macwan A., 2005, Optimal Module Selection for Preliminary Design of Reconfigurable Machine Tools. Transactions of ASME, J. Manufacturing Science & Eng., p 127

Cunha P., Dionisio J., Henriques E., 2003, An architecture to support the manufacturing system design and planning. International Journal of Computer Integrated Manufacturing 17:605–612

Dallery Y., Frein Y., 1988, An Efficient Method to Determine the Optimal Configuration of a Flexible Manufacturing System. Annals of Operations Research 15:207–225

Darwin C.R., 1859, On the Origin of Species, 1st edn. John Murray, London

Deif A.M. and ElMaraghy W., 2004, Reconfigurable Manufacturing Systems Design Architecture 14th CIRP Design Seminar, Cairo, Egypt, May 16–18

Deif A.M. and ElMaraghy W., 2006, Investigating Optimal Capacity Scalability Scheduling in Reconfigurable Manufacturing Systems. International Journal of Advanced Manufacturing Technology, pp 1–6

ElMaraghy H.A., 2006, Reconfigurable Process Plans for Responsive Manufacturing Systems. Keynote Paper, Proceedings of the CIRP International Design Enterprise Technology (DET) Conference. Portugal

ElMaraghy H.A., 2007, Reconfigurable Process Plans for Responsive Manufacturing Systems. Digital Enterprise Technology: Perspectives & Future Challenges Editors: P.F. Cunha and P.G. Maropoulos, Springer Science, ISBN: 978-0-387-49863-8, pp 35–44

ElMaraghy H.A., AlGeddawy T., Azab A., 2008, Modelling Evolution in Manufacturing: A Biological Analogy. CIRP Annals 57/1:467–472

Fleischer J., Denkena B., Winfough B., Mori M., 2006, Workpiece and Tool Handling in Metal Cutting Machines. CIRP Annals 55/2:817–839

Galan R., Racero J., Eguia I., Garcia J.M., 2007, A systematic approach for product families formation in Reconfigurable Manufacturing Systems. Robotics and Computer-Integrated Manufacturing 23:489–502

Gu P., Rao H.A., Tseng M.M., 2001, Systematic Design of Manufacturing Systems Based on Axiomatic Design Approach. CIRP Annals, Manufacturing Technology 50:299–304

Hu S.J., Koren Y., 2005, Reconsider Machine Layout to Optimize Production. Manufacturing Engineering 134:81–85

Jose A., Tollenaere M., 2005, Modular and platform methods for product family design: literature analysis. Journal of International Manufacturing 16:371–390

Kochhar J.S., Heragu S.S., 1999, Facility layout design in a changing environment. International Journal of Production Research 37:2429–2446

Kulak O., Durmusoglu M.B. and Tufekci S., 2005, A complete cellular manufacturing system design methodology based on axiomatic design principles. Computers & Industrial Engineering, p 48

Lee H.F., Srinivasan M.M. and Yano, C.A., 1989, An Algorithm for Minimum Cost Configuration Problem in Flexible Manufacturing Systems. Proceedings of the 3rd ORSA/TIMS

Liberopoulos G., 2002, Production Capacity Modeling of Alternative, Non-identical, Flexible Machines. International Journal of Flexible Manufacturing Systems 14:345–359

Liu J.P., 2004, Manufacturing system design with optimal diagnosability. International Journal of Production Research 42:1695–1714

Liu W., Liang M., 2008, Multi-objective design optimization of reconfigurable machine tools: A modified fuzzy-Chebyshev programming approach. International Journal of Production Research 46:1587–1618

Macedo J., 2004, Unified structural – procedural approach for designing integrated manufacturing systems. International Journal of Production Research 42:3565–3588

Mellichamp J.M., Kwon O., Wahab A.F., 1990, FMS Designer: An Expert System for Flexible Manufacturing System Design. International Journal of Production Research 28:2013–2024

Meng G., Heragu S.S., Zijm H., 2004, Reconfigurable Layout Problem. International Journal of Production Research 42:4709–4729

Moon Y., Kota S., 2002, Design of Reconfigurable Machine Tools. Transactions of the ASME 124:480–488

Murata S., et al., 1998, A 3-D Self-Reconfigurable Structure. Proceedings of the 1998 IEEE, Int. Conf. on Robotics & Automation, pp 432–439

Perez R.R., et al., 2004, A Modularity Framework for Concurrent Design of Reconfigurable Machine Tools. Y. Luo (ed): CDVE, pp 87–95

Ridely M., 2004, Evolution, Blackwell Publishing

Simpson T.W., 1998, A Concept Exploration Method for Product Family Design. PhD Thesis, Georgia Institute of Technology

Sluga A., Butala P., Peklenik J., 2005, A Conceptual Framework for Collaborative Design and Operations of Manufacturing Work Systems. CIRP Annals, Manufacturing Technology 54:437–440

Spicer P., Carlo H.J., 2007, Integrating reconfiguration cost into the design of multi-period scalable reconfigurable manufacturing systems. Journal of Manufacturing Science and Engineering. Transactions of the ASME 129:202–210

Spicer P., Yip-Hoi D., Koren Y., 2005, Scalable reconfigurable equipment design principles. International Journal of Production Research 43:4839–4852

Suh N.P., Cochran D.S., Lima P.C., 1998, Manufacturing System Design. Annals of the CIRP 47/2:627–639

Tang Y., Qiu R.G., 2004, Integrated design approach for virtual production line-based reconfigurable manufacturing systems. International Journal of Production Research 42:3803–3822

Ueda K., 2001, Synthesis and Emergence-Research Overview. Artificial Intelligence in Engineering, p 15

Ueda, K. (2007) Emergent Synthesis Approaches to Biological Manufacturing Systems. Digital Enterprise Technology, Springer-Verlag, New York inc., 25–34

Vaughn A., F. and Shields J.T., 2002, A Holistic Approach to Manufacturing System Design in The Defense Aerospace Industry. ICAS 2002 Congress, 622:1–10

Vinod B., Solberg J.J., 1985, The optimal Design of Flexible Manufacturing Systems. International Journal of Production Research 23:1141–1151

Xiaobo Z., Jiancai W., Zhenbi L., 2000, A stochastic model of a reconfigurable manufacturing system Part 1: A framework. International Journal of Production Research 38:2273–2285

Yamada, Y., Ookoudo, K. and Komura, Y. (2003) Layout Optimization of Manufacturing Cells and Allocation Optimization of Transport Robots in Reconfigurable Manufacturing Systems Using Particle Swarm Optimization. Proceedings of the 2003 IEEE/RSJ, Intl. Conf. on Intelligent Robots and Systems, 2049–2054

Yang T., Peters B.A., 1998, Flexible Machine Layout Design for Dynamic and uncertain Production Environments. European Journal of Operation Research 108:49–64

Yigit A.S., Allahverdi A., 2003, Optimal selection of module instances for modular products in reconfigurable manufacturing systems. International Journal of Production Research 41:4063–4074

Youssef A.M.A., ElMaraghy H.A., 2006, Assessment Manufacturing Systems Reconfiguration Smoothness. International Journal of Advanced Manufacturing Technology 30:174–193

Youssef A.M.A., ElMaraghy H.A., 2007, Availability consideration in the optimal selection of multiple-aspect RMS configurations, International Journal of Production Research (IJPR) 46/21:5849–5882

Internet recourses: Mori Seiki Co. (http://www.moriseiki.com)

Chapter 16
Managing Change and Reconfigurations of CNC Machine Tools

R. Hedrick[1] **and J. Urbanic**[2]

Abstract Several factors must be considered to effectively manage process change. However, when changing or reconfiguring a system, care should be taken to assure that the changes minimize deviations from the original manufacturing process plans in a controlled manner. The new system configuration should utilize much of the original programs, tooling, fixturing, material handling, and inspection equipment in order to minimize the chances of introducing new quality or logistics issues. A systematic, spreadsheet based methodology for assessing configurations or reconfigurations for CNC machine tools is presented. This methodology can be used in the initial planning stages, or when a change is introduced into the system. Heuristics are utilized that consider the physical and functional characteristics to determine a candidate machine's suitability. The application of this methodology is demonstrated using practical examples.

Keywords CNC Machines, Reconfiguration Management, Process Planning, Change Management

16.1 Introduction

As technologies become more complex and product lives are shortened, manufacturing systems that balance flexibility, specialization, and performance in order to meet market demands are essential for any manufacturer wishing to remain competitive. The requirements for fast, efficient production processes conflict with the need to quickly change or modify these processes due to the ever-shortening life cycle of modern products (Skinner, 1978). Also important is the ability to keep a process running if some part of the process experiences an equipment failure. Once an efficient

[1] CNC Software Canada, Windsor, Canada

[2] Industrial and Manufacturing Systems Engineering Department, University of Windsor, Canada

H.A. ElMaraghy (ed.), *Changeable and Reconfigurable Manufacturing Systems*,

manufacturing strategy has been engineered, changes should be evolutionary, not revolutionary; therefore, a controlled change or reconfiguration management strategy is required. The overall strategy should consider both hard and soft reconfiguration issues (ElMaraghy, 2005) at multiple levels of resolution. This work focuses on pinpointing hard and logical reconfiguration issues at the machine and operational levels.

Hard reconfiguration deals with changes to the physical structure of the machine or system. In the context of this research, hard reconfiguration issues consist of physical changes to the:

1. Cutting tools and tooling systems used to mount the tools into the machine- this may affect the tool setups, tool/tool life control, the part program and possibly introduce quality issues.
2. Part location in its fixture: this introduces re-tooling costs, part re-programming, new inspection procedures, and potential quality issues.
3. Part orientation on the machine: this alters the part program, and may affect chip/coolant flow from the part.
4. Machine topology: this may require re-programming, or depending on the reconfiguration, a totally different programming strategy may be required. The machine work envelope may be affected, and operator training is required with a new program.

Soft reconfiguration deals with changes to the process plan and part flow/routing through a given physical configuration. Reconfiguration issues consist of:

1. Process plan changes (order and type of operations),
2. Machine program, work and tool offsets changes, and
3. Changes to the control functionality/language.

If it becomes necessary to reconfigure a manufacturing system, the required modules must be purchased, manufactured, or taken from other processes. The expense, lead-time or impact on other processes must be considered before applying a reconfiguration strategy. Authors such as Bruccoleri et al., (2006) have shown that reconfigurable technologies can be used to adapt to dynamic situations in zones where routing flexibility is not feasible (i.e. work cell or transfer line). However, the specific detailed nature of the reconfiguration technologies is not defined. Katz (2007) discusses the design principles of reconfigurable machines, but does not discuss specific design details or the operational impacts of the configuration changes. A more comprehensive analysis is required to support process planning, reconfiguration strategies and effectively manage changes due to shifting product or production requirements. Such a methodology should complement existing CNC machine tools and computer-aided manufacturing (CAM) tools as well as provide a foundation for future technological developments at both the process planning and machine levels. The presented approach complements the work developed by Youssef and ElMaraghy (2006) and Shabaka and ElMaraghy (2007).

In an idealized reconfigurable manufacturing system (RMS), the individual modules of a given scale are totally interchangeable; however in actual practice this is rarely true. Examples of this are: (i) machine tools axes have travel limits which may be imposed by collisions of the part or tooling with other machine elements or the machine structure, rather than the soft/hard limits of the axes themselves, and (ii) different machine configurations have different programming requirements due to functionality limitations, rigidity issues, and so forth. In addition the human interface aspect must be taken into account. Radically changing a manufacturing process by reconfiguring the system means downtime to train the actors (engineers, technicians, managers and/or other key production support personnel) for the new process. During the learning curve period product quality may suffer.

16.1.1 Reconfiguration Considerations

Hard reconfiguration almost always results in soft reconfiguration. Here an operation is defined by the tool, the program parameters (feed, speed, lead-in, lead-out approaches, coolant parameters and so forth) and the resulting tool path set. A process plan is defined as a group of operations to produce a product.

A form of machine configuration control exists in commercially available CAD/CAM software. In such systems, when the user decides to move a manufacturing process to a new machine tool, the configuration of the new machine is checked against the operations in the process plan and the results are displayed to the user. Machine compatibility for each operation in the process is rated using levels of compatibility, as illustrated in Table 16.1.

The 'Not compatible' state (0) indicates that the process parameters for the operation cannot be changed to make the tool path run on the new machine. For example, if the tool path utilizes rotary axes, it is not compatible with a machine that only has linear axes (Fig. 16.1).

The 'Compatible with major operation changes' state (1) requires the user to make major changes to the operation parameters and/or associated geometry, or utilize different tooling/fixturing so the operation can run on the new machine. This

Table 16.1 Compatibility states matrix

Compatibility State	Value
Not compatible	0
Compatible with major process changes	1
Compatible with minor process changes	2
Compatible with automatic process changes	3
Fully compatible	4

Fig. 16.1 '0' Compatibility state

scenario can occur with topology changes. For example, if operations are moved from a 'Table/Table' to a 'Table/Head' 5 axis machine (Fig. 16.2) the part, machine positions and orientation will be different. This has to be taken into consideration when editing the operation parameters.

The 'Compatible with minor operation changes' state (2) requires the user to manually edit some operation parameters or make minor process changes so that the operation can run on the new machine. This would occur if the coolant control commands or peripheral automation commands in the operation are not valid on the new machine. The 'Compatible with automatic operation changes' state (3) indicates that the CAD/CAM software can automatically update operation parameters so that the tool path can run on the new machine. For example, if the new machine has a lower maximum spindle speed, the spindle speed for the operation can be automatically adjusted to the maximum for the new machine and the feed rates adjusted accordingly to maintain a consistent chip load on the tool. The 'Fully compatible' state (4) indicates that the operation can run on the new machine as-is, without any changes. A methodology needs to be developed to recognize these compatibility states for potential machine candidates (configurations or reconfigurations) for a set of operations. Once these compatibility states for each operation for each machine are defined, the proposed machine configuration control methodology can be applied.

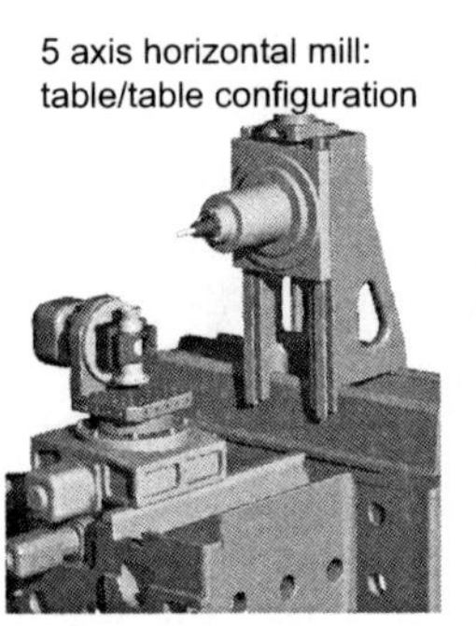

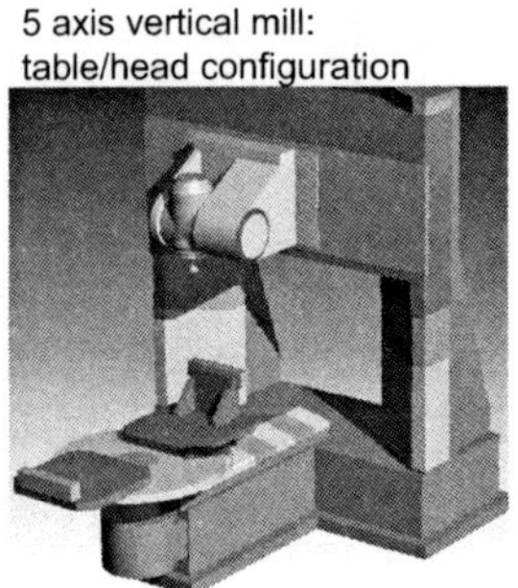

Fig. 16.2 '1' Compatibility state

16.2 The Change or Reconfiguration Management Methodology

There are generally a fixed number of machine configurations that are practical for manufacturing a given product. An optimum reconfiguration strategy is dictated by the actors' perceptions and the operating environment. 'Reconfiguration optimization' for a process could include: minimizing the number of machines that must be reconfigured, minimizing the total number of machines required, minimizing the number of manual setup changes or manual operations, and/or minimizing the total cycle time. The actual criteria are situation dependent; hence, the output from this methodology would serve as input data into other optimization tools that consider the total operating strategy and costs. Insight into other issues such as minimization of the reconfiguration costs, machine downtime, and operator training can be derived from the data, but are beyond the focus of this work. The spreadsheet based change management methodology presented here consists of: (i) seeding the system with machine configurations represented in a manner that enables the process planner to make comparisons, (ii) creation of a detailed process plan, and (iii) assessing the machine compatibility using sorting and filtering criteria with the developed process plan (illustrated in Fig. 16.3).

The first step in applying a controlled reconfiguration strategy is to define a domain of valid machine configurations for manufacturing the product to seed the system with viable alternatives. The candidate machines consist of existing machines and virtual potential reconfigurations of these machines. The key characteristics of these candidate machines must be extracted as the goals are to: (i) select a machine that is capable of performing an operation or set of similar operations within a process plan (Lei, 2005), (ii) select a machine or set of machines capable of executing a complete process plan, and (iii) provide a basis to manage changes to an

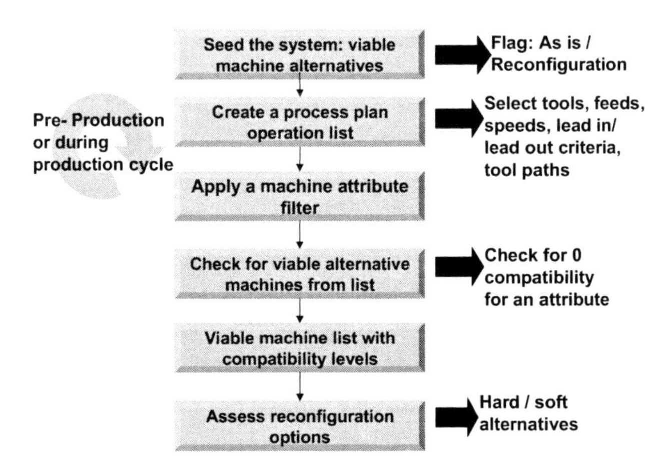

Fig. 16.3 The reconfiguration management methodology flow

existing configuration. The key characteristics being considered are partitioned into four areas, which are labeled the *universal, structural, functional* and the *peripheral interface* characteristics. The universal characteristics focus on the machine family and types and general structural information. The machine family definition is based on the processes of which the machine is capable, such as milling, drilling, turning, and multi-tasking (a machine family that can perform both turning and milling operations). The machine type is a specific subset of the machine family. The machine type is based on key general structural attributes within a family, which would introduce both physical and logical reconfiguration issues. For example, there may be issues with tool and chip management and program compatibility when moving from a 3 axis horizontal to a 3 axis vertical mill or moving from a standard to a sliding headstock lathe. The machine type is also used to cluster or identify potential configuration sets. A 3 axis horizontal mill can be reconfigured into a 4 axis machine by adding an A axis or B axis. There are also several potential 5 axis configurations from a base 3 axis machine (Fig. 16.4). Modules may be interchangeable between families of machines, but functionality issues may exist.

The basic universal operating parameter attributes are machine family and type dependent. The operating parameters being considered are: the maximum recommended part size and weight, the maximum spindle speed, the maximum feed rate, and the number of tools for the tool magazine.

The detailed structural characteristics are determined by assessing the machine's kinematic chains (Shabaka and ElMaraghy 2007). Each axis within a chain has a type (translational or rotational) affiliated with a label (X, Y, Z, A, B, C), and a stroke distance. Multiple axes for a given kinematic chain are stacked in a specific order, which must also be identified. It is assumed that each kinematic chain within a machine ends with a fixture or a tool/end-effector. An axis combination is defined as a pair of kinematic chains combined to form a closed loop chain or a kinematic circuit (note, one chain can be fixed). Each machine has at least one station and tool kinematic chain (high volume production machines may have several). There can be multiple stations, tool chains and valid axis combination sets. These attributes are more extensive for multi-tasking flexible machines.

The control characteristics are described in the machine functional record. It is assumed that the control mechanisms are through a CNC based interface. The controller attributes being considered are rapid motion control, arc moves, helical inter-

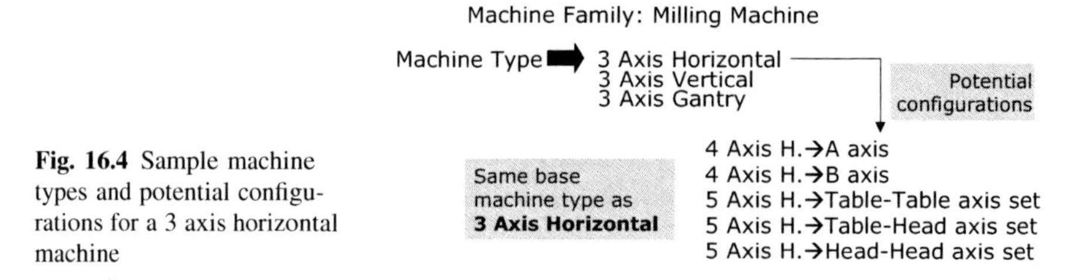

Fig. 16.4 Sample machine types and potential configurations for a 3 axis horizontal machine

polation, multiple work offsets, canned cycles, tool and wear offset compensation. There can be multiple programs or program streams being executed by the same controller. This is standard for multi-tasking lathes, which are mill-lathe hybrids and are becoming widely available in many different configurations (Hedrick et al., 2004). The same tool kinematic chain can work into two separate stations – this is the dynamic machining mode. Multiple tool kinematic chains can engage the same part (typical for dedicated manufacturing systems and multi-tasking lathes). Hence, it is necessary to check the specific axis combinations for each station and operation.

The key characteristics for the peripheral systems must also be considered, as these aspects are required for full functionality of the equipment for the machine type and processes. Application specific peripheral controls being considered are various coolant controls (mist, flood and so forth), and certain machine specific material handling components, as illustrated in the example in Sect. 16.3.

After developing a detailed process plan, the next step is to define the filtering and sorting criteria necessary for the particular application. The list of candidate machines will be assessed for each operation based on these criteria. Depending on the desired output, the criteria could either be used as a filter or a sort option during the assessment process. These criteria can be integrated with the machine/operation compatibility indices described in the introduction. A sample of common filtering criteria for a tool path is listed in Table 16.2, where the filter only criteria are shaded.

Some criteria will not apply to all operation types and must be selectively ignored by the system (e.g. the lathe spindle speed criterion is not required for parts that don't require any operations on a lathe). Once the process planner has defined the list of candidate machines and the filtering/sorting criteria, viable machine alternatives are extracted while they are developing the process plan. As each operation in the process is created, the list of machines is first reduced, using the filtering options.

Table 16.2 Selected criteria for the filter/sort process (U-universal, S-structure, F-functional, P-peripheral)

Type	Sample Criterion	Filter	Sort Option	Compatibility States
U	General compatibility	Y	N	0, 4
S	Work envelope	Y	N	0, 4
S	Number of axes required to perform the operation	Y	N	0, 4
P	Controller auxiliary function support	Y	N	0, 4
U	Power requirements	Y	Y	0, 3, 4
U	Axis feed rates	Y	Y	0, 3, 4
U	Spindle speed (tools)	Y	Y	0, 3, 4
U	Spindle speed (lathe part)	Y	Y	0, 3, 4
S	Axis configuration (topology)	Y	Y	0, 1, 4
F	Controller language	Y	Y	0, 3, 4
F	Controller canned cycle support	Y	Y	0, 2, 3, 4
P	Controller coolant support	Y	Y	0, 2, 4

The remaining machines in the list are then sorted according to the sort options. If filtering removes too many machines from the list, some of the filtering options can be changed to sort options.

Once the process plan is complete, the compatibility state data for each machine configuration needs to be transformed into relevant knowledge in order to provide a platform for making system management decisions. Various comparison techniques are applied to the data in order to accomplish this. Each operation is analyzed for each machine, and the results are summarized, as follows:

1. For each machine, multiply the compatibility states for all criteria. Any machine that has a value of zero as the result cannot be used for the operation.
2. For each machine, sum the compatibility states for all criteria. Sorting by descending value on the result of these operations will rank the machines by the impact of the machine change on the process plan tool path operations.
3. For each criterion, sum the compatibility states for all of the machines. The higher the number, the more flexibility there is for this operation.
4. Create a worst case compatibility 'operation versus machine' summary matrix (Fig. 16.5), where:

   - Black cells represent where there are 1 or more '0 compatibility' conditions for a machine/operation set,
   - Grey cells with a number represent the number of '1 compatibility' criteria for a machine/operation set (note, any 2 compatibility criteria intermingled with 1 compatibility criterion is not illustrated), and
   - White cells with a number represent the number of '2 compatibility' criteria for a machine/operation set. 3 and 4 compatibility states are not considered, as any changes are transparent. These states are represented by blank white cells.
   - Note: An 'all black' row will flag any operation that cannot be performed by any candidate machine. A gray cell without a number is an invalid condition.

Applying the first procedure will eliminate machines that do not meet the specific criterion from the list. This is distinctly highlighted for any machine that contains

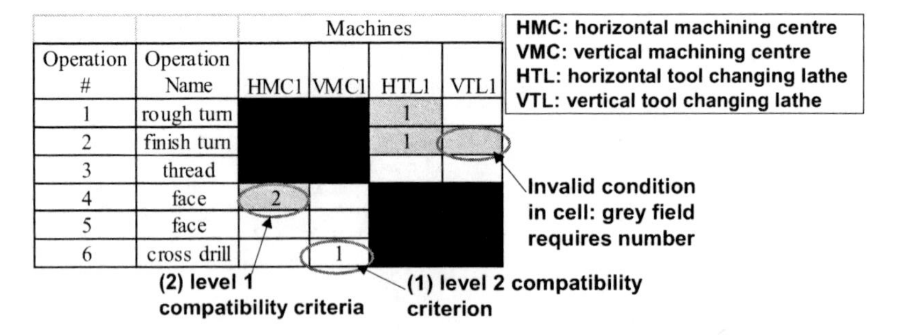

Fig. 16.5 Typical 'operation versus machine' summary matrix

a black cell for any operation. The compatibility 'operation versus machine' summary matrix provides insight into compatible machine configurations, and the potential operational challenges. More sophisticated analysis techniques can then be used to determine configurations with the least number of machines, minimum number of setups or changes from the current process, minimum cycle time, etc.

A straightforward example is used to illustrate a sample set of compatibility states for two vertical 5 axis general purpose machining centers (Fig. 16.6), labeled M/C A and M/C B. Machine 'M/C A' incorporates orthogonal rotary axes, while Machine 'M/C B' has offset rotary axes. There is only one workstation and spindle, and axes exist for both the station and tool kinematic chains. There is one axis combination. The maximum spindle speed for M/C A is 8000 rpm, and for M/C B is it 18 000 rpm. The maximum feed rate for M/C A is 22 m/min (900 ipm) rapid feed rate, for M/C B is 50 m/min (1968 ipm). The axis stack topology is the same for both machines (BC stack). The stroke lengths and power requirements are presented in Fig. 16.6.

When assessing a potential product transfer from M/C A to M/C B, the ratio B:A is used to compare attributes, and vice versa. Machine M/C A is larger, slower and less powerful than M/C B, as illustrated in Fig. 16.7.

If products are to be moved from M/C B to M/C A, there are no issues with the work envelope attributes (compatibility state of 4); however, there are significant differences in the power, speed and feed attributes, which lead to a compatibility state of 1. Analysis must be done to adjust the process parameters appropriately when moving a product between these two machines.

The converse is true when moving from M/C A to M/C B. The work envelope for M/C B is slightly smaller (within 85%). For this example, the XY work envelope may still contain a component processed on M/C A; however, it may require positioning at an angle. Consequently, the NC program will have to be edited, using specific program rotation codes, if this setup strategy is required. This physical, structural difference between machines may require a logical reconfiguration

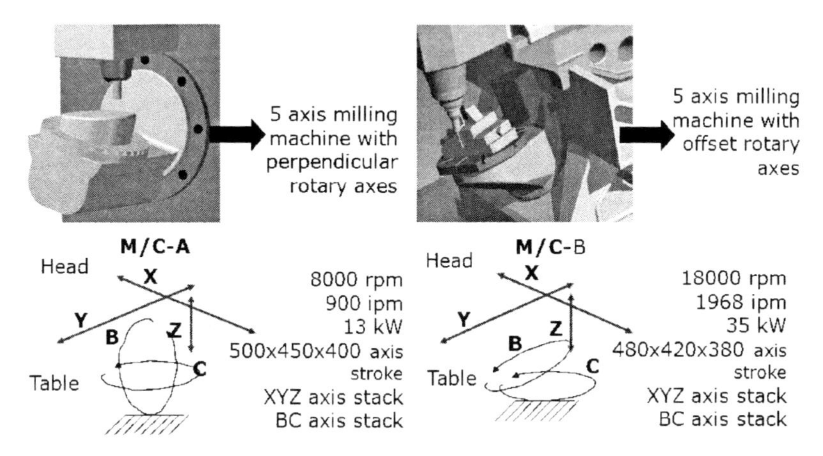

Fig. 16.6 Selected attributes for two CNC machine configurations

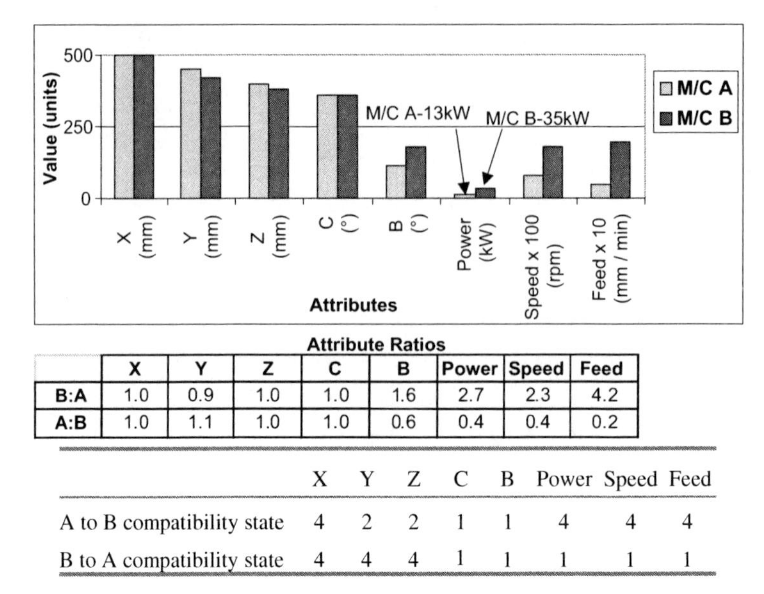

Attribute Ratios

	X	Y	Z	C	B	Power	Speed	Feed
B:A	1.0	0.9	1.0	1.0	1.6	2.7	2.3	4.2
A:B	1.0	1.1	1.0	1.0	0.6	0.4	0.4	0.2

	X	Y	Z	C	B	Power	Speed	Feed
A to B compatibility state	4	2	2	1	1	4	4	4
B to A compatibility state	4	4	4	1	1	1	1	1

Fig. 16.7 Selected attributes comparison

(compatibility state of 2). Fixture clearances also require assessment. There are no issues with power, speed and feed (compatibility state 4). Without assessing the rotary axis type, the rotational axis information is misleading with a direct ratio analysis. M/C B has a larger sweep angle for the B axis; however, the axis types are distinctly different. For M/C B, conventional 4 axis machining must utilize a 5-axis configuration. Hence, the compatibility state is 1 for both the B and C axes. The differences in axis type as well as the motion characteristics need to be captured. In the next section, these analysis techniques are applied to an industrial case study.

16.3 Pneumatic Flow Control Valve Case Study

A typical brass pneumatic flow control valve along with a list of the tool-path operations and a representative tool path verification illustration is presented in Fig. 16.8. There are 17 operations, which consist of facing, rough and finish ID and OD turning, milling and drilling operations, part cutoff and transfer. Details with respect to the tool selection (12 different tools are used), the feeds, speeds and depth of cut, the lead-in/lead-out approach angles and the resulting program are compressed.

The operations have been developed for a specific machine: the Tool Changer Lathe (TCL). TCL is the current configuration and is similar to commercially avail-

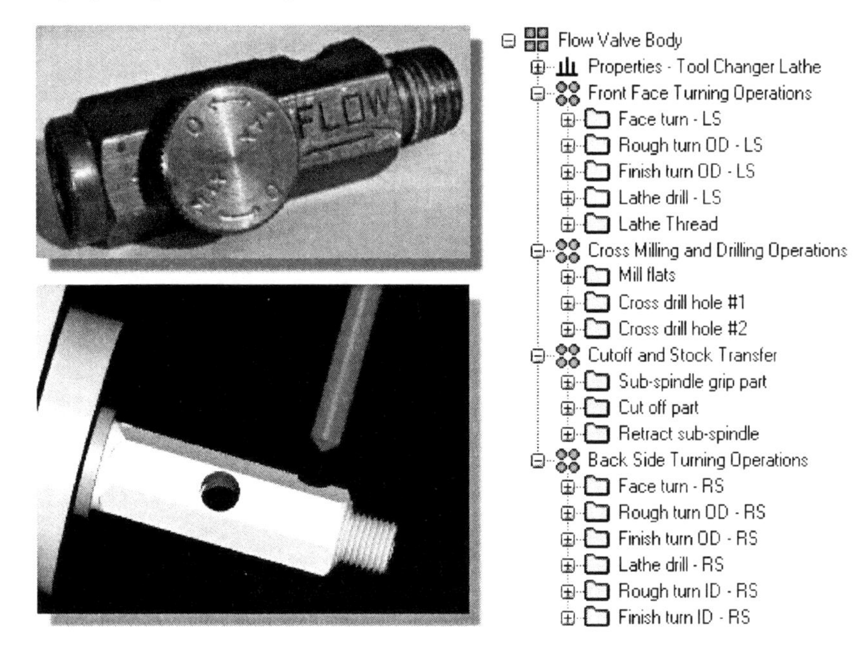

Fig. 16.8 Pneumatic flow control valve

able B-axis head machines. VMC1 (Vertical Machining Center) is configured as a 3 axis mill with an extra rotary axis. VMC2 is a 5 axis mill table-table configuration with a C rotary axis ($\pm 360^{\circ}$) stacked on an A rotary axis ($\pm 90^{\circ}$). HTL1 (Horizontal Turret Lathe) is a C-axis lathe with live tooling. HTL2 is the same configuration as HTL1 with the addition of a second spindle and related Z and C axes. HTL3 is a simple 2-axis lathe. The kinematic configurations for these machines are illustrated in Fig. 16.9.

For this case study, the following initial conditions are assumed:

- The process was running successfully on TCL until the right spindle failed, and suitable alternatives must be found.
- Machines HTL1, HTL3 and VMC1 are available for running the process.
- Lathes HTL1, HTL2, HTL3 are not as sophisticated as TCL. They do not support CNC canned cycles. VMC1 and VMC2, having mill-based controls, do not support lathe operations at all.
- Machine HTL2 represents a reconfiguration of machine HTL1 using the secondary spindle components (Z2 axis, C2 axis and Fixture 2) from TCL.
- HTL3, being an older machine, has a maximum spindle speed and feed rate that are below those used in the existing machining processes.
- Machine VMC2 is another machine available to transfer the process to, if feasible.

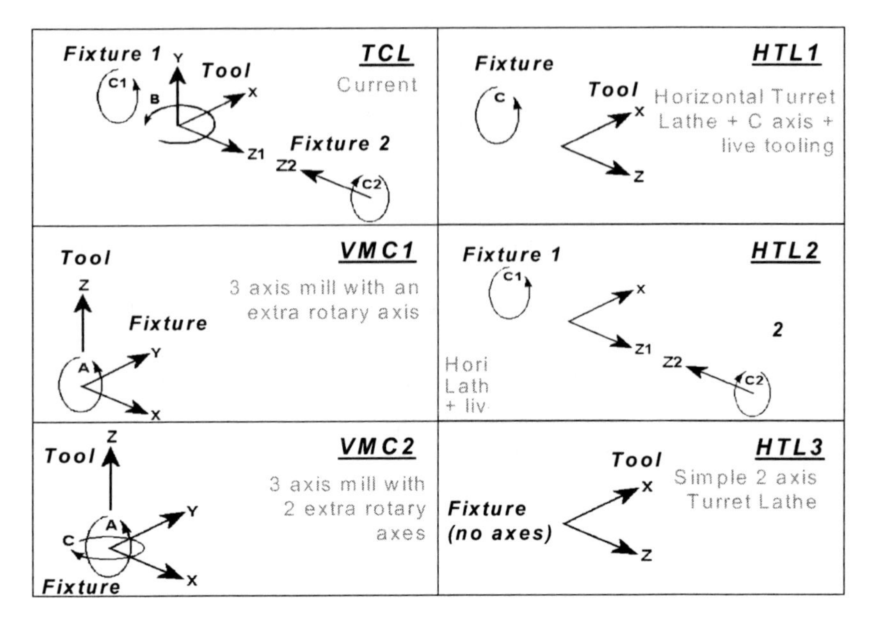

Fig. 16.9 Machine configurations

Based on these conditions, the machines were assessed for each operation, and a worst-case compatibility 'operation versus machine' summary matrix was created (Table 16.3). The values in the summary matrix were derived as follows:

- Although it can be argued that some lathe operations can be performed on a milling machine, (i.e. face turning, and centerline drilling operations), because of the set up and other procedural changes, it is determined not to be feasible.
- C-Axis Contour Milling/drilling (operations 6, 7, 8) – These operations require 3 axes of motion. HTL3, the 2-axis lathe configuration, has been assigned a ranking of 0 for the 'number of axes' criteria. VMC2, the 5-axis mill has a ranking of 0 for the work envelope. It cannot perform this operation, because the C-axis stacked on the A-axis would not allow the tool to reach all of the flats machined by this operation. This illustrates that the machine topology, is as critical as the number of linear and rotary axes when assessing a machine's compatibility with an operation.
- Stock Transfer (automated) – This operation is treated as a pass-fail filter for each machine. Machine configuration HTL2, with the secondary C and Z axes, make it the only other alternate configuration that supports this operation.

For the 'Lathe Thread', 'Mill Flats' and stock transfer operations (op. 5, 6, 9 and 11 respectively), a detailed compatibility assessment is presented in Tables 16.4 to 16.6. Within the tables, the double asterisk (**) indicates the original machine, a single asterisk (*) highlights a reconfigured machine.

Table 16.3 Pneumatic flow control valve operation versus machine summary

		Machines					
Op. #	**Operation Name**	**TCL****	**HTL1**	**HTL2***	**HTL3**	**VMC1**	**VMC2**
1	Face turn - LS						
2	Rough turn OD - LS						
3	Finish turn OD - LS						
4	Lathe drill - LS						
5	**Lathe Thread**						
6	**Mill flats**						
7	Cross drill hole #1						
8	Cross drill hole #2						
9	Sub-spindle grip part						
10	Cut off part						
11	Retract sub-spindle						
12	Face turn - RS						
13	Rough turn OD - RS						
14	Finish turn OD - RS						
15	Lathe drill - RS						
16	Rough turn ID - RS						
17	Finish turn ID - RS						

Machines HTL1, HTL2 and HTL3 have a controller that can support threading, but not in the compact canned cycle format of TCL. The canned cycle compatibility ranking of 3 represents the conversion of the NC code from canned cycle to long-hand format. VMC1 and VMC2 do not support lathe threading at all. Threads can be cut on a mill using a thread milling cycle, but this is a different machining process that requires new fixturing, specialized tooling and programming. Due to the extreme process differences, a general compatibility ranking of 0 has been assigned.

By assessing the machine operation compatibility and the compatibility rankings for each operation, an appropriate reconfiguration strategy can be developed. From these observations, 3 general reconfiguration strategies can be developed:

- Move the Z2, C2 and Fixture2 components from TCL to HTL2 and run the process on this machine. There are clearly no process issues based on the compatibility criteria.
- Move the process to HTL1 and manually flip the part to do the back work.
- Move the turning processes to HTL3 and the milling process to VMC1.

The landscape of this solution will change with different sorting and filtering criteria, which is shop or situation specific. Recognizing that certain lathe operations

Table 16.4 Threading compatibility analysis

Criteria	Machines					
	TCL**	HTL1	HTL2*	HTL3	VMC1	VMC2
General compatibility	4	4	4	4	0	0
Work envelope	4	4	4	4	4	4
No. axes required for the oper.	4	4	4	4	4	4
Axis configuration (topology)	4	4	4	4	4	4
Power requirements	4	4	4	4	4	4
Axis feed rates	4	4	4	3	4	4
Spindle speed (tools)	4	4	4	4	4	4
Spindle speed (lathe part)	4	4	4	3	4	4
Cntlr canned cycle support	4	3	3	3	1	1
Cntlr auxiliary function support	4	4	4	4	4	4
Machine - Operation Compatibility	T	T	T	T	F	F
Compatibility ranking	40	39	39	37	33	33

Table 16.5 Mill Flats compatibility analysis

Criteria	Machines					
	TCL**	HTL1	HTL2*	HTL3	VMC1	VMC2
General compatibility	4	4	4	0	3	0
Work envelope	4	4	4	4	4	4
No. axes required for the oper.	4	4	4	0	4	4
Axis configuration (topology)	4	4	4	0	4	0
Power requirements	4	4	4	4	4	4
Axis feed rates	4	4	4	4	4	4
Spindle speed (tools)	4	4	4	4	4	4
Spindle speed (lathe part)	4	4	4	4	4	4
Cntlr canned cycle support	4	4	4	4	4	4
Cntlr auxiliary function support	4	4	4	4	4	4
Machine - Operation Compatibility	T	T	T	T	F	F
Compatibility ranking	40	40	40	28	39	32

could be replaced with rotary axis milling operations, the pneumatic flow control valve case study is reassessed using the machine general compatibility criterion as a filter, with values 0, 1, and 4, for demonstration purposes. The resulting summary matrix for op. 1–8 is illustrated in Table 16.7.

Using this alternative assessment scenario, VMC1 could be used to perform several of the turning operations using the rotary A axis; however, with consider-able effort. It must be noted that right angle heads or multiple setups are required to perform both face (center drilling) and cross-drilling operations for VMC1.

Table 16.6 Stock transfer (material handling) compatibility analysis

Criteria	TCL**	HTL1	HTL2*	HTL3	VMC1	VMC2
General compatibility	4	0	4	0	0	0
Work envelope	4	4	4	4	4	4
No. axes required for the oper.	4	1	4	1	1	1
Axis configuration (topology)	4	1	4	1	1	1
Power requirements	4	4	4	4	4	4
Axis feed rates	4	4	4	4	4	4
Spindle speed (tools)	4	4	4	4	4	4
Spindle speed (lathe part)	4	4	4	4	4	4
Cntlr canned cycle support	4	4	4	4	4	4
Cntlr auxiliary function support	4	4	4	4	4	4
Machine - Operation Compatibility	T	F	T	F	F	F
Compatibility ranking	40	30	40	30	30	30

Table 16.7 Pneumatic flow control valve alternative operation versus machine summary

		Machines					
Op. #	Operation Name	TCL**	HTL1	HTL2*	HTL3	VMC1	VMC2
1	Face turn - LS					1	1
2	Rough turn OD - LS					1	
3	Finish turn OD - LS					1	
4	Lathe drill - LS					1	1
5	Lathe Thread						
6	Mill flats						
7	Cross drill hole #1						
8	Cross drill hole #2						

16.4 Summary and Conclusions

A new set of management tools must be introduced to assist the process planners in developing relevant manufacturing process strategies. This work assesses the configuration implications at a detail level that has not been rigorously addressed in the academic or industrial environments. The reconfigurable manufacturing paradigm adds a new dimension to the process planning problem. The proposed configuration management strategy can be used when responding to changes in production scenarios, and can also be applied to determine preferred machine configurations from a group of candidate machines – real or virtual – for longer term planning. Machine compatibility heuristics and a spreadsheet based methodology are presented that can quickly isolate configuration issues. The compatibility rankings for each operation for each machine can be assessed to determine a set of valid alternatives. The results are displayed in a simple and concise manner that can be understood by the various actors within a facility. This work can be extended to include specific tool related attributes (e.g., spindle taper), cutting tool axis (to determine number of required setups), and other machine attributes such end-effector controls (i.e. spindle speed and orientation, draw bar mechanism for a boring head). Comprehensive data with respect to the machines and the process plan are required as input for this assessment process. While this low-level tool is not fully automated, the resulting output data can be used as input into a formal optimization model that considers product volumes, routing, layout, set up times, human factors and other system related issues.

References

Bruccoleri M., Pasek Z., Koren Y., 2006, Operation management in reconfigurable manufac-turing systems: Reconfiguration for error handling. Int J Production Economics 100:87–100

ElMaraghy H.A., 2005, Flexible and reconfigurable manufacturing systems paradigms. International Journal of Flexible Manufacturing Systems, Special Issue on Reconfigurable Manufacturing Systems 17:261–276

Hedrick B., Urbanic R.J., ElMaraghy H.A., 2004, Multi-tasking machine tools and their impact on process planning. CIRP 2004 Int. Design Conference, Cairo, Egypt

Katz R., 2007, Design principles of reconfigurable machines. Int J Adv Manuf Technology 34:430–439

Lei W (2005) Dynamic reconfigurable machine tool controller. PhD dissertation, Brigham Young University

Shabaka A.I., ElMaraghy H.A., 2007, Generation of Machine Configurations based on Product Features. IJCIM 24/4:355–369, Special Issue

Skinner W., 1978, Manufacturing in the Corporate Strategy. John Wiley and Sons

Wiendahl H.-.P, Heger C.L., 2003, Justifying changeability: A methodical approach to achieving cost effectiveness. In: Proceedings of the CIRP 2nd International Conference on Reconfigurable Manufacturing, Michigan

Youssef A.M.A., ElMaraghy H.A., 2006, Assessment of manufacturing systems reconfiguration smoothness. Int J Ad Manuf Technol (IJAMT) 30:174–193

Cost and Quality Management

Chapter 17
Economic and Strategic Justification of Changeable, Reconfigurable and Flexible Manufacturing

O. Kuzgunkaya and H.A. ElMaraghy[1]

Abstract The evolving characteristic of changeable manufacturing systems requires design and assessment techniques that consider both the strategic and financial criteria and incorporate the reconfiguration aspects as well as fluctuations in the demand over the planned system life cycle. The economic evaluation approaches to reconfigurable and flexible manufacturing systems have been reviewed. A fuzzy multi-objective mixed integer optimization model for evaluating investments in reconfigurable manufacturing systems used in a multiple product demand environment is presented. The model incorporates in-house production and outsourcing options, machine acquisition and disposal costs, operational costs, and re-configuration cost and duration for modular machines. The resulting configurations are optimized by considering life-cycle costs, responsiveness performance, and system structural complexity simultaneously. The overall model is illustrated with a case study where FMS and RMS implementations were compared. System configurations generated from the proposed model are simulated to compare the life-cycle costs of FMS and RMS. The suitable conditions for RMS investments have been discussed.

Keywords Reconfiguration, Economic Justification, Multi-criteria Decision Making, Complexity, Responsiveness

17.1 Introduction

Due to the increased competition in today's manufacturing environment, companies are trying to cope by producing a wide range of products and adapting quickly to market variations. The changing manufacturing environment requires creating production systems that are themselves easily upgradeable to incorporate new technologies and new functions.

---

[1] Intelligent Manufacturing Systems (IMS) Center, University of Windsor, Windsor, Canada

In order to meet the need for agility and responsiveness, Reconfigurable Manufacturing Systems (RMS) has been proposed by Koren et al., (1999). The USA's National Research Council has identified reconfigurable manufacturing as first priority among six grand challenges for the future of manufacturing (1998). Unlike traditional manufacturing systems, RMS can be achieved by using reconfigurable hardware and software, such that its capacity and/or functionality can be changed over time. The reconfigurable components include machines and material handling systems, mechanisms, modules and sensors for individual machines, as well as process plans, production plans, and system control algorithms for the entire production system. In order to analyze the capital investments of RMS, a tool is needed that considers both the reconfiguration aspects and the strategic benefits of having an agile manufacturing system (ElMaraghy, 2005).

The changeability of manufacturing systems has been recently introduced as an arching concept that incorporates the reconfiguration and flexibility paradigms. Its implementation at various levels of the enterprise and factories must also be justified both economically and strategically (Wiendahl et al., 2007).

17.2 Literature Review

The main difference between RMS and conventional manufacturing systems lies in its ability to evolve over time. In relation to life-cycle modeling and justification of changeability in manufacturing systems, Wiendahl and Heger (2003) stated that a cost-effective manufacturing system alternative can be found between a conventional inflexible system and an extremely transformable system. Wiendahl et al. (2007) point out that the benefits of changeability are usually felt in the long term hence it is more difficult to economically justify the additional cost that has to be incurred in order to achieve its benefits. There are several research areas in literature that can be related to the life-cycle cost modeling and economic justification of changeable manufacturing systems including FMS and RMS:

1. FMS selection considering the machines technological obsolescence
2. Equipment replacement due to technological change
3. RMS capacity expansion modeling using real options analysis
4. RMS configuration selection

Abdel-Malek and Wolf (1994) developed a methodology that ranks candidate FMS designs based on strategic financial and technological criteria. Yan et al. (2000) applied a modified integrated product and process development (IPPD) approach to the design of an FMS, including the modeling of machines upgrades that are necessary due to technological obsolescence. Rajagopalan et al. (1998) considered a problem where sequences of technological breakthroughs are anticipated but their magnitude and timing are uncertain. Lotfi (1995) developed a multi-period flexible automation selection model where both financial and strategic aspects are considered. Amico et al. (2003) applied real options theory to RMS investment evaluation to quantify the characteristics of RMS such as scalability, and convertibility. Spicer (2002) addressed the principles of designing scalable machining systems and reconfigurable machines. The methodology developed by Spicer is purely based on economic evaluation and did not include the potential strategic benefits of RMS. Zhang and Glardon (2001) compared four types of manufacturing systems empirically with respect to several criteria such as adaptability, complexity, production rate, reconfiguration time, ramp-up time and life-cycle cost whose values were assigned subjectively. Abdi and Labib (2004) presented a Fuzzy Analytic Hierarchy Process (AHP) tool for tactical design justification of RMS and focused on the first step of tactical design in which the feasibility of manufacturing operations and economic requirements are evaluated.

Previous studies concerned with the life-cycle modeling of manufacturing systems don't fully capture the reconfiguration process and associated cost. In the studies related with FMS selection, both strategic and financial performance of the alternatives was considered. The studies also included determining the number of necessary FMS system upgrades; however, they fell short of capturing the uncertain nature of future investments, and did not include the reconfiguration costs. In the equipment replacement models under technological change, the demand is considered deterministic with the objective of minimizing the overall lifetime system cost.

It should be noted that advanced manufacturing technologies should be evaluated by including not only their financial performance but also their strategic benefits. The real options analysis captures this strategic value by converting it into an option value and it has the benefit of using a stochastic market demand; however, there is room for improvement by including multiple options/reconfigurations in the analysis.

Due to the uncertain nature of future investments in RMS, the anticipated costs related with its operation can only be estimated. Including additional criteria, which are expressed by the system's features such as flexibility, reconfigurability, and responsiveness would compensate for the typical inaccuracy of the cost figures. In addition, the ability to easily reconfigure the system should be included in the analysis to fully express the benefits of such systems. Otherwise, the investment analysis of RMS technologies would not be realistic.

17.3 Proposed RMS Justification Model

The proposed model includes three criteria: economic considerations, structural complexity of manufacturing systems and their responsiveness.

The model incorporates uncertainty and the decision maker's preferences by converting the objective functions into fuzzy membership functions. Fuzzy membership functions are also important in expressing the degree of satisfaction with the obtained solution. Furthermore, having each objective function's value within [0, 1] interval overcomes the difficulty of having variables with different scales and units.

Fuzzy linear programming can be used to formulate the vagueness inherent in decision-making problems in an efficient way. When the objective function and the constraints are fuzzy, the corresponding fuzzy linear programming model is expressed as follows:

$$\begin{aligned} f(\mathbf{x}) &\tilde{\leq} \mathbf{f_{min}} \\ g(\mathbf{x}) &\tilde{\geq} \mathbf{g_{max}} \\ \mathbf{Ax} &\tilde{\leq} \mathbf{b}, \quad \mathbf{x} \geq 0\,, \end{aligned} \tag{17.1}$$

where f_{min} and g_{max} define the level to be achieved by the objective, and imply the fuzziness of the objective function. In other words, an achievement level is determined for each objective function and the decision-maker allows for the violation of these levels.

$$\mu_{\text{D}}(x^*) = \max_{\text{X}} \min_{\text{i}} [\mu_1(x), \ldots, \mu_{\text{m}}(x)]\ . \tag{17.2}$$

Bellman and Zadeh (1970) defined maximizing a decision in a fuzzy environment by using the following principle. Suppose there are a fuzzy objective function f and a fuzzy constraint C in a decision space X, which are characterized by their membership functions $\mu_{\text{f}}(X)$ and $\mu_{\text{C}}(X)$, respectively, where X_{M} represents the optimal solution. Their combined effect can be represented by the intersection of the membership functions as shown in Fig. 17.1 and Eq. 17.2.

A fuzzy linear program can be transformed to a classical linear programming formulation as represented in Eq. 17.3. A fuzzy multiple objective optimization model allows incorporating several objectives along with constraints. The max-min approach allows satisfying each objective with an overall satisfaction degree of λ. In addition, the fuzzy membership functions are used to represent various types of objectives with different scale units. This approach is also incorporates the decision maker's preferences regarding the desired performance levels for each objective.

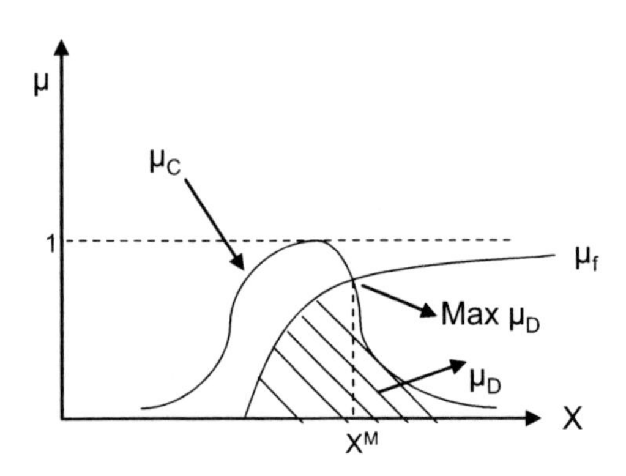

Fig. 17.1 The relationship of μ_{f}, μ_{C} and μ_{D} in fuzzy decision-making

A mathematical model of the life-cycle cost of a reconfigurable manufacturing system is presented and its use is illustrated with a case study. A reconfigurable manufacturing system that consists of modular multi-spindle machine tools is considered. Each machine consists of a base structure to which several modules can be added or removed as capacity requirements change similar to those proposed by Spicer (2002).

$$\begin{aligned}
&\text{Max } \lambda \\
&\text{Subject to:} \\
&\lambda \leq \frac{f_{\max} - f(x)}{f_{\max} - f_{\min}} \\
&\lambda \leq \frac{g(x) - g_{\min}}{g_{\max} - g_{\min}} \\
&Ax \leq b \\
&0 \leq \lambda \leq 1,\ x \geq 0
\end{aligned} \tag{17.3}$$

It is assumed that a candidate part family to be produced has been identified for a planned time horizon of T periods. These part types, denoted by index set i, are to be manufactured in a multi-machine system, involving a set of modular CNC machines. The operations required for a part type i are denoted by the set j. These operations are to be performed by a machine type set m having k configurations. The operation capabilities are represented by an incidence matrix z_{ijmk} the elements of which assume a value of one if operation j of part type i (i.e. operation (i,j)) can be processed by machine type m at configuration k (i.e. machine (m,k)), and zero otherwise. Table 17.1 and Table 17.2 include the required data and decision variables to be used in the optimization model.

The machines types, with different capabilities, capacities, and cost parameters, are selected by optimizing three objectives simultaneously over the planning hori-

Table 17.1 Decision Variables

Description	Notation
Number of machine type m at configuration k in period t	X_{mkt}
Production quantity for part type i in period t	M_{it}
Number of products i outsourced in period t	Q_{it}
Production quantity for operation (i, j) on machine (m,k) in period t	Y_{ijmkt}
Number of machine bases of type m in period t	B_{mt}
Number of modules for machine type m of configuration k in period t	MD_{mkt}
Depreciation charge for machine type (m, k) in period t	DP_{mkt}
Book value of the assets at the end of period t	BV_{t}
Reconfiguration task in period t	RT_{t}
Reconfiguration cost in period t	RC_{t}
Reconfiguration duration in period t	RD_{t}

Table 17.2 Model parameters

Description	Notation
Product Index (i = 1,. . . ,I)	I
Operation index (j = 1,. . . , J)	J
Machine Type Index (m = 1,. . . ,M)	M
Configuration state of a machine type (k = 1,. . . ,K)	K
Period index (t=1,. . . ,T) (e.g. week, month, year)	T
Cash Flow discounting factor	(P/F,I,t)
Tax rate	TR
Reliability of machine type (m,k) n = 1 up-time, n = 2 down-time	p_{mkn}
Demand of part type i in period t	D_{it}
Unit Sales price of part type i in period t	P_{it}
Unit Material cost of part type i in period t	MC_{it}
Unit outsourcing cost of product i	OC_{it}
Outsourcing percentage	OL%
Equals 1, if operation (i, j) can be performed on a machine type (m, k)	z_{ijmk}
Process time of operation (i, j) on a machine type m. k	p_{ijmk}
Efficiency of machine (m, k) with respect to operation (i, j)	e_{ijmk}
Response ability of machine type (m, k) with respect to product i	RA_{imk}
Demand ratio of product i	P_i
Probability of assigning product i to machine type (m, k)	P_{imk}
Lot size in period t	L_t
Setup time of operation (i, j) on a machine type (m, k)	ST_{ijmk}
Setup cost of operation (i, j) on a machine type (m, k)	SC_{ijmk}
Fixed operation cost of operation (i, j) on a machine type (m, k)	FC_{ijmk}
Variable operation cost of operation (i, j) on a machine type (m, k)	VC_{ijmk}
Investment cost of a machine type m, at configuration k in period t	IC_{mkt}
Salvage value of a machine type m at configuration k in period t	S_{mkt}
Straight line depreciation factor for machine type (m, k)	d_{mk}
Available time of a machine type m at configuration k in one period	AH_{mk}
Labor rate	LR
Available workforce in period t	W_t
Time required to install/remove a machine base	t_B
Time required to install/remove a machine module	t_{MD}

zon. These objectives are to maximize the net present value of after-tax cash flows, minimize average system inherent structural complexity, and maximizing its average responsiveness.

17.3.1 Financial Objective

The financial objective, as shown in Eq. 17.4, consists of the present value of several cash flows that occur during the life-cycle of a system. These are the sales profit, machine initial investment costs, operational costs such as variable, fixed, and setup costs, reconfiguration costs, cash flows occurring from machine disposal and acquisition, tax savings from the depreciation of machines, and the book value of machines at the end of the planning horizon.

$$
\begin{aligned}
f_1 = {} & \text{Maximize } NPV(Aftertaxcashflows) = \\
& + \sum_{t=1}^{T} \sum_{i=1}^{I} (P_{it} - MC_{it}) M_{it} (1 - TR)(\mathrm{P/F,\ I,\ t}) \\
& + \sum_{t=1}^{T} \sum_{i=1}^{I} (P_{it} - OC_{it}) Q_{it} (1 - TR)(\mathrm{P/F,\ I,\ t}) \\
& - \sum_{m=1}^{M} \sum_{k=1}^{K} IC_{mk1} X_{mk1} - \sum_{t=2}^{T} \sum_{m=1}^{M} \sum_{k=1}^{K} IC_{mkt} \max(0, X_{mkt} - X_{mk(t-1)})(\mathrm{P/F,\ I,\ t}) \\
& - \sum_{t=1}^{T} \sum_{i=1}^{I} \sum_{j=1}^{J} \sum_{m=1}^{M} \sum_{k=1}^{K} (VC_{ijmk} Y_{ijmkt} + FC_{ijmk} X_{mkt})(1 - TR)(\mathrm{P/F,I,t}) \qquad (17.4) \\
& - \sum_{t=1}^{T} \sum_{i=1}^{I} \sum_{j=1}^{J} \sum_{m=1}^{M} \sum_{k=1}^{K} \frac{SC_{ijmk} Y_{ijmkt}}{L_t} (1 - TR)(\mathrm{P/F,I,t}) - \sum_{t}^{T} RC_t (1 - TR)(\mathrm{P/F,I,t}) \\
& + \sum_{t=2}^{T} \sum_{m=1}^{M} \sum_{k=1}^{K} S_{mkt} \max(0, X_{mk(t-1)} - X_{mkt})(\mathrm{P/F,\ I,\ t}) \\
& + \sum_{t=1}^{T} \sum_{m=1}^{M} \sum_{k=1}^{K} DP_{mkt}(TR)(\mathrm{P/F,I,t}) + BV_T(\mathrm{P/F,I,T})
\end{aligned}
$$

17.3.2 System Complexity

In addition to the economic considerations in comparing various systems configurations, one of the objectives is to opt for those with the least complexity. The structural complexity metric of a manufacturing system configuration is used to capture this criterion, in order to differentiate between otherwise comparable system configurations. The complexity measure used in this work is an entropy-based index that uses the reliability of each machine to describe its state in the manufacturing system, combined with an equipment type code index, a_{mk}, to incorporate the inherent structural complexity of the various hardware and technologies used. Detailed explanation of Eq. 17.5, which expresses the complexity due to the machines, is found in Kuzgunkaya and ElMaraghy (2006).

$$
f_2 = \text{Minimize Average Complexity} = \frac{\sum_{t=1}^{T} \sum_{m=1}^{M} \sum_{k=1}^{K} a_{mk} X_{mkt} \sum_{n=1}^{2} p_{mkn} \log_2 \left(\frac{1}{p_{mkn}} \right)}{T} \qquad (17.5)
$$

17.3.3 System Responsiveness

Responsiveness is the ability of a production system to respond to internal and external disturbances, which impact upon production goals as defined by Matson and McFarlane (1999). The responsiveness metric defined in Eq. 17.7 captures the ability of a system to change from one product to the next without changing its configuration. This metric is based on the mix response flexibility metric developed by Van Hop (2004). A definition of the efficiency of each machine with respect to an operation has been added to Van Hop's formulation as shown in Eq. 17.6:

$$e_{\mathrm{ijmk}} = \frac{\min\limits_{\mathrm{m,k}}(ST_{\mathrm{ijmk}})}{ST_{\mathrm{ijmk}}} \times \frac{\min\limits_{\mathrm{m,k}}(p_{\mathrm{ijmk}})}{p_{\mathrm{ijmk}}} \tag{17.6}$$

$$f_3 = \text{Maximize Average Responsiveness} = \frac{\sum\limits_{\mathrm{t=1}}^{T}\sum\limits_{\mathrm{m=1}}^{M}\sum\limits_{\mathrm{k=1}}^{K}\sum\limits_{\mathrm{i=1}}^{I} P_{\mathrm{imk}} RA_{\mathrm{imk}} X_{\mathrm{mkt}}}{T} \tag{17.7}$$

Where, P_{imk} represents the probability of assigning product i to machine (m, k), and RA_{imk} represents the response ability of machine (m, k) relative to product i. These terms are calculated as follows:

$$RA_{\mathrm{imk}} = \frac{\sum\limits_{\mathrm{j}}^{J} z_{\mathrm{ijmk}} e_{\mathrm{ijmk}}}{J} \tag{17.8}$$

$$P_{\mathrm{imk}} = P_{\mathrm{i}} \frac{RA_{\mathrm{imk}}}{\max\limits_{\mathrm{m,k}} \{RA_{\mathrm{imk}}\}} \tag{17.9}$$

17.3.4 Overall Model

The objective functions are expressed as constraints in Eqs. 17.10, 17.11, and 17.12. The maximum and minimum values of the objective functions have been determined by maximizing and minimizing each objective function subject to the problem constraint. These three objectives are subject to several constraints.

Max λ

Subject to:

$$\mu_1 = \frac{NPV(ATCF) - \min NPV}{\max NPV - \min NPV} \geq \lambda \tag{17.10}$$

$$\mu_2 = \frac{\max C - Complexity}{\max C - \min C} \geq \lambda \tag{17.11}$$

$$\mu_3 = \frac{Responsiveness - \min R}{\max R - \min R} \geq \lambda \tag{17.12}$$

$$0 \leq \lambda \leq 1 \tag{17.13}$$

Total demand in any period must be equal to the sum of the in-house production and outsourced amount.

$$M_{\text{it}} + Q_{\text{it}} = D_{\text{it}} \tag{17.14}$$

In addition, it will be assumed that the outsourced amount should not exceed a specified percentage (OL) of the total annual demand as expressed in Equation (17.15). The upper limit on the outsourcing amount can be decided by the decision maker depending on the specific conditions of the case.

$$Q_{\text{it}} \leq (OL)D_{\text{it}} \tag{17.15}$$

Production for an operation (i, j) in period t, can be assigned to a machine only if it is capable of performing the operation:

$$Y_{\text{ijmkt}} \leq z_{\text{ijmk}} M_{\text{it}} \tag{17.16}$$

A given operation (i, j) may be assigned to different machine types, but the total quantity produced should be equal to M_{it}.

$$\sum_{\text{m}=1}^{M} \sum_{\text{k}=1}^{K} Y_{\text{ijmkt}} = M_{\text{it}} \quad \forall i, j, t \tag{17.17}$$

The capacity on each machine should be available to meet the demand within the available time in one period.

$$\sum_{\text{i}=1}^{I} \sum_{j=1}^{J} \left(p_{\text{ijmk}} Y_{\text{ijmkt}} + \left(\frac{ST_{\text{ijmk}}}{L_t} \right) Y_{\text{ijmkt}} \right) \leq AH_{\text{mk}} X_{\text{mkt}} - RD_t \tag{17.18}$$

Machine utilization must be higher than 85%.

$$\sum_{\text{i}=1}^{I} \sum_{\text{j}=1}^{J} \left(p_{\text{ijmk}} Y_{\text{ijmkt}} + \left(\frac{ST_{\text{ijmk}}}{L_{\text{t}}} \right) Y_{\text{ijmkt}} \right) \geq 0.85 (AH_{\text{mk}} X_{\text{mkt}} - RD_{\text{t}}) \tag{17.19}$$

In order to calculate the book value of assets at each period, the depreciation of each machine in the system in each period should be determined. Assuming a straight line depreciation method, the depreciation of each machine type in one period is expressed as follows:

$$DP_{\text{mkt}} = DP_{\text{mk(t-1)}} + RX_{\text{mkt}} IC_{\text{mkt}} d_{\text{mk}} \tag{17.20}$$

The book value of assets at each period is equal to the book value of the previous period less the depreciation, salvage value of disposed assets, and plus the value of purchased assets in each period. Book value at each period is calculated using the following equation:

$$BV_{\mathrm{t}} = BV_{\mathrm{t-1}} + \sum_{\mathrm{m=1}}^{M} \sum_{\mathrm{k=1}}^{K} (RX_{\mathrm{mkt}}^{+} IC_{\mathrm{mkt}} - RX_{\mathrm{mkt}}^{-} SV_{\mathrm{mkt}} - DP_{\mathrm{mkt}}) \tag{17.21}$$

Reconfiguration task time is equal to the number of machine bases installed or removed and the number of modules purchased or removed during reconfiguration as expressed in Eq. 17.24. Reconfiguration cost is then calculated by multiplying the reconfiguration task time with hourly labor rate, and its duration is determined by dividing the reconfiguration task to the available workforce.

$$\mathrm{B}_{\mathrm{mt}} = \sum_{\mathrm{k=1}}^{K} X_{\mathrm{mkt}} \tag{17.22}$$

$$\mathrm{MD}_{\mathrm{mkt}} = kX_{\mathrm{mkt}} \tag{17.23}$$

$$RT_{\mathrm{t}} = \sum_{\mathrm{m=1}}^{M} t_B \left(RB_{\mathrm{mt}}^{+} + RB_{\mathrm{mt}}^{-}\right) + \sum_{\mathrm{m=1}}^{M} \sum_{\mathrm{k=1}}^{K} t_{MD} \left(RMD_{\mathrm{mkt}}^{+} + RMD_{\mathrm{mkt}}^{-}\right) \tag{17.24}$$

$$Cost_{\text{ Reconfiguration}} = RC_{\mathrm{t}} = LR(RT_{\mathrm{t}}) \tag{17.25}$$

$$Time_{\text{ Reconfiguration}} = RD_t = \frac{RT_t}{W_t} \tag{17.26}$$

The constraints from (17.27) to (17.30) are used in order to linearize the nonlinear terms in (17.4) such as $\max(0, X_{\mathrm{mkt}} - X_{\mathrm{mk(t-1)}})$. Constraint (17.27) calculates RX_{mkt}, which represents the difference in the number of machines of (m, k) between period t and (t − 1). Since RX_{mkt} is a real number, constraint (17.28) separates them into two positive variables where RX_{mkt}^{+} represents the positive difference and RX_{mkt}^{-} represents the negative difference. Constraints (17.29) and (17.30) ensures that either RX_{mkt}^{+} or RX_{mkt}^{-} is positive by using the binary variable δ_{mkt} and M, a positive large number. The terms $\max(0, X_{\mathrm{mk}(t-1)} - X_{\mathrm{mkt}})$ in and $Max(0, X_{\mathrm{mkt}} - X_{\mathrm{mk(t-1)}})$ in (17.4) can be replaced by RX_{mkt}^{-} and RX_{mkt}^{+} respectively.

$$RX_{\mathrm{mkt}} = X_{\mathrm{mkt}}^{-} X_{\mathrm{mk(t-1)}} \tag{17.27}$$

$$RX_{\mathrm{mkt}} = RX_{\mathrm{mkt}}^{+} - RX_{\mathrm{mkt}}^{-} \tag{17.28}$$

$$RX_{\mathrm{mkt}}^{+} \leq \delta_{\mathrm{mkt}} M \tag{17.29}$$

$$RX_{\mathrm{mkt}}^{-} \leq (1 - \delta_{\mathrm{mkt}}) M \tag{17.30}$$

17.4 Illustrative Example

Consider the simultaneous production of two potential parts throughout the life-cycle of a manufacturing system (Suresh, 1992). In order to manufacture these parts, three types of machines need to be installed: Drill, mill, and lathe. All of these machine types have numerical control and a modular structure that allows adding/removing modules (e.g. spindles or axes of motion). It is assumed that, based on these changeable modules, each machine type can have three different configurations with different capability and/or capacity as shown in Table 17.4. A planning horizon of 8 years is assumed. The prices are assumed to decrease while the material costs are expected to rise. A demand scenario that represents changeable market conditions is used.

By following the fluctuating demand requirements, shown in Table 17.3, the available machine candidates and their cost structures, the developed model selects the right machine configuration and the acquisition strategy, and determines the optimal production schedules. Since this is a multi-objective optimization based on the satisfaction degree of each objective, it will generate results that accomplish both the financial and strategic objectives.

The following assumptions and parameters are used in this case study:

- There are three types of machine bases each of which can be in three different configuration states, i.e. m = 1, 2, 3 and k = 1, 2, 3.
- 8 years of planning horizon is considered.
- For each part, a maximum of 20% outsourcing is allowed.

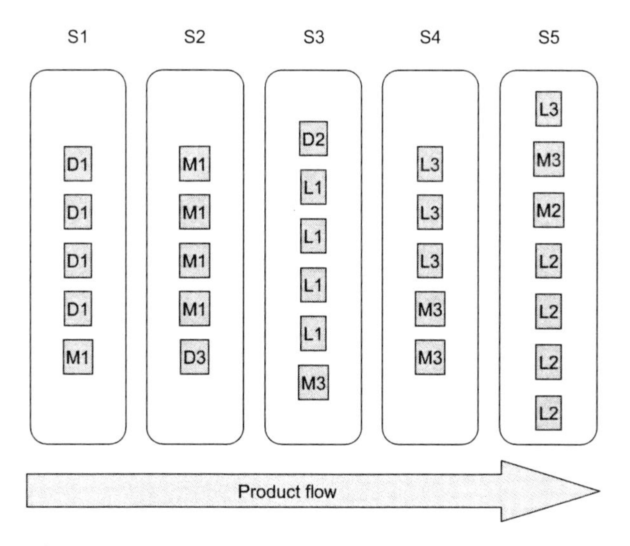

Fig. 17.2 System layout

Table 17.3 Demand scenario

Year	1	2	3	4	5	6	7	8
Part1	1 750	1 750	850	850	2 500	2 500	1 500	1 500
Part2	1 300	1 300	500	500	1 900	1 900	900	900

Table 17.4 Machine processing capabilities

					Z_{ijmk}					
Part	Operation		k = 1			k = 2			k = 3	
I	J	D	M	L	D	M	L	D	M	L
1	1	1						1		
	2		1	1				1	1	
	3				1				1	
	4				1	1			1	1
	5					1	1			1
2	1	1	1					1		
	2		1					1	1	
	3			1					1	
	4				1	1			1	1
	5					1	1		1	1

- Each period consists of one production year, which consists of 250 days, and 7.5 hours/day production time.
- Each machine configuration has an availability value depending on the number of modules attached to the base. We assume 0.92, 0.9, and 0.88 availability for configuration states 1, 2 and 3 respectively.
- Time required to install a machine base, tb, is 300 man-hours, and time to install a machine module, t_{MD}, is 150 man-hours.
- Available workforce for reconfiguration, Wt, is 50 workers.
- Interest rate for each period is 12% and the tax rate is 40%.

The model has been implemented in GAMS software package and solved using CPLEX solver algorithm on SunBlade 2000 Unix computer. Each objective function has been maximized and minimized subject to the case studies constraints in order to define the range of each objective. These values have been used to determine the fuzzy membership functions of each objective, followed by the multiple objective optimization run. Each run required 22 hours CPU time on average with a solution obtained within 2% of the relaxed solution.

Table 17.5 shows the results of the illustrative example. The satisfaction degree results for NPV, complexity, and responsiveness are 0.867, 0.862, and 0.872 respectively. The number of machine configurations follows the demand trend. As a result of dynamically following the demand changes, some reconfiguration activities are performed with an average cost of $34 050. Reconfiguration cost and capital outlay values at the end of the fourth period indicate a major reconfiguration. This investment in reconfiguration was required in order to meet the demand increase from 1.35M units/year to 4.4M units/year. Due to this re-investment, period 4 results in a negative cash flow; however, the RMS investment generates an after tax value of \$13.6M throughout its life-cycle. The payback period for this RMS investment is approximately 1.31 years.

Complexity and responsiveness metrics for the generated configurations are shown in Table 17.5. The complexity level follows the demand trend in both financial and multiple objective evaluations since it is dependent on the number of machines in the system. The complexity, responsiveness and financial performance of RMS configurations were satisfied in each period and the utilization of each period's configuration remained above 89%.

Table 17.5 RMS justification results for considered example

Year	0	1	2	3	4	5	6	7	8
Complexity		10.64	10.65	5.26	5.26	15.00	15.06	8.50	8.50
Responsiveness		4.06	3.80	2.26	2.26	4.08	4.29	4.10	4.10
Utilization		0.98	0.99	0.89	0.89	0.97	0.98	0.99	0.99
Outsourcing Level		0.2	0.2	0.2	0.2	0.2	0.2	0.2	0.2
Reconfiguration Cost Actual value (\$K)		19	57.6	0	99.6	32.4	63.6	0	0
Capital Outlays Actual value (\$K)		1 520	0	0	7 940	1 315	0	0	0
ATCF Present value (\$K)	−12 080	9 097	9 522	2 086	−3 486	2 038	2 003	1 220	3 246
Cumulative ATCF Present value (\$K)	−12 080	−2 982	6 540	8 626	5 140	7 178	9 182	10 403	13 649
NPV(ATCF) (\$K)	13 649								

17.4.1 Comparison of Reconfigurable and Flexible Scenarios over the System Life Cycle

In order to compare the performance of Flexible and Reconfigurable systems operating under the same demand scenario and all other conditions as well as the planning period being identical, the developed assessment model has been modified to generate FMS configurations to satisfy these demands and conditions. In the FMS case, the reconfiguration aspect of configurations evaluation has been disabled, and the candidate machines have been replaced by FMS machine types capable of various operations through out the considered periods, i.e. the whole system life cycle. Table 17.6 includes the processing capabilities of those FMS machines.

Table 17.6 FMS machine capabilities

Part i	Operation j	CNC Drill	CNC Mill	CNC Lathe
1	1	1		
	2	1	1	
	3		1	
	4		1	1
	5			1
2	1	1		
	2	1	1	
	3		1	
	4		1	1
	5		1	1

Table 17.7 FMS configurations

Year	1	2	3	4	5	6	7	8
CNC drill	9	9	9	9	9	9	9	9
CNC mill	19	19	19	19	19	19	19	19
CNC lathe	4	4	4	4	4	4	4	4
TOTAL	32	32	32	32	32	32	32	32

Table 17.8 FMS performance results

Year	0	1	2	3	4	5	6	7	8
Complexity		19.51	19.51	19.51	19.51	19.51	19.51	19.51	19.51
Responsiveness		15.52	15.52	15.52	15.52	14.80	14.67	14.60	14.49
Utilization		0.78	0.78	0.38	0.38	1.00	1.00	0.68	0.68
Outsourcing Level		20%	20%	1%	1%	20%	20%	20%	20%
Capital Outlays Actual value($K)	−27 740								
Cash Flows Actual value ($K)	−27 740	12 137	10 759	3 461	2 910	9 373	7 767	4 570	3 815
Cumulative Cash flows Actual value($K)	−27 740	−15 603	−4 844	−1 383	1 528	10 901	18 668	23 238	27 052
Cash Flows Present value($K)	−27 740	10 837	8 577	2 464	1 850	5 319	3 935	2 067	1 541
Cumulative Cash flows Present value($K)	−27 740	−16 903	−8 326	−5 863	−4 013	1 305	5 240	7 308	8 848
NPV(ATCF) ($K)	8 848								

Table 17.7 and Table 17.8 include the results from FMS implementation that satisfies the same demand scenario. The satisfaction degrees for NPV, complexity, and responsiveness objectives are 0.953, 0.962, and 0.930 respectively. The complexity level of FMS configuration is at 19.51 bits, which is 98% more complex on average than the RMS implementation. The responsiveness level is 15.08/system or 0.47/machine for the FMS implementation. This result shows that the FMS system is more complex due to the use of complex machine structures with redundant modules for additional capabilities. In contrast with complexity levels, the average responsiveness level of 15.08/system shows that the FMS system better than an RMS system whose average responsiveness is 3.62/system. Since the responsiveness metric used in this methodology tries to capture the ability to changeover the production from one product to another within the same configuration, the FMS system was found to be more responsive considering that its machines are more flexible and have more built-in capabilities.

The financial results of the FMS implementation shows that it requires 44% more total investment compared to an RMS implementation to meet the same demand requirements over the examined period. The FMS system generates an NPV of $8.8M compared to an NPV of $13.6M of an equivalent FMS implementation. This can be explained by the fewer outsourced products in the FMS case compared to the results of RMS implementation. The 15% average level of outsourcing in FMS case versus the 20% outsourcing level in RMS case is mainly due to the initial built-in excess capacity of FMS configuration.

The utilization of both systems shows that FMS is underutilized compared to the RMS implementation throughout the planning horizon. Since FMS is designed to meet the anticipated demand increases, it is expected that it will be underutilized in the periods where lower demand levels occur. While the built-in capacity and capability allow better responsiveness in FMS, the RMS configurations are used more efficiently.

17.4.2 FMS and RMS Comparison Through Life-Cycle Simulation

In order to examine the performance of the generated configurations and validate the results from the developed model, each period's configuration is simulated using ARENA software package for both the generated FMS and RMS configurations.

Table 17.9 Simulated demand scenario

Year	1	2	3	4	5	6	7	8
Part1					1 000	1 300	1 500	1 850
Part2	1 250	1 500	1 700	2 000	2 000	2 000	2 000	2 000

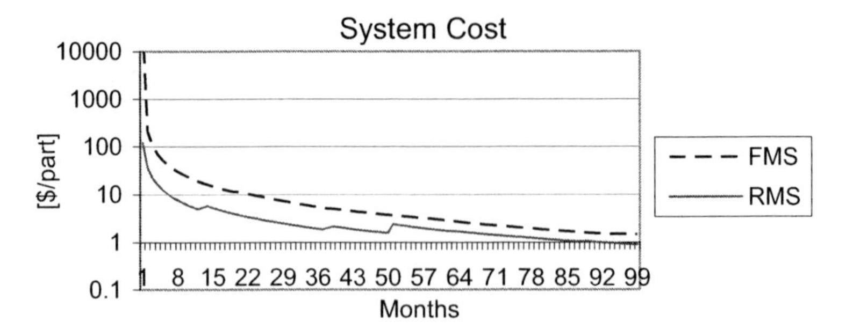

Fig. 17.3 Average system cost

Figure 17.3 represents the average cost per part throughout the life-cycle. The average cost/part for RMS increases in periods where a reconfiguration task is performed, whereas FMS starts with higher average cost/part that decreases as the production increases. The difference between RMS and FMS's average cost/part is due to the high initial investment in FMS and efficient reconfiguration of RMS by only adding the necessary capacity and capability when needed.

17.5 Conclusions

A fuzzy multi-objective evaluation model was developed in order to analyze, assess and justify the RMS investments. The model takes into account both financial and strategic objectives simultaneously, in order to generate manufacturing systems configurations that meet the demand forecast. The model considers in-house and outsourcing options, operational costs, reconfiguration costs and effective utilization of machines while minimizing the system complexity and maximizing its responsiveness.

The use of the model has been illustrated by studying a reconfigurable manufacturing system operating under fluctuating market demands. The results indicate that reconfiguration provides the means to use the acquired equipment effectively.

In addition, the developed model has been used to compare investments in both RMS and FMS as potential alternatives for meeting the same demand requirements over the same time period. The RMS implementation had the ability to reconfigure depending on the market conditions whereas the FMS configuration consisted of machines selected to carry out all anticipated processes due to their built-in versatile capabilities. A fluctuating demand scenario has been applied to both types of systems. For this example, the results showed that the higher investment levels required for the FMS configuration could not be justified since RMS performed better in terms of utilization, complexity and financial performance levels.

The developed model can help assess the trade-off between high initial capital investments in FMS vs. investment as needed for RMS. Life-cycle simulation analysis of the FMS and RMS configurations generated by the optimization tool allowed comparing their life-cycle cost performance. The simulation results, for this particular case study, illustrated the advantage of incremental investing as opposed to committing the total investment a priory. The responsiveness metric demonstrated that FMS responds better to demand changes within the same configuration.

The advantages of including the strategic benefits coupled with the financial objectives were demonstrated. Adding strategic criteria such as complexity and responsiveness generated manufacturing systems configurations that are less complex and more responsive while maintaining acceptable financial performance. The sensitivity analysis for unit reconfiguration time showed that the time required for reconfiguration is an important factor in justifying investing in RMS compared with FMS. Other critical parameters such as investment cost, variable costs and set-up time can also affect the system performance and consequently the ultimate investment decisions and should be studied further.

The developed model can support system designers by applying what-if scenarios when designing new systems and/or reconfiguring existing ones, and help system managers justify the investments in either FMS or RMS for given scenarios and market conditions.

Acknowledgements This research was conducted at the Intelligent Manufacturing Systems (IMS) Center, while the first author was a Ph.D. Candidate and Post Doctoral Fellow. The support from The Canada Research Chairs (CRC) Program and the Natural Science and Engineering Research Council (NSERC) of Canada is greatly appreciated.

References

Abdi M.R., Labib A.W., 2004, Feasibility study of the tactical design justification for reconfigurable manufacturing systems using the fuzzy analytical hierarchical process. Int J Prod Res 42:3055–3076

Abdel-Malek L., Wolf C., 1994, Measuring the impact of lifecycle costs, technological obsolescence, and flexibility in the selection of FMS design. J Manuf Syst 13/1:37–47

Amico M., Asl F., Pasek Z., Perrone G., 2003, Real Options: an application to RMS Investment Evaluation. CIRP 2nd Conference on RMS, Ann Arbor, MI, USA

Bellman R.E., Zadeh L.A., 1970, Decision-making in a fuzzy environment. Manage Sci 17:141–164

ElMaraghy H.A., 2006, A complexity code for manufacturing systems. In: Proceedings of 2006 ASME Int. Conf. on Manufacturing Science and Engineering (MSEC), Symposium on Advances in Process & System Planning, Ypsilani, MI, USA, 8–11 October 2006

ElMaraghy H.A., 2005, Flexible and reconfigurable manufacturing systems paradigms. International Journal of Flexible Manufacturing Systems 17/4:261–276

ElMaraghy H.A., Kuzgunkaya O., Urbanic R.J., 2005, Manufacturing systems configuration complexity. CIRP Annals 54/1:445–450

Gindy N.N., Saad S.M., 1998, Flexibility and responsiveness of machining environments. Integrated Manufacturing Systems 9/4:218–227

Koren Y., Heisel U., Jovane F., Moriwaki T., Pritschow G., Ulsoy G., Van Brussel H., 1999, Reconfigurable manufacturing systems. Ann CIRP 48/2:527–540

Kuzgunkaya O., ElMaraghy H.A., 2006, Assessing the structural complexity of manufacturing systems configurations. Int J Flex Manuf Syst 18/2:145–171

Lotfi V., 1995, Implementing flexible automation: a multiple criteria decision making approach. Int J Prod Economics 38:255–268

Matson J.B., McFarlane D., 1999, Assessing the responsiveness of existing production operations. Int J Oper Prod Manage 19/8:765–784

Rajagopalan S., Singh M.R., Morton T.E., 1998, Capacity expansion and replacement in growing markets with uncertain technological breakthroughs. Manage Sci 44/1:12–30

Spicer JP (2002) A design methodology for scalable machining systems. PhD dissertation, University of Michigan

Suresh N.C., 1992, A Generalized multimachine replacement model for flexible automation investments. IIE Trans 24/2:131–143

Van Hop N., 2004, Approach to measure the mix response flexibility of manufacturing systems. Int J Prod Res 42/7:1407–1418

Visionary Manufacturing Challenges For 2020 (1998) National Academy Press Washington, DC

Wiendahl H.P., ElMaraghy H.A., Nyhuis P., Zah M.F., Wiendahl H.-H., Duffie N., Kolakowski M., 2007, Changeable manufacturing-classification, design and operation. CIRP Annals, Manuf Technol 56/2:783–809

Wiendahl H.P., Heger C.L., 2003, Justifying Changeability. A Methodical approach to Achieving Cost Effectiveness. CIRP 2nd Conf on RMS, Ann Arbor, MI, USA

Yan P., Zhou M., Caudill R., 2000, A lifecycle engineering approach to FMS development. In: Proc. of IEEE Int. Conf. on Robotics and Automation. San Francisco, CA, pp 395–400

Zhang G., Glardon R., 2001, An Analytical Comparison on Cost and Performance among DMS, AMS, FMS, and RMS. 1st Conf. on Agile Reconfigurable Manufacturing 21.–22. May Ann Arbor, MI, USA

Chapter 18
Quality and Maintainability Frameworks for Changeable and Reconfigurable Manufacturing

W.H. ElMaraghy and K.T. Meselhy[1]

Abstract Despite the existence of many tools for assessing the product quality in manufacturing systems, there is limited research and/or tools that are concerned with studying the impact of manufacturing system design on the resulting product quality; especially, at the system development stage. The methodologies that are used for designing the product for quality, especially when considering form, function and variations and their interaction with the manufacturing system design, are rather limited. In the context of reconfigurable manufacturing systems, the designer will be faced with many configuration alternatives, and other changes. From the quality point of view, the designer should have an insight, and most preferably mathematical models, of how design decisions could affect the product quality. Except for the research work that was devoted to investigate the impact of the system layout on quality, until recently the relationship between the quality and the different system parameters were not well defined and quantified.

Manufacturing system changeability affects product quality in two respects: 1) manufacturing system design, and 2) maintenance of the manufacturing system equipment. Concerning the first aspect, the changeability in a manufacturing system affects many dimensions of the product quality. Some of these effects are positive and others are negative. A framework is presented for the complex relationship between quality, and the changes in reconfigurable manufacturing parameters. Details are also in the first Author's publications and other publications referred to in this Chapter. With regards to maintainability, it is an important concern in choosing the manufacturing system parameters. A maintainability strategy based on axiomatic design and complexity reduction is presented using the relationships between the manufacturing system parameters and the multi-objectives for optimizing quality, cost and availability. This should lead to maintenance systems that are less complex and adaptive to the changes in manufacturing.

Keywords Quality, Reliability, Maintainability, Changeability, Reconfigurable Manufacturing Systems, Complexity

---

[1] Intelligent Manufacturing Systems Center, University of Windsor, Ontario, Canada

18.1 Introduction

With the rapid evolution of today's products and the frequent changes in the global manufacturing environment; technological, economical, geographical and even political, the manufacturing system should have a high degree of changeability to cope with these variations and to remain competitive. Nevertheless, worldwide competition and the pace of technological innovation should not lead to distraction from industries' primary task, which is to produce quality products at competitive prices (Chen and Adam, 1991).

Achieving effective changeable manufacturing system has become a goal for both academia and industry. The targeted manufacturing system should be able to change or evolve quickly and adapt easily to new changes. Wiendahl et al. (2007) and ElMaraghy (2007) presented a comprehensive study of both the external and internal factors driving the need for manufacturing systems changeability. The effect of both physical/hard changes on all levels, as well as the logical/soft changes required to achieve them, on the different aspects of manufacturing systems performance including the all important product quality aspects of manufacturing system performance is yet to be fully investigated.

The importance of product quality in today's competitive industrial environment has been addressed by many researchers from different disciplines. Ben-Daya and Duffuaa (1995) stated that quality is becoming a business strategy leading to success, growth, and enhanced competitive position. Organizations with successful quality improvement programs can enjoy significant competitive advantages. Liu et al. (2004) stated that manufacturing enterprises are faced with unpredictable and rapidly changing market competition by customer demands all over the world. Companies must put tremendous emphasis on improving their product quality in order to survive and remain competitive in such environments. Elsayed (2000) explained the intensity of the global competition to develop new products in shorter time with higher reliability and overall quality.

Now, with the introduction of the new paradigm of changeable manufacturing systems as an umbrella for all the flexibilities, it is ever more critical to study the relationship between changeability in the manufacturing system and the expected product quality.

18.2 Quality and the Manufacturing System Design

The perceived product quality by the customer has always been a main concern for any business. The perceived product quality is a final result of many phases through which the products pass, starting from the product design to manufacturing system design to manufacturing operations on the shop floor. As a result of economic globalization and intensified competition, quality and cost have become crucial factors to the success of any manufacturing industry.

Despite of the importance and the wide applicability of the quality measures available in the literature, most cannot predict the quality level of the manufactured product in terms of the manufacturing system parameters at the design stage. Yet this is what must be used for a proactive strategic approach to quality. In this Chapter, a complexity approach is used to determine the relationships between product quality and the manufacturing system design. Quality in the changeable manufacturing environment is studied from the perspective that quality is an integrated result of product design, manufacturing system design and the applied maintenance strategy. The effects of different changeability enablers on system design, and hence product quality, are presented.

Shibata (2002) developed a method for predicting assembly defect rates at the pre-production stages. In his development of the Global Assembly Quality Methodology, he highlighted strong correlation between the occurrence of defects and the complexities in the assembly process. Hence, he developed metrics for assembly complexity using two engineering measures; these are: assembly time estimates and a rating for ease-of-assembly. The main issue in this developed tool is that the prediction of the defect rate is assessed only in terms of the assembly complexity and it does not take into account any other factors in the manufacturing environment that might affect the defect rate.

Variation risk management (VRM) is a systematic method to identify, assess, and mitigate variation throughout the product development process (Thornton et al., 2000). VRM can be applied either proactively, during product development, or to an existing product being manufactured. Variation risk management integrates all functional groups impacting product quality including design engineering, manufacturing, quality engineering, system engineering, customers, procurement, and suppliers. The variation risk framework developed by (Jay, 1998 and Thornton, 1999) is described in a three-step process as follows:

1. Risk Identification (I): Identify variation sensitive system requirements and latitudes; and Identify system, sub-system, feature and process characteristics that contribute to the system variation.
2. Risk Assessment (A): Quantify the probability of variation; and quantify the cost of variation.
3. Risk Mitigation (M): Select mitigation strategy based on costs, schedule and strategic impact; and Execute the strategy.

Figure 18.1 indicates how the Thornton (1999) framework may be applied in the product development stages. Thornton et al. (2000) concluded that industry typically applies VRM practices late in the design process when the product is about to be transitioned into manufacturing. The problems associated with the industry implementation are due to a lack of qualitative and quantitative models that enable designers to make quick and accurate decisions.

The manufacturing system affects product quality by several means including the manufacturing system design, manufacturing processes and manufacturing system operation. In general, product quality has a complex relationship with the condi-

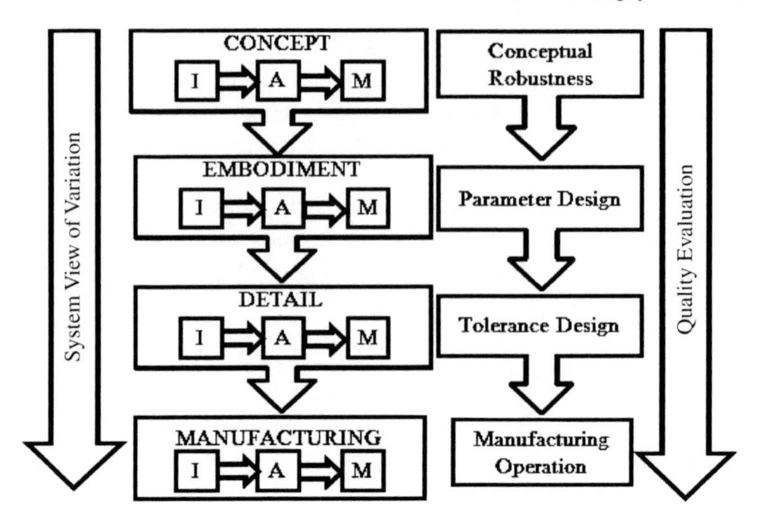

Fig. 18.1 The VRM Framework in the various product/ process development phases

tions, layouts, and interaction of the manufacturing system components. This relationship is more complex for changeable and reconfigurable manufacturing systems because of the continuously changing relationships between the manufacturing systems components and its configuration. There are many tools developed in the literature that enhance quality in the product design (first phase), such as the quality function deployment, robust design, poka yoke, and fault tree analysis to mention only a few.

The second phase is concerned with the realization of the designed product quality by the manufacturing processes and system. This relationship is depicted throughout the life cycle of the manufacturing system in ElMaraghy (2005), as illustrated in Fig. 18.2. Initially, the manufacturing system is designed to fulfill certain requirements and constraints, which affect product quality. This fact has been investigated by Inman (2003) and Nada (2006). The resulting product quality is ultimately determined on the shop floor by the manufacturing processes and their capability to meet the design requirements. The quality in this final phase is affected by the state of the machine tools, which depends partly on the applied maintenance policy. Thus, product quality is a combined outcome of product design, manufacturing systems components interaction and the applied maintenance system.

After some time in operation, and as indicated in Fig. 18.2, new needs appear which require the manufacturing system to reconfigure or to change. For the manufacturing system to be able to respond effectively to these new requirements, it should be changeable. Wiendahl et al. (2007) and ElMaraghy (2008) have defined enablers required for realization of that changeability such that the manufacturing system ability to change is proportionate to the degree of existence of these enablers. These enablers should be incorporated early into the design of the system and its

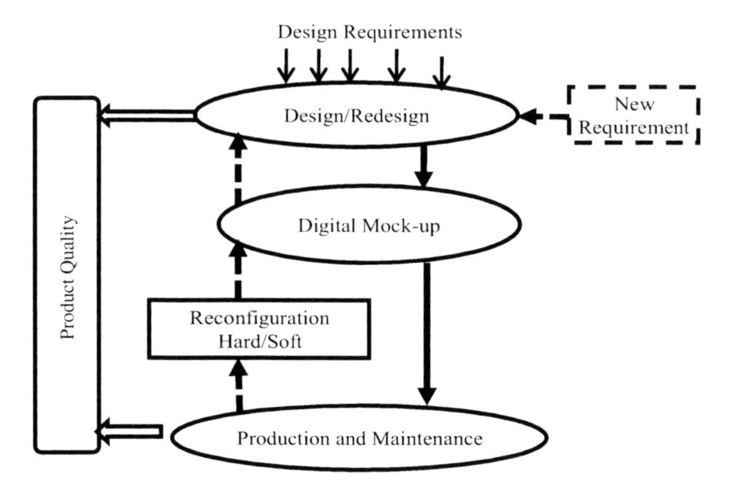

Fig. 18.2 Reconfiguration and product quality in manufacturing systems

operation strategies. This will influence the production methods and maintenance strategies, which will consequently reflect on the product quality.

In the following Section, the effect of incorporating the changeability enablers in manufacturing systems design on the product quality will be discussed.

18.3 Changeable Manufacturing and Quality

The effect of changeability on product quality can be investigated by exploring the effect of changeability enablers on the product quality. ElMaraghy (2007) defined two groups of changeability enablers: a logical group and a physical group. The physical enablers will be emphasized since they are the ones expected to affect manufacturing system design and hence product quality. There are some important common enablers in all modules, which are modularity, scalability and convertability as shown in Fig. 18.3.

Modularity is defined by Gershenson et al. (2003) as the building of complex product or process from smaller sub-systems that can be designed independently yet function together as a whole. The effect of modularity in the design of the manufacturing system on product quality is scarcely investigated in the literature. Arnheiter and Harren (2006) conducted qualitative research to study the effect of modularity on quality. They discussed the effect of manufacturing systems modularity on various aspect of product quality. They adopted the quality aspects stated by Garvin (1996), which are: aesthetics, perceived quality, performance, conformance, features, serviceability, reliability, and durability. Since only the effect of modularity of manufacturing system on the quality of produced products is of interest, the relevant quality aspects will be expanded upon.

	Enablers		*Effect on Product Quality*	
Changeability Enablers	*Modularity*	Conformance	?	**Product Quality**
		Features	+	
		Performance	-	
		servicability	+	
		Reliability	+	
	Scalability	Variability	-	
	Convertability	Configuration	-	
		Machine Tools	?	
		MHS	?	

Fig. 18.3 Manufacturing System Changeability and Product Quality

It is of interest now to discuss the relationship between ***modularity*** and the expected performance of the manufacturing system, as product quality is affected by both the performance of each module of a manufacturing system and by the integration between these modules. Thus, the manufacturing system designer chooses between a limited number of on the shelf modules that will lead to non-optimum performance or at least a lesser performance than the customized system design. In addition to that, if the modules are not fully integrable, this will lead to a decrease in system performance.

Conformance is traditionally managed using on-line quality management systems, such as control charts. There is no direct practical proof that modularity affects conformance in manufacturing systems (Dhafr et al., 2006).

Modularity and Serviceability in manufacturing systems affect the adopted maintenance policies because the modular structure enhances the replacement choices as it would be easy and inexpensive to replace the defective module compared to the non-modular structure, which promotes the repair choices and reduces replacement and repair cost. Therefore, a modular manufacturing system structure is easier to be maintained in a near as good as new state, which will reflect on the produced quality.

The ***reliability*** of a manufacturing system and its components and modules affects product quality. Yong and Jionghua (2005) developed a model to describe the complex relationship between product quality and the manufacturing system components reliability. This kind of quality and reliability interaction characteristics can be observed in many manufacturing processes such as machining, assembly, and stamping, etc. The effect of modularity on system reliability was discussed by Nepal et al. (2007). They used the failure potential metric to explain that system *reliabil-*

ity is improved by the modular structure. Another possible effect of modularity on system reliability is the ease of adding redundant modules such as a machine tool or a redundant sub-module, which enhances the whole system reliability and availability.

The manufacturing system production volume/capacity *scalability* problem, which addresses when, where, and by how much the capacity of the manufacturing system should be scaled. ***Scalability*** is defined by Deif and ElMaraghy (2007) as the ability to adapt to changing demand. Manufacturing systems may be scaled up or down to adjust their production capacity. A traditional technique of introducing capacity scalability into the manufacturing system is by parallelism as indicated by Sung-Yong et al. (2001). They stated that stage paralleling is an approach to scalability for RMSs and showed the economic feasibility of a parallel scalable manufacturing system compared to a balanced transfer line. Nada et al. (2006) studied the effect of a manufacturing system configuration on product quality. They explained that as the number of flow paths or the number of parallel lines increases, the product variability increases and hence quality decreases. Therefore, it can be concluded that scalability, through implementing parallelism, has a negative effect on the product quality.

Convertability was defined by Maier-Speredelozzi et al. (2003) as the capability of a system to rapidly adjust production functionality, or change from one product to another. They addressed three main contributions to manufacturing system convertability: configuration, machine and material handling system. They developed a metric to quantitatively compare the different alternatives from the convertability point of view. The effect of each one of these manufacturing system convertability contributions on product quality will be briefly discussed here.

Configuration convertability is defined by Maier-Speredelozzi et al. (2003) to be dependent upon the minimum increment of conversion, the routing connections, and the number of replicated machines. The minimum increment of conversion for a parallel system is less than that for a serial system because introducing a new product in a serial system requires shutting off the entire line, making the required changes, and then restarting it. In a parallel system however, the system can keep working while making changes in any of the parallel production segments. The other parameter in the configuration that enhances the convertability is the routing with a greater number of routing connections. It can be noticed that a parallel structure has a greater number of connections than a serial line of equivalent number of machines. The third parameter is the number of replicated machines, which is coincident with the concept of configuration width (Spicer et al. 2002). This factor indicates the system ability to produce more than one part type at the same time. Since there is one flow path for a serial line; the system can only produce one product. However, as the machine replications increase, more flow paths exist and more process plans can co-exist in the system, yielding more ability to produce different products. Therefore, convertability is enhanced by the parallel structure, which negatively affects the product quality.

18.4 Effect of Reconfigurable Manufacturing System Design on Quality

Introducing high quality products to the customers involves two important aspects. One aspect is related to the quality of the product design; which necessitates designing the product with all the quality features that satisfy the customer requirements. The other aspect that should be considered involves the development of a manufacturing system that is capable of producing products with minimal deviation from the design targets. Assessing the capability of a system configuration is a challenging task, especially, at the early stages of manufacturing system design. Yet it is obvious that the prediction of the resulting quality level at the early stages of system development is important.

Inman et al. (2003) have recently explored the intersection of two important fields of research: Quality and manufacturing system design. They stated that the production system used to manufacture a product does indeed affect its quality. They also provided evidence from the automotive Industry in order to support their argument. They pointed out that there is a lack of attention in literature to the impact of production system design on product quality and suggested several future research issues, related to the manufacturing system design and quality, which are important to industry. Some of these research issues are studying the impact of each of the following on quality: ergonomics, line or machine speed, plant layout, number and location of inspection stations, buffer location and size, batch size, level of automation, and flexibility. Some of the proposed issues are partially explored, however others are largely unexplored.

In the case of reconfigurable manufacturing systems, not only the cost of the system design modifications is the main concern, but also the time needed to modify the system, particularly that they are intended to be responsive systems. Therefore, the ramp-up time for such systems should be minimal. With traditional manufacturing systems, the effort to design the system and assess the expected quality was to be done once. With changeable and reconfigurable manufacturing systems, the system needs to cope with the changeable requirements. This will make the assessment of the resulting quality an ongoing activity with the changing of the manufacturing system configuration. At the early stages of system configuration or reconfiguration, the designer will be faced with many configuration alternatives. It is critical, at these stages to have the tools that can give the designer an insight of whether or not the proposed system configuration is capable of manufacturing products conforming to their design specifications.

A ***Conceptual Quality Framework*** for the prediction of quality of changeable and reconfigurable manufacturing systems has been developed (Nada et al., 2006 and ElMaraghy et.al., 2008), and is illustrated in Fig. 18.4.

In the proposed framework, it is illustrated that there are two possible relationships between the configuration parameters and the resulting quality. One is the direct relation between each configuration parameter and the resulting quality level.

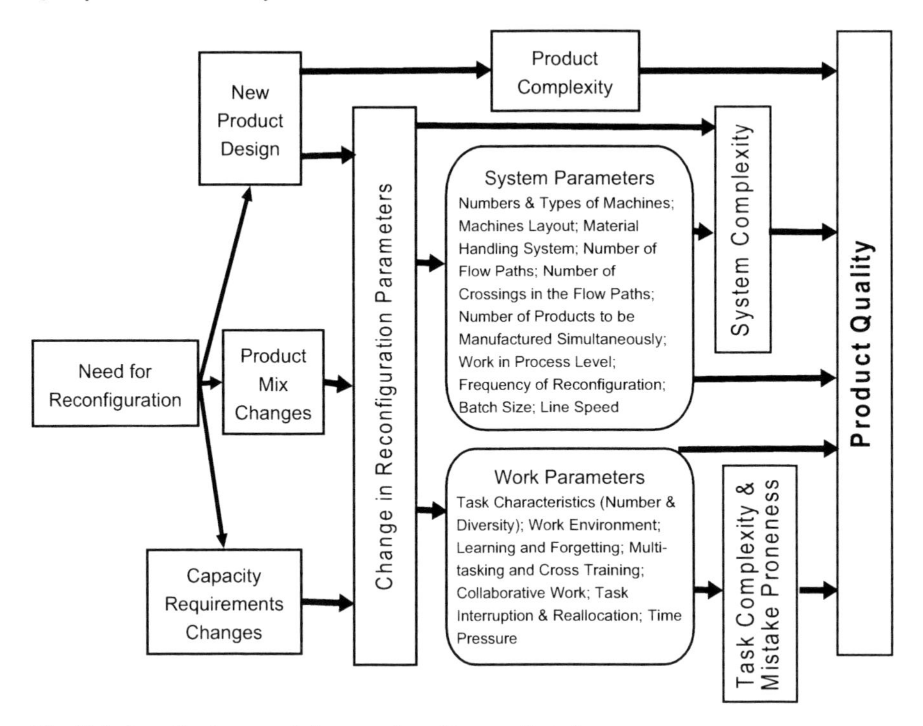

Fig. 18.4 A quality framework for reconfigurable manufacturing

Considering that path, in order to predict changes in quality level due to reconfiguration, the relations between each configuration parameter and quality should be investigated and quantified. The other relation is an indirect one; in which changes in complexity will be used as an intermediate relation between configuration parameters and quality. Considering that path, the configuration parameters are classified as system related parameters and worker related parameters.

Changes in product design, system related configuration parameters, and worker related parameters could be assessed by changes in product complexity, system complexity and task complexity respectively. The changes can be used as indicators to changes in the resulting quality. This is the approach used in this research which is discussed in detail in ElMaraghy et al. (2008) and Nada et al. (2006). This research discussed qualitatively and quantitatively how the quality is affected by the manufacturing system configuration and reconfiguration.

At *the system level*, the impact of different system configurations on the process quality was considered from the variation propagation point of view. At *the machine and component level*, the impact of modular design of machines components on the process quality was investigated. This requires studying how the machine capability could be affected by its modular design.

It is important to study the limitations on worker training time and adaptation to a newly assigned task that might not be completely different from their current task, which affects the product quality. At this *human (worker) performance level*, the impact of dynamic task allocation on product quality was investigated and a novel model, based on physical as well as cognitive ergonomics, was developed. The details and results of this are reported by ElMaraghy et al. (2008). For this research it was essential to study the effect of learning and forgetting rates and their impact on the system performance, in terms of product quality, especially with cross-trained workers in a reconfigurable environment with highly dynamic task allocation.

18.5 The Changeability and Maintainability Relationship

The relationship between product quality and machine operational status has been emphasized by several researchers. Decisions in the process design phase, such as tolerance assignment and maintenance policies play a substantial role for the overall manufacturing quality and costs. Chen et al. (2006) presented a new framework to integrate tolerance design and maintenance planning for multi-station manufacturing processes. When compared to other non-integrated approaches, this integrated design methodology leads to more desirable system performance with a significant reduction in production cost.

Maintenance of manufacturing systems is a multidisciplinary subject that combines reliability, scheduling and optimization. Traditionally, the maintenance of manufacturing systems is planned to optimize one of the performance criteria; cost, availability, reliability or quality. Figure 18.5 depicts the maintenance actions classifications (Aurich et al., 2006 and Wu and Clements-Croome, 2005).

The maintenance is traditionally performed according to a maintenance policy, which is the concept or strategy that describes what events (failure, passing of time or certain item condition) trigger what type of action (inspection, repair, maintenance or replacement).

The effect of maintenance on product quality has been addressed in several research publications, e.g.: Cassady et al. (2000), Linderman et al. (2005) and Chen et al. (2006). All agree that the quality of products is affected by the adopted mainte-

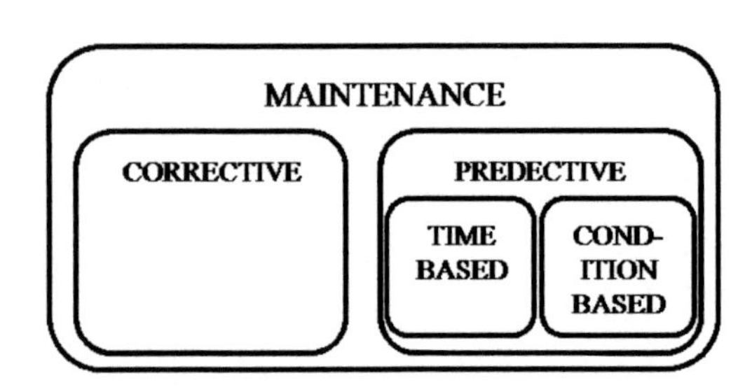

Fig. 18.5 Maintenance actions classification

nance policy for the manufacturing system, although the formulation of this causal relationship is not yet fully defined.

In a changeable manufacturing system, the elements of the system, either physical (machine tools, material handling equipments, etc.) or logical (process plans, production plans, controls, etc.), are changeable. Therefore, the maintenance policy should be changeable as well to adapt to the variations in the manufacturing system components. For example, assuming a new product is introduced into the family of products, this may need adding a new module to a machine(s) tool. This new configuration of the machine would cause a change in the machine structural complexity (ElMaraghy, 2006) and machine reliability, which needs a corresponding change in the maintenance policy.

The problem with the existing policies is that they are pre-planned to optimize one or more of the performance criteria. Therefore, when changes occur, the maintenance policy is no longer optimal and it may even negatively affect the system performance. Therefore, after introducing new product and accordingly changing the manufacturing system configuration, the maintenance policy in place is not guaranteed to satisfy the optimum reliability for the new system configuration. A possible solution for this problem is to re-plan the maintenance policy with every change in the system. Unfortunately, with the large number of parameters to be considered, this solution may not be practical. Another alternative would be to remove or reduce the sources of complexity in the maintenance system to simplify it and hence make it more agile and adaptable. The relationship between the system complexity and agility has been emphasized by Arteta and Giachetti (2004) as well as others.

Based on the literature survey of the vast number of developed preventive maintenance policies, it can be concluded that any maintenance policy can be described by three main parameters:

1. Level of failure repair (ranging from minimal to perfect)
2. Frequency of performing the preventive maintenance
3. Level of preventive maintenance (ranging from minimal to perfect).

Furthermore, the general main goals associated with any maintenance system are: 1) Quality, 2) Cost, and 3) Availability.

Therefore, according to the axiomatic design formulation of .Suh (1990), the maintenance policy can be formulated as a design matrix (Eq. 18.1):

$$\begin{bmatrix} \text{Quality} \\ \text{Cost} \\ \text{Availability} \end{bmatrix} = \begin{bmatrix} X_{11} & X_{12} & X_{13} \\ X_{21} & X_{22} & X_{23} \\ X_{31} & X_{32} & X_{33} \end{bmatrix} \begin{bmatrix} \text{Failure Repair level} \\ \text{Preventive Maintenance Frequency} \\ \text{Preventive maintenance Level} \end{bmatrix} \quad (18.1)$$

The goal is to reduce the complexity of the maintenance system in order to be easily changeable. Therefore, it is required to investigate the sources of complexity in the maintenance system and the different alternatives for mitigating them.

A novel model is suggested here, based on the complexity minimization approach, and the axiomatic design matrix (Xij), which relates the maintenance policy goals and parameters. Based on a comprehensive literature review and from the earlier discussion and understanding of the role of quality and maintainability in manufacturing systems, it can be stated that:

- The matrix elements are functions of manufacturing system parameters such as machine structural complexity, machine failure rate, process plan, etc.
- Each one of the requirements; quality, cost and availability is a function in all the three maintenance policy parameters. Therefore, all the matrix elements are non-zero, which leads to a coupled design matrix.

Thus, this coupling is the apparent source of the complexity in maintenance systems. Hence, the goal is to transform this coupled design to one of the simpler forms; un-coupled or de-coupled design. In cases of un-coupled design, the design matrix will take the following form, as shown in Equation 18.2:

$$\begin{bmatrix} \text{Quality} \\ \text{Cost} \\ \text{Availability} \end{bmatrix} = \begin{bmatrix} X_{11} & 0 & 0 \\ 0 & X_{22} & 0 \\ 0 & 0 & X_{33} \end{bmatrix} \begin{bmatrix} \text{Failure Repair level} \\ \text{Preventive Maintenance Frequency} \\ \text{Preventive maintenance Level} \end{bmatrix} \tag{18.2}$$

Therefore, when any of the manufacturing system parameters change, the matrix element/s would also change. However, it would be a simpler task to change the maintenance policy parameters to keep the functional requirements at the required levels. If de-coupled design can be achieved, the design matrix will assume the following form (or its diagonal mirror image), as shown in Equation 18.3:

$$\begin{bmatrix} \text{Quality} \\ \text{Cost} \\ \text{Availability} \end{bmatrix} = \begin{bmatrix} X_{11} & 0 & 0 \\ X_{21} & X_{22} & 0 \\ X_{31} & X_{32} & X_{33} \end{bmatrix} \begin{bmatrix} \text{Failure Repair level} \\ \text{Preventive Maintenance Frequency} \\ \text{Preventive maintenance Level} \end{bmatrix} \tag{18.3}$$

For resolving this coupling and its associated complexity, the off-diagonal elements should ideally be zero. In applying such a solution, the following relationship formulas need to be developed:

$$\begin{aligned} \text{Quality} &= f(\text{RL, PMF, PML, Manufacturing system parameters}) \\ \text{Cost} &= g(\text{RL, PMF, PML, Manufacturing system parameters}) \\ \text{Availability} &= h(\text{RL, PMF, PML, Manufacturing system parameters}) \end{aligned}$$

where:

RL = Repair Level
PMF = Preventive Maintenance Frequency
PML = Preventive Maintenance Level

The next step is to perform an experimental design for each function; f, g, and h, in order to generate the effect of maintenance policy parameters on the functional requirements. Then the design matrix in Eq. 18.1 would be rearranged to place the main (most important) factors on the diagonal.

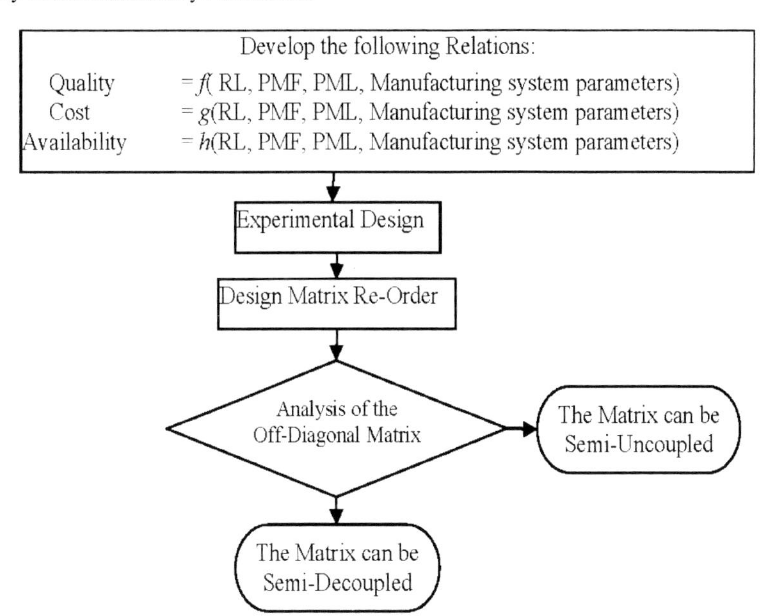

Fig. 18.6 Framework for Maintenance System Complexity Reduction

The relationship between the manufacturing system parameters and the off-diagonal elements is defined to highlight their effect on the complexity of the maintenance system. This approach is explained in the framework illustrated in Fig. 18.6. The goals are to develop a less complex maintenance strategy that optimizes Quality, Cost and Availability. Note that in this figure the terms semi-uncoupled and semi-decoupled are used to differentiate between the ideal forms of design matrices stated by axiomatic design theory, where the off-diagonal elements are exactly zero. In reality the off-diagonal elements may not be zeros, but rather small values compared to the diagonal elements which indicates weak coupling.

18.6 Conclusion

Quality is assessed based on whether a product possesses certain characteristics that satisfy the customer and the functional requirements. Quality measurement at the manufacturing level is a critical activity as it allows the assessment of the degree of conformance to specifications. Quality is inversely proportional to variability resulting from the manufacturing system as designed and/or as a result of changing capability. Hence, one relies on reducing the manufacturing variation in addition to the utilization of techniques mitigating the negative effects, or reducing the sensitivity to variations.

Quality approaches can be classified as: a) Passive/Reactive and b) Proactive/Preventive. The Proactive strategy, which is of interest here, emphasizes the design of quality into products and processes, identifies possible sources of variation and uses mathematical models to evaluate and monitor the process.

Despite the many existing tools for assessing the product quality in manufacturing systems, there is limited research that is concerned with studying the impact of manufacturing system design on the resulting product quality; especially, at the system development stage. Much of the measures used in the literature are applicable to the "on-line" quality control activities. Even the methodologies that are used for designing the product for quality, especially when considering form, function and variations and their interaction with the manufacturing system design, are rather limited.

In the context of Reconfigurable Manufacturing Systems, the designer will be faced with many configuration alternatives, and other changes. From the quality point of view, the designer should have an insight, and most preferably mathematical models, of how design decisions could affect the product quality. Except for the research work that was devoted to investigate the impact of the system layout on quality, until recently the relationships between the quality and the different system parameters were not well defined and quantified.

Manufacturing system changeability affects product quality in two respects: 1) manufacturing system design, and 2) maintenance of the manufacturing system equipment. Concerning the first aspect, the changeability in a manufacturing system affects many dimensions of the product quality. Some of these effects are positive and others are negative, as discussed in this Chapter. A framework is presented for the complex relationship between quality, and the reconfigurable manufacturing parameters changes. Details are in the first Author's publications and other referenced publications.

Concerning the second aspect, maintenance should be included as an important dimension and concern in choosing the manufacturing system parameters. Maintenance strategy, based on axiomatic design and complexity reduction was presented using the relationships between the manufacturing system parameters and the multi-objectives of optimizing quality, cost and availability. This should lead to maintenance systems that are less complex and adaptive to the changes in the manufacturing system.

Acknowledgements The support from The Canada Research Chairs (CRC) Program and The Natural Science and Engineering Research Council (NSERC) of Canada is greatly appreciated.

References

Arnheiter E.D., Harren H., 2006, Quality Management in a Modular World. TQM Magazine 18/1:87–96

Arteta B.M., Giachetti R.E., 2004, A Measure Of Agility As The Complexity Of The Enterprise System. Robotics and Computer-Integrated Manufacturing 20/6:495–503

Aurich J.C., Siener M., Wagenknecht C., 2006, Quality Oriented Productive Maintenance Within The Life Cycle Of A Manufacturing System, In: Duflou J.R.; Dewulf W., Willems B., Devoldere T. (ed): Towards a Closed Loop Economy – Proceedings 13th CIRP International Conference on Life Cycle Engineering, Leuven University, pp 669–673

Ben-Daya M., Duffuaa S.O., 1995, Maintenance and Quality: The Missing Link. Journal of Quality in Maintenance Engineering 1/1:20–26

Cassady C.R., Bowden R.O., Leemin Liew, Leemin Liew Pohl E.A., 2000, Combining Preventive Maintenance and Statistical Process Control: A Preliminary Investigation. IIE Transactions 32/6:471–8

Chen F.F., Adam E.E. Jr., 1991, The Impact of Flexible Manufacturing Systems on Productivity and Quality. IEEE Transactions on Engineering Management 38/1:33–45

Chen Y., Ding Y., Jin J., Ceglarek D., 2006, Integration of Process-Oriented Tolerancing and Maintenance Planning in Design of Multistation Manufacturing Processes. IEEE Transactions on Automation Science and Engineering 3/4:440–453

Deif A.M., ElMaraghy W., 2007, Investigating Optimal Capacity Scalability Scheduling in a Reconfigurable Manufacturing System. International Journal of Advanced Manufacturing Technology 32/5-6:557–62

Dhafr N., Ahmad M., Burgess B., Canagassababady N., 2006, Improvement of Quality Performance in Manufacturing Organizations by Minimization of Production Defects. Robotics and Computer-Integrated Manufacturing 22/5-6:536–42

ElMaraghy H.A., 2005, Flexible and Reconfigurable Manufacturing Systems Paradigms. International Journal of Flexible Manufacturing Systems 17/4:261–276

ElMaraghy H., 2007, Enabling Change in Manufacturing Systems. International Journal of Flexible Manufacturing Systems 19/3:125–127

ElMaraghy W.H., Nada O.A., ElMaraghy H.A., 2008, Quality Prediction for Reconfigurable Manufacturing Systems via Human Error Modelling. International Journal of Computer Integrated Manufacturing (IJCIM) 21/5:584–598

Elsayed E.A., 2000, Perspectives and Challenges of Research in Quality and Reliability Engineering. International Journal of Production Research 38/9:1953–1976

Jay D.J., 1998, Evaluation and Assessment of Variation Risk Management and the Supporting Tools and Techniques. MIT, MA, USA, M. Sc. Thesis

Inman R.R., Blumenfeld D.B., Huang N., Li J., 2003, Designing Production Systems for Quality: Research Opportunities from an Automotive Industry Perspective. International Journal of Production Research 41/9:1953–71

Garvin D.A., 1996, Competing on the Eight Dimensions of Quality. IEEE Engineering Management Review 24/1:15–23

Gershenson J.K., Prasad G.J., Zhang Y., 2003, Product Modularity: Definitions and Benefits. Journal of Engineering Design 14/3:295–313

Linderman K., McKone-Sweet K.E., Anderson J.C., 2005, An Integrated Systems Approach To Process Control And Maintenance. European Journal of Operational Research 164/2:324–40, 16 July 2005

Liu P., Luo Z.B., Chu L.K., Chen Y.L., 2004, Manufacturing System Design with Optimal Diagnosability. International Journal of Production Research 42/9:1695–1714

Maier-Speredelozzi V., Koren Y., Hu S.J., 2003, Convertability Measures For Manufacturing Systems. CIRP Annals 52/1:367–370

Nada O., 2006, Quality Prediction in Manufacturing System Design, Ph.D. Dissertation, University of Windsor

Nada O.A., ElMaraghy H.A., ElMaraghy W.H., 2006, Quality Prediction in Manufacturing System Design. Journal of Manufacturing Systems (JMS) 25/3:153–171

Nepal B., Monplaisir L., Singh N., 2007, A Framework to Integrate Design for Reliability and Maintainability in Modular Product Design. International Journal of Product Development 4/5:459–84

Shibata, H., 2002, Assembly Quality Methodology: A New for Evaluating Assembly Complexities in Globally Distributed Manufacturing, Ph.D. Dissertation, Stanford University

Spicer P., Koren Y., Shpitalni M., Yip-Hoi D., 2002, Design Principles for Machining System Configurations. CIRP Annals 51/1:275–280

Suh N., 1990, Axiomatic Design: Advances and Applications, The Oxford Series on Advanced Manufacturing, Oxford University Press

Sung-Yong S., Olsen T.L., Yip-Hoi D., 2001, An Approach to Scalability and Line Balancing for Reconfigurable Manufacturing Systems. Integrated Manufacturing Systems 12/7:500–11

Thornton A.C., 1999, Variation Risk Management Using Modeling and Simulation. ASME Journal of Mechanical Design 121:297–304

Thornton A.C., Donnelly S., Ertan B., 2000, More Than Robust Design: Why Product Development Organizations Still Contend with Variation and Its Impact on Quality. Research in Engineering Design 12:127–143

Wiendahl H.-P., ElMaraghy H.A., Nyhuis P., Zäh M.F., Wiendahl H.-H., Duffie N.A., Brieke M., 2007, Changeable Manufacturing – Classification, Design and Operation. CIRP Annals 56/2:783–809

Wu S., Clements-Croome D., 2005, Optimal Maintenance Policies under Different Operational Schedules. IEEE Transactions on Reliability 54/2:338–346, SCI

Yong C., Jionghua J., 2005, Quality-Reliability Chain Modeling For System-Reliability Analysis of Complex Manufacturing Processes. IEEE Transactions on Reliability 54/3:475–88

Chapter 19
Maintenance Strategies for Changeable Manufacturing

A.W. Labib[1] and M.N. Yuniarto[2]

Abstract This chapter includes a review of the recent developments in this field motivated by the need for changeability and reconfiguration and its consequences on the maintenance strategies. It also includes technical details about the author's own work and obtained scientific results in the field of fuzzy adaptive preventive maintenance in manufacturing control systems and self-maintenance as well as industrial application, future challenges and new directions.

Keywords Maintenance strategies, intelligent manufacturing system, fuzzy logic

19.1 Introduction

Maintenance is becoming ever more important under the current trends in changeable and reconfigurable manufacturing systems and the ever increasing complexity of manufacturing equipment. Machines are required to be available and ready to produce in response to variation in demand. Among the main sources of operational uncertainty in such an environment is maintenance in terms of severity of downtime and frequency of failures. Hence, preventive maintenance is crucial for efficient operations. This leads to the fundamental trade-off in maintenance theory between maximizing planned maintenance, which is usually unproductive downtime, versus the risk of unplanned and much costlier failures.

[1] University of Portsmouth, UK

[2] ITS-Surabaya, Indonesia

19.2 Recent Developments

Up until recently, the problems of maintenance, production, and quality have been addressed in isolation. We focus on recent research developments that attempted to take into consideration both the integration and interaction between the maintenance policies and the production planning and control.

Takata et al. (2004) provide a discussion of the changing role of maintenance from the perspective of life cycle management.

Yao et al. (2004) consider semiconductor manufacturing and shows how information on workloads of machines and work-in-process levels plays a critical role in preventive maintenance (PM) decisions.

Boukas and Liu (2001) propose a stochastic control approach for manufacturing flow control and preventive maintenance of failure-prone manufacturing systems. They propose a model whose transitions are governed by a continuous-time Markov chain.

Iravani and Duenyas (2002) considered an integrated maintenance and production control policy using a semi-Markov decision process model, where a heuristic policy with simple structure is proposed and analyzed. Other related work includes the work of Sloan and Shanthikumar (2000, 2002), in which they address the problem of joint equipment maintenance scheduling and production dispatching. The work of Sloan (2004) studies a joint production-maintenance problem, and explores some structural properties of optimal polices.

Chelbi and Ait-Kadi (2002) developed an analytical model to determine both the buffer stock size and the preventive maintenance period for an unreliable production unit, which is submitted to regular preventive maintenance of random duration.

Rezg et al. (2004, 2005), considered the optimization of production lines operating with a given maintenance policy and a given inventory control policy.

19.3 Current Research and Trends

This section is based on recent research (Yuniarto and Labib, 2006), and (Yuniarto 2004). In this work, a framework of reconfiguring preventive maintenance (PM) and manufacturing control system is proposed. Fuzzy logic control is used to enable an intelligent approach of integrating PM and manufacturing control system. It is thought that this contributes to the novel development of an integrated and intelligent framework in those two fields that are sometimes difficult to achieve.

This idea is based on combining work on intelligent real-time controller for failure prone manufacturing system using fuzzy logic approach (Yuniarto and Labib, 2004 and 2005) and the work on PM proposed by Labib (2004). The aim of the research is to control a failure prone manufacturing system and at the same time propose which PM method is applicable to a specific failure prone manufacturing system. The mean time to repair (MTTR) and mean time between failures (MTBF) of the system are used as integrator agents, by using them to couple the two areas to be integrated (i.e. maintenance system and manufacturing system).

The proposed reconfiguration is based on the ability to specify and control what production mode (make-to-order or make-to-stock) should be chosen and at what production rate a product should be produced. Optimal operation in here means that the chosen production mode and rate is able to compensate for breakdown that has occurred by minimizing the total cost incurred. From the results that have been reported, there is an opportunity (time) that can be used to do preventive maintenance. This opportunity occurred when the intelligent controller tells the manufacturing system to change the production mode, i.e. from make-to-stock to make-to-order mode or vice versa. Other previous work by Labib et al., (1998) on PM was only able to specify what type of preventive maintenance should be applied. They did not explicitly specify when the suggested PM has to be implemented. Based on the previous work capability, we designed a combined fuzzy logic controller that is able to determine when and what type of preventive maintenance should be implemented to the system in any given condition. This is the novel contribution of the research reported in this chapter.

19.3.1 Model of Integration Between Intelligent Manufacturing Control System and Intelligent Maintenance System

As it has been elaborated in previous section, the aim of integration/reconfiguration of manufacturing system is to have a control system which is able to specify optimum rate of production and at the same time propose what type and when a maintenance strategy should be performed on the system. The two objectives are performed intelligently and autonomously using a model of integration, which is depicted in Fig. 19.1.

The model depicted in Fig. 19.1 is able to reconfigure itself as a response to uncertain condition (i.e. machine breakdown). It can specify whether at any time to produce part at certain rate or to do maintenance (Preventive Maintenance) task. It is developed with three intelligent subsystems, which are the fuzzy logic controller I, fuzzy logic controller II, and fuzzy maintenance (Decision Making Grid/DMG) system. Fuzzy Logic Controllers I and II are developed based on hedging point theory to control a failure prone manufacturing system (Kimemia and Gershwin, 1987), while fuzzy maintenance and DMG are based on the work on maintenance specification process (Labib, 1998). As it has been outlined previously, the fuzzy logic controller for failure prone manufacturing system has the capability to determine the production mode of the system. It could be make-to-stock mode or make-to-order mode. It is also capable of determining when the machine should produce part or not to produce part when the machine is in operational state. Based on this fact,

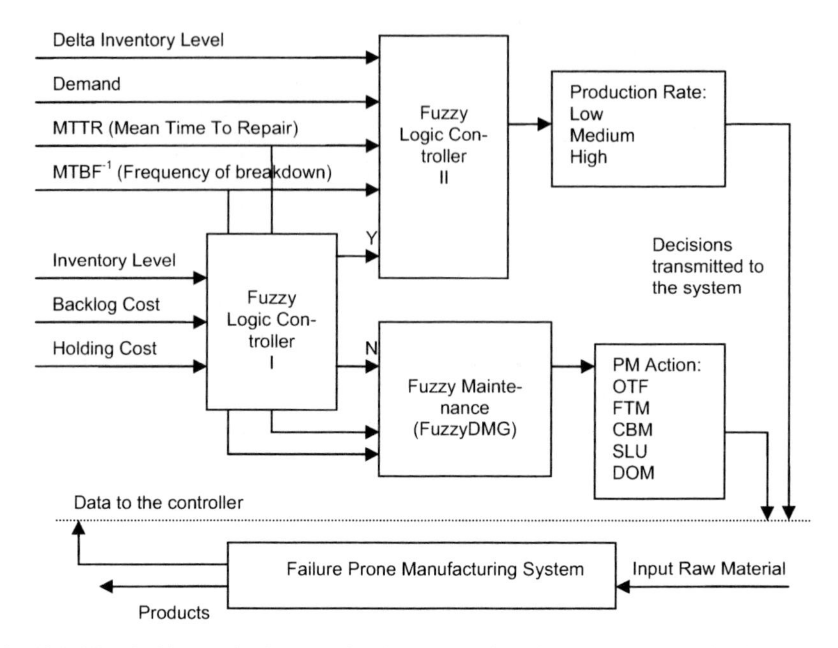

Fig. 19.1 Model of integration between intelligent manufacturing system and the intelligent maintenance system

we propose to make use of the time when the system is operational but not producing parts, as the time for the maintenance should take place. This approach is often called 'opportunity maintenance'.

The proposed model of integration in Fig. 19.1 can be explained as follows. There are two types of data for the proposed fuzzy logic controller, which are production system data (Delta inventory level, demand, inventory level, backlog cost and holding cost) and maintenance system data (MTTR and MTBF-1). The first fuzzy logic controller (FLC I) evaluates its inputs (backlog cost, inventory level and holding cost) and makes decision whether the system should produce the requested part or not. If the decision is to produce part (YES to produce part) then it will trigger the second fuzzy logic controller (FLC II). The FLC II then specifies at what rate the requested part should be produced. When the decision of the FLC I is NO to produce part then it will trigger the fuzzy logic maintenance and DMG controller to prescribe what the appropriate maintenance action should be taken based on the current value of MTTR (downtime) and $MTBF^{-1}$ (frequency of breakdown). Then the decision is transmitted to the shop floor (production system) to be executed while the machine is still in operational state but not producing requested part; In other words, when the machine is idle.

When the maintenance action is successful in reducing the downtime and the frequency of breakdown (the objective of all maintenance activities) then the new value of downtime and frequency of breakdown are fed back to the fuzzy logic controller and a new cycle of controlling and maintenance is started. This model is a closed loop system and it enables the failure prone manufacturing system to continuously improve its performance. In addition, it aims to integrate data from CMMS and ERP systems. This fact is also a significant step into what it is called a self maintenance manufacturing system due to the fact that the failure prone manufacturing system is able to specify when and which PM action should take place without affecting the operational optimality of the system, as the PM is suggested to take place when the machine is operational but not producing part (idle). This is particularly useful in changeable manufacturing where response to changes on the shop floor, including machine failure, can be made efficiently. More detail explanation about the three sub-systems of fuzzy logic is presented as follows.

19.3.2 Fuzzy Logic Controller I and II (FLC I and II)

The fuzzy logic controllers I and II in Fig. 19.1 are based on the hedging point method for controlling a failure prone manufacturing system. The idea of the hedging point method is to find a certain level of inventory, in which it is safe to build up products to anticipate or compensate for demand when the manufacturing system is down. Safe here means that the building up inventory at this level will not make unnecessary additional (minimize) total cost, which is a combination between holding cost (penalty for being in surplus of the demand) and backlog cost (penalty for being in shortage).

Based on the objectives of hedging point method, a control law has been proposed to achieve that goal. The control law can be expressed as follows:

Control Law: Define $H \geq 0$ to be an important value of x (inventory level).

$$\begin{array}{lll} \text{If} \quad x < H, & \text{then} & u = \mu\gamma \\ \text{If} \quad x = H, & \text{then} & u = d\gamma \\ \text{If} \quad x > H, & \text{then} & u = 0 \end{array} \tag{19.1}$$

Where u is the controlled production rate, μ is the maximum production rate, $d$is the demand, $H$is the optimal threshold value or the Hedging point and γ is the state of the system or machine; it could be 0 if the machine is down (under repair) and 1 if the machine is in operational condition. The value of H has been approximated by Gershwin et al. (1984) and can be expressed as:

$$H = \frac{dT_{\mathrm{r}}}{2} \tag{19.2}$$

Where d is the demand and T_{r} is time to repair and in most cases it could be the value of Mean Time To Repair (MTTR) of the system or machine.

The drawback of the hedging point method has been outlined by Yuniarto and Labib (2005), which is stated that such a control policy is not able to give the characteristic of what they called as gradual production rate reduction if the inventory level is approaching its hedging point level. This is similar to the braking phenomenon if one is driving a car, where if we want to stop the car at a designated point, one must apply the brake so that the speed of the car is gradually reduced, where eventually the car will stop at exactly the designated point. If we apply the brake at the designated point (this is what the hedging point suggested) then the car will overshoot stopping point, and it has to reverse back to stop at the designated point. This incapability leads to the chattering phenomenon, where the production rate oscillates around its hedging point level (similar to the reverse back in the driving car illustration). Other drawbacks have also been identified by Gharbi and Kenne (2003) who argued that the approach is time consuming in terms of computation time and the results are not tractable.

To overcome those problems, Yuniarto and Labib (2005) proposed a method of controlling a failure prone manufacturing system by using fuzzy logic control I and II (Fig. 19.1). The FLC I and II give the optimal solution by controlling the production rate of the manufacturing system. They are designed to improve the control policy provided by the classical hedging point method. They also managed to reduce the tendency for chattering phenomenon to occur by providing an extra capability that the classical hedging point does not have, which is the capability of gradually reducing the production rate of the system when the inventory level is approaching its hedging point. This controller has been applied for controlling a single part single machine manufacturing system with characteristics shown in Table 19.1 (Yuniarto and Labib, 2005).

Based on Table 19.1, the membership functions for each input and output of the controller are then defined. In this case, triangular values are chosen as the fuzzy logic membership functions. The complete triangular values for the inputs and outputs of the FLC I are shown in Table 19.2 for the FLC I and Table 19.3 for the inputs and outputs of the FLC II, respectively.

Table 19.1 Manufacturing system data

Parameter	Values
Maximum production rate (μ)	2.5 parts/unit time
MTBF (Mean Time Between Failures)	Low limit = 0 unit time Upper limit = 200 unit time
MTTR (Mean Time To Repair)	Low limit = 0 unit time Upper limit = 9 unit time
Maximum demand (d)	2 parts/unit time
Holding cost	0–25 unit cost/part
Backlog cost	0–25 unit cost/part

In order to obtain robust and optimum results, the fuzzy rule-base has to be determined. The fuzzy rule-base determination is based on the expert knowledge of the failure prone manufacturing system. IF-THEN rules are used in this case. Below is a sample of fuzzy rule base.

19.3.2.1 FLC I Fuzzy Rule-Base

This fuzzy rule base is defined based on the rationale that: production of a part is required whenever there is a demand and the current level of inventory is less than zero. If the current level of inventory is more than zero, the system should check whether to produce a part or not based on the holding and backlog cost. The objective is to find the appropriate decision that minimizes the total cost (holding cost + backlog cost).

Fuzzy rule example:

1. IF Inventory Level is Low and Holding cost is Low and Backlog cost is Low THEN the decision is to produce part (YES)
2. IF Inventory Level is Positive Big and Holding cost is Medium and Backlog cost is Low THEN the decision is not to produce part (NO)

⋮

45. IF Inventory Level is Positive Big and Holding cost is High and Backlog cost is High THEN the decision is not to produce part (NO)

Numerical example:

Inventory Level = 4, Holding Cost = 5 and Backlog Cost = 5. Based on the Inventory Level membership functions, fuzzy value for Inventory Level = 4 is Positive Small. For Holding Cost and Backlog Cost = 5, the fuzzy value are Low. These fuzzy values will fire fuzzy rules, where the output is to produce part (YES). This decision means that the failure prone manufacturing system should continue to produce part if there is demand to do so.

19.3.2.2 FLC II Fuzzy Rule-Base

The FLC II fuzzy rule-base is defined based on the logic that high production rate is required when the current level of inventory is far below its hedging point, or the demand is quite high on a production system that is less reliable. Low production rate is required when the production system is reliable enough; i.e. low MTTR and MTBF-1, the demand is low and the current inventory level is quite close to its hedging point. If there is no condition that meets the two production-rates stated before, the medium production rate is chosen. This idea will enable the controller to switch from one production rate to another smoothly and also reduce the possibility of chattering to occur.

Table 19.2 Input and output variables of fuzzy-logic controller I

Variables	Linguistic Values	Triangular Fuzzy Values
Inputs		
Inventory Level	Negative Big (NB)	(−9, −9, −6)
	Negative Small (NS)	(−9, −6, 0)
	Zero (ZO)	(−6, 0, 4)
	Positive Small (PS)	(0, 4, 9)
	Positive Big (PB)	(4, 9, 9)
Holding Cost	Low (L)	(0, 0, 15)
	Medium (M)	(0, 15, 25)
	High (H)	(15, 25, 25)
Backlog Cost	Low (L)	(0, 0, 10)
	Medium (M)	(0, 10, 25)
	High (H)	(10, 25, 25)
Outputs		
Decision Output	YES	(1, 1, 1)
	NO	(0, 0, 1)

Fuzzy rule example:

1 IF Demand is Low and MTTR is Low and MTBF-1 is L and Delta inventory is Low THEN Production rate is Low
2 IF Demand is Medium and MTTR is Medium and MTBF-1 is Medium and Delta inventory is Medium THEN Production rate is Medium

⋮

81. IF Demand is High and MTTR is High and MTBF-1 is High and Delta inventory is High THEN Production rate is High

Numerical example:

Delta Inventory Level = 5, Demand = 1, MTTR = 4.5 and MTBF-1 = 100. Based on the FLC II inputs membership functions fuzzy value for:

Delta Inventory Level = 4 is Low and Medium
Demand = 1 is Low and Medium
MTTR = 4.5 is Low
MTBF-1 = 100 is Medium

These input values will fire fuzzy rules, where its output is to set the production rate of the failure prone manufacturing system to Medium Production Rate.

19.3.3 Fuzzy Maintenance and Decision Making Grid

The fuzzy maintenance and DMG is used to specify what maintenance action to be done on the system based on the equipment criticality and reliability properties.

This work has been proposed by Labib et al. (1998) to address a problem in maintenance from not having clear criteria and not having robust decision criteria with which to maintain failing equipment. This work was then extended in Labib (2004). A two-step method is used. The first step is to obtain a prioritized criterion for maintenance and hence identify the most critical machines and their related faults using the Analytical Hierarchy Process (AHP) analysis. The second step is to use weights obtained from the first step as crisp inputs to a fuzzy logic controller in order to obtain a prescriptive model for maintenance action.

In this chapter, we are only interested in the second step of Labib et al. (1998) work. Their proposed fuzzy logic controller is depicted in Fig. 19.2.

Table 19.3 Input and output variables of the second fuzzy-logic controller

Variables	Linguistic Values	Triangular Fuzzy Values
Inputs		
Delta Inventory Level	Low (L)	(0 0 10)
	Medium (M)	(0 10 20)
	High (H)	(10 20 20)
Demand	Low (L)	(0 0 1)
	Medium (M)	(0 1 2)
	High (H)	(1 1 2)
MTTR	Low (L)	(0 0 4.5)
	Medium (M)	(0 4.5 9)
	High (H)	(4.5 9 9)
$MTBF^{-1}$	Low (L)	(0 0 100)
	Medium (M)	(0 100 200)
	High (H)	(100 200 200)
Outputs		
Production Rate	Low (L)	(0 0 0)
	Medium (M)	(1.25 1.25 1.25)
	High (H)	(2.5 2.5 2.5)

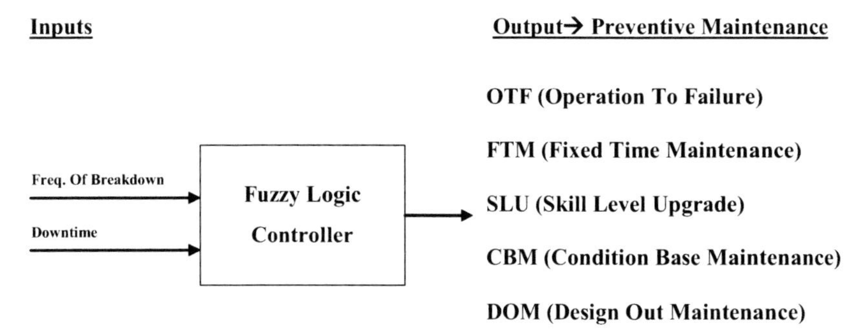

Fig. 19.2 Fuzzy-logic controller for a maintenance system

The inputs for the fuzzy logic controller are the frequency of breakdown (MTBF-1) and the time spent to repair (MTTR or downtime). These two inputs can be initiated from a computerized maintenance managements system (CMMS) if applicable or from manually data inputted from the system. Based on the membership functions of the inputs (Low, Medium and High) and certain rule-based developed in the controller, the controller then proposed an output, which is a prescriptive method of maintenance. This output could be; OTF, FTM, SLU, CBM and DOM.

1. OTF: Operation To Failure
This action is suitable to the machine if the frequency of breakdown and the downtime of the system/machine are low. This is the ideal condition of a system, where preventive maintenance is not applicable. The machine or system is so reliable such that it can be operated continuously. It means the only parameter that affect the availability of the machine in this region is the scheduled maintenance not the random breakdown caused by poor maintenance.

2. FTM: Fixed Time Maintenance
This maintenance action is prescribed when the frequency of breakdown and the downtime of the machine/system are medium.

3. SLU: Skill Level Upgrade
When the frequency of breakdown of the machine is high and the downtime of the machine is low then the Skill Level Upgrade is prescribed to the system. This condition means that the breakdown occurres frequently, but the time spent for repairing the machine or the system is short. This could be a simple breakdown that could be easily repaired or in some cases it is just a breakdown that is caused by an operator who does not know how to operate the machine/system properly

4. CBM: Condition Base Maintenance
This action is prescribed when the frequency of breakdown (MTBF-1) of the machine is low and the downtime of the machine (MTTR) is high. This condition means that the breakdown rarely occurs, but when it occurs it will take a long time to repair. For the machine/system with this condition, the condition base monitoring action is the most suitable one.

5. DOM: Design Out Maintenance
This action is the worst-case scenario of the maintenance function, when the breakdown is frequently occurring and it takes a long time to repair it. The machine/system with this condition is uneconomic to be operated. The only thing that is appropriate to do when dealing with such machine/system is by replacing it or redesigning it during the shutdown phase of production. That is why the DOM action is implemented for such cases.

The five outputs criteria are often called as the Decision-Making Grid (DMG) for machine/system, and it can be seen in Fig. 19.3. The membership functions for inputs and output in the FuzzyDMG are presented in Table 19.4.

The triangular fuzzy values in Table 19.4 (Output FuzzyDMG are associated with cost code function of the correlated PM action, for instance Operation To Failure (OTF) action has cost code function of 0, as there is no money or time spent for doing this PM action. Whereas Design Out Maintenance (DOM) costs 50 unit cost,

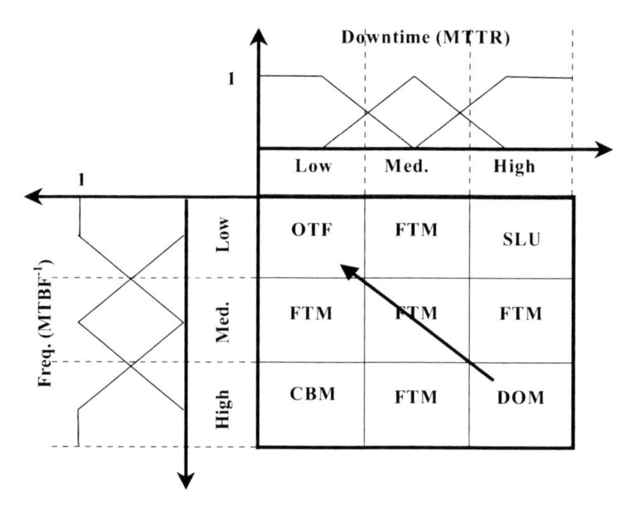

Fig. 19.3 Decision-making grid (DMG) of the fuzzy maintenance system

which is the highest. This is due to the fact that when this maintenance action is taken, it will take a lot of time or money.

Table 19.4 Input and output variables of FuzzyDMG

Variables	Linguistic Values	Triangular Fuzzy Values
Inputs		
MTTR	Low (L)	(0 0 4.5)
	Medium (M)	(0 4.5 9)
	High (H)	(4.5 9 9)
$MTBF^{-1}$	Low (L)	(0 0 100)
	Medium (M)	(0 100 200)
	High (H)	(100 200 200)
Outputs		
PM Action	OTF	(0 0 20)
	FTM	(20 20 20)
	SLU	(30 30 30)
	CBM	(40 40 40)
	DOM	(40 50 50)

Based on the membership functions of the inputs and output of the Fuzzy DMG, then the fuzzy rule-based is developed. The developed rule base is presented in Table 19.5.

This fuzzy rule-based has been simulated using Fuzzy Logic Toolbox Matlab, and the result is shown in Fig. 19.4.

The idea of the DMG is as a map or indicator of the machine performance. Based on the performance of the machine, an appropriate maintenance action is then sug-

Table 19.5 FuzzyDMG rule-base

MTTR	$MTBF^{-1}$	PM ACTION
Low	Low	OTF
Low	Medium	FTM
Low	High	CBM
Medium	Low	FTM
Medium	Medium	FTM
Medium	High	FTM
High	Low	SLU
High	Medium	FTM
High	High	DOM

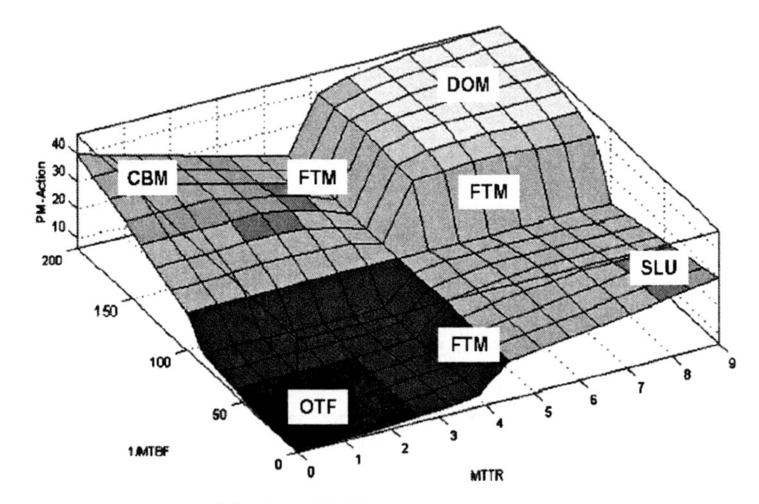

Fig. 19.4 Response surface of the fuzzy DMG

gested with the objective that it will move to the top left area (OTF) of the DMG. One limitation that the fuzzy logic maintenance and DMG have is that it cannot specify when the suggested maintenance action should be executed. This is the key of the effectiveness of the maintenance system. In the next section, we will try to address this problem by integrating it into the production system controller; so that the exact time when the maintenance should take place can be specified.

19.4 Case Study

The robustness of the proposed FLC I and FLC II has been tested on real time environment of a failure prone manufacturing system. Please consult Yuniarto and Labib (2005) for more detail explanation about the case study. In this chapter, the authors would like to illustrate how the developed framework works in controlling and at the same time specifying what maintenance action should be performed in a failure prone manufacturing system. Suppose at one stage of its operation, a failure prone manufacturing system has properties as follows; demand = 1 part/unit-time, inventory level = 5 part, MTTR = 7 hours, $\text{MTBF}^{-1} = 60/200$ hours, Holding cost = backlog cost = 5 unit cost/part, delta inventory level = 4 part.

If the developed fuzzy logic controller is applied to the system at current state, it will provide decision to produce part at speed 2 part/unit-time at simulation time of 5 unit time. Now if the holding cost is changed to the high value, say 20 unit-cost/part. The controller will tell the system to stop production until the inventory level reaches zero (the new optimal inventory level). During this time the machine is in operational condition but has been told not to produce part. Then the FLC I will trigger the Fuzzy DMG to suggest PM action, which in this case it produces a code cost function equal to 24.3 (based on the value of the MTTR and MTBF-1). This cost code function value lies between FTM and SLU action, then using simple percentage calculation, the recommended PM action need to be implemented on the system while system is idle is 42% SLU and 58% FTM. It means that 42% of resources available in terms of effort, money, or time has to be allocated to the machine or system to perform SLU, and 58% of the resources on FTM (Fixed Time Maintenance).

From the case study above, it is shown that the system is capable to perform an autonomous maintenance and autonomous control of its production rate, by specifying exactly when and what type of PM should be applied on the manufacturing system.

19.5 Conclusions and Future Research

As it has been elaborated in the previous section, the proposed approach (fuzzy logic control) has made the manufacturing system more adaptable to uncertain environment, i.e. demand and machine reliability variation. Furthermore, with the addition of FuzzyDMG, it made the manufacturing has special ability which it is able to determine what type and when a maintenance strategy has to be implemented. With those two control system, the manufacturing system will be more responsive to the environment changing (changeability) and at the same time, it is able to maintain its level of reliability (due to right maintenance method) and optimality.

Future research directions that could be conducted based on the proposed framework are:

- Development and implementation of the controller into a real failure prone machine/system.
- Test the scalability of the proposed framework, whether it is only suitable for one machine and one product manufacturing system or can be applied in a more general failure prone system.

- If the controller is to be developed in modular concept then it would be interesting to propose a communication method between one machine to another using the current intelligent manufacturing system paradigm, e.g. holonic manufacturing system concepts or multi agents concepts.
- It could be argued that the human body, when in state of relaxation, sleep or in other words 'idle', the body performs some self-maintenance tasks. Our approach in maintaining a machine when it is idle can be considered as a step towards self-maintenance. This needs to be further developed and compared with human activities in self-maintenance.

References

Akella. R., Kumar P.R., 1986, Optimal Control of Production Rate in a Failure Prone Manufacturing System. IEEE Transactions on Automatic Control AC-31/2:116–126

Boukas E.K., Liu Z.K., 2001, Production and maintenance control for manufacturing systems. IEEE Transactions on Automatic Control 46:1455–1460

Chelbi A., Ait-Kadi D., 2002, Joint Optimal Buffer Inventory and Preventive Maintenance Strategy for a Randomly Failing Production Unit. Journal of Decision Systems 11/1:91–108

Gershwin S.B., Akella R. and Choong Y., 1984, Short-term Production Scheduling of an Automated Manufacturing Facility, Massachusetts Institute of Technology Laboratory for Information and Decision Systems Report, LIDS-FR-1356

Iravani M.R., Duenyas I., 2002, Integrated maintenance and production control of a deteriorating production system. IIE Transactions 34/5:423–435

Kimemia J.G. and Gershwin S.B., 1983, An Algorithm for the Computer Control of Production in Flexible Manufacturing Systems, Massachusetts Institute of Technology Laboratory for Information and Decision Systems Report, LIDS-P-1134

Labib A.W., 2004, A Decision Analysis Model for Maintenance Policy Selection Using a CMMS. Journal of Quality in Maintenance Engineering (JQME) 10/3:191–202

Labib A.W., Yuniarto M.N., 2005, Intelligent Real time Control of Disturbances in manufacturing Systems. Journal of Manufacturing Technology Management 16/8:864–889

Labib A.W., Williams G.B., O'Connor R.F., 1998, An Intelligent Maintenance Model (System): An Application of Analytic Hierarchy Process and A Fuzzy Logic Rule-Based Controller. Journal of the Operational Research Society 49:745–757

Rezg N., Chelbi A., Xie X.-L., 2005, Modeling and optimizing a joint buffer inventory and preventive maintenance strategy for a randomly failing production unit: Analytical and simulation approaches. International Journal of Computer Integrated Manufacturing 18/2-3:225–235

Rezg N., Xie X.-L., Mati Y., 2004, Joint Optimization of Preventive Maintenance and Inventory Control in A Production Line Using Simulation. International Journal on Production Research 42/10:2029–2046

Sloan T.W., 2004, A periodic review production and maintenance model with random demand, deteriorating equipment, and binomial yield. Journal of the Operational Research Society 55:647–656

Sloan T.W., Shanthikumar J.G., 2000, Combined production and maintenance scheduling for a multiple-product, single machine production system. Production and Operations Management 9:379–399

Sloan T.W., Shanthikumar J.G., 2002, Using in-line equipment condition and yield information for maintenance scheduling and dispatching in semiconductor wafer fabs. IIE Transactions 34:191–209

Takata S., Kimura F., van Houten F., Westkamper E., Shpitalni M., Ceglarek D., Lee, 2004, Maintenance: Changing Role in Life Cycle Management. CIRP Annals - Manufacturing Technology 53/2:643–655

Yao Fernandez-Gaucherand X.E., Fu M., Marcus S.I., 2004, Optimal preventive maintenance scheduling in semiconductor manufacturing. IEEE Transactions on Semiconductor Manufacturing 17:345–356

Yuniarto M.N., 2004, Intelligent Real-time Control and Monitoring of a Failure Prone Manufacturing System: A Fuzzy Logic Approach, PhD Thesis, The University of Manchester

Yuniarto M.N., Labib A.W., 2006, Fuzzy Adaptive Preventive Maintenance in a Manufacturing Control System: A Step Towards Self-Maintenance. International Journal of Production Research (IJPR 44/1:159–180

Yuniarto M.N., Labib A.W., 2005, Optimal Control of an Unreliable Machine Using Fuzzy Logic Control: From Design to Implementation. International Journal of Production Research (IJPR) 43/21:4509–4537

Part V
Future Directions

Chapter 20
The Cognitive Factory

M.F. Zäh[1], M. Beetz[2], K. Shea[3], G. Reinhart[1], K. Bender[4], C. Lau[1], M. Ostgathe[1], W. Vogl[1], M. Wiesbeck[1], M. Engelhard[3], C. Ertelt[3], T. Rühr[2], M. Friedrich[4] and S. Herle[5]

Abstract The automation of processes and production steps is one of the key factors for a cost effective production. Fully automated production systems can reach lead times and quality levels exceeding by far those of human workers. These systems are widely spread in industries of mass production where the efforts needed for set-up and programming are amortized by the large number of manufactured products. In the production of prototypes or small lot sizes, however, human workers with their problem solving abilities, dexterity and cognitive capabilities are still the single way to provide the required flexibility, adaptability and reliability. The reason is that humans have brains, computational mechanisms that are capable of acting competently under uncertainty, reliably handling unpredicted events and situations and quickly adapting to changing tasks, capabilities, and environments. The realization of comparable cognitive capabilities in technical systems, therefore, bears an immense potential for the creation of industrial automation systems that are able to overcome today's boundaries. This chapter presents a new paradigm of production engineering research and outlines the way to reach the Cognitive Factory, where machines and processes are equipped with cognitive capabilities in order to allow them to assess and increase their scope of operation autonomously.

Keywords Cognition, Flexibility, Intelligent Automation

---

[1] Institute for Machine Tools and Industrial Management (*iwb*), Technical University of Munich, Germany

[2] Computer Science Department, Technical University of Munich, Germany

[3] Institute of Product Development, Technical University of Munich, Germany

[4] Institute of Information Technology in Mechanical Engineering, Technical University of Munich, Germany

[5] Department of Automation, Technical University of Cluj-Napoca, Romania

20.1 Introduction

Some decades ago, mass production with a high degree of automation seemed to be the silver bullet to reach an economical production of products. With the shift from seller markets to buyer markets and increasing dynamics, such as rising customer demands, increasing number and variety of products, and changing market demands, flexibility and changeability became main enablers for an efficient production (Wiendahl et al., 2007).

In order to reach this changeability, several concepts have been proposed for the physical system, the control system and the organization of production systems. Even though remarkable results can be achieved with existing concepts, they disregard the immense cognitive capabilities that humans possess and which enable them to react to unpredictable situations, to plan their further actions, to learn and gain experience and to communicate with others. Hence, the most flexible and changeable production system remains the skilled and experienced human worker.

To reach the next level of changeability, it is therefore necessary to combine the advantages of automated systems with the cognitive capabilities of common human workers. Future work in production science will thus include research on mimicking human behavior to enable the Cognitive Factory.

20.2 Intelligence in Automated Systems

Many researchers have identified the necessity to develop novel manufacturing paradigms in order to achieve higher degrees of flexibility, adaptability, autonomy and intelligence of production systems (Scholz-Reiter and Freitag, 2007; Monostori et al., 2006; Valckenaers and Van Brussel, 2005; Koren et al., 1999; Van Brussel, 1990). The quasi-standard of rigid, hierarchical control architectures in today's industry has been unable to cope with the new challenges successfully, since the production schedules and plans are known to become ineffective after a short time on the shop floor. Established production planning and control systems are therefore vulnerable to abrupt changes and unforeseen events in production processes and do not allow a real-time computation of sophisticated decision models (Scholz-Reiter and Freitag, 2007; Monostori et al., 2006; Valckenaers and Van Brussel, 2005). Furthermore, with the increasing size and scope of central-planning-based manufacturing execution systems, the structural complexity of these systems is growing rapidly (Monostori et al., 2006). The emphasis for future research is thereby put on the development of new organizational methods as well as new paradigms in manufacturing and automation technology (Feldmann and Rottbauer, 1999).

In order to overcome the aforementioned issues and to increase the productivity of production processes, several authors have proposed clustering of manufacturing systems into subsystems and modules. Reconfigurable Manufacturing Systems (RMS) could be reconfigured both on the overall system's structure level and on the machine level (e.g. machine hardware and control software) (Scholz-Reiter and Freitag, 2007; Koren et al., 1999). Other research approaches went even further and propagated decentralized or heterarchical manufacturing systems, where intelligent and autonomous products control the production in cooperation with intelligent resources (Scholz-Reiter and Freitag, 2007; Monostori et al., 2006; Valckenaers and Van Brussel, 2005). Among these solutions, isolated approaches as well as high level concepts such as agent-based manufacturing systems, Holonic Manufacturing Systems (HMS) and Biological Manufacturing Systems (BMS) can be found. The common aspect of these approaches is the application of Artificial Intelligence (AI) methods and techniques.

In agent-based manufacturing systems and Holonic Manufacturing Execution Systems, centralized, hierarchical control architectures are replaced by a group of loosely connected agents. However, these agents operate solely on a software basis. The term "holon" was originally coined by Koestler (1967) from the Greek word *holos* = whole and the suffix *on* as in proton or neutron, indicating a particle. As shown in Gou et al. (1994), HMS is a manufacturing system, where the key elements such as resources, products and orders are represented by distributed, autonomous, cooperative holons. Figure 20.1 shows the three types of basic holons, defined in the PROSA reference architecture (Van Brussel et al., 1998). The agents are able to communicate with each other, to reason about received messages and to learn from experiences. The intelligent agents use planning and optimization heuristics from the known methods and tools of artificial intelligence, such as genetic algorithms, neural networks or fuzzy logic.

An architecture similar to PROSA, called ADACOR was introduced by Leitao and Restivo (2006). The structure comprises a supervisor holon that enables coordination, group formation and global optimization in the decentralized control. The authors state that the system increases agility and re-configurability of a production system. Approaches using two basic building blocks can be found in Tseng et al.,

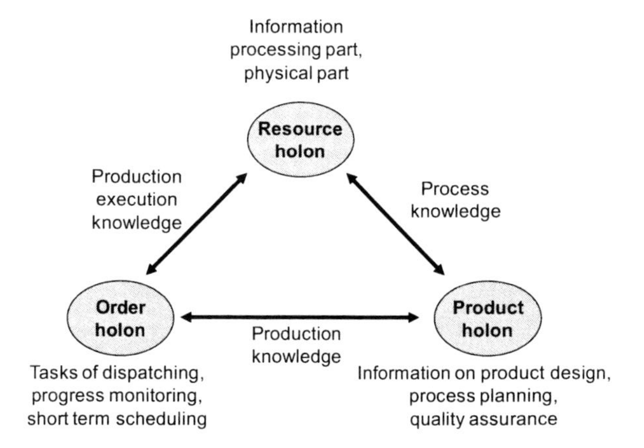

Fig. 20.1 Reference architecture for Holonic Manufacturing Systems (Van Brussel et al., 1998)

(1997) (resource and job agents) as well as in Kádár et al., (1998) and Wiendahl and Ahrens (1997) (order and machine agents).

The approach of HMS represents a good trade-off between fully hierarchical and heterarchical systems (Bongaerts et al., 2000; Valckenaers et al., 1994). A holonic architecture may hereby include temporary as well as permanent hierarchies (Valckenaers et al., 1994). The aspects of holonic architectures and their industrial applications were extensively discussed (Monostori et al., 2006; Valckenaers and Van Brussel, 2005; Valckenaers, 2001; Van Brussel et al., 1998; Teti and Kumara, 1997; Márkus et al., 1996). However, the concept of HMS mainly focuses on production scheduling and control, thus enabling a system only to react to changes in terms of re-sequencing predefined tasks.

In the field of dynamic reconfiguration of manufacturing systems, the concept of Biological Manufacturing Systems was proposed in the second half of the 1990ies (Ueda et al., 2006; Ueda et al., 2001; Ueda et al., 1997). BMS uses biologically inspired ideas such as self-growth, self-organization, adaptation, and evolution. The single elements in a BMS, such as work materials, machine tools, transporters and robots are considered as autonomous organisms. The characteristics of each component within the BMS are represented by genetic information evolving through generation (called DNA type) and individually acquired experience during the lifetime of a system's element (called BN type) (Fig. 20.2). The main focus of this concept is to deal autonomously with dynamic and unpredictable changes in internal and external production environments by changing a system configuration. In order to implement the concept of a Biological Manufacturing System, methods of evolutionary computation, self-organization and reinforcement learning were developed (Ueda et al., 2001, 2000; Ueda et al., 1997).

In addition to these high level concepts, different AI techniques have been applied to several manufacturing problems. Teti and Kumara (1997) identified Knowledge Based Systems and Expert Systems, Neural Networks, Fuzzy Logic, Multi-Agent Systems, Genetic Algorithms and Simulated Annealing as the most promising ones.

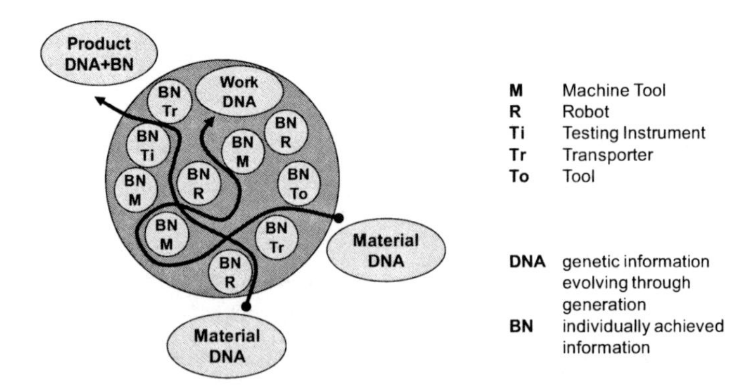

Fig. 20.2 Concept of Biological Manufacturing Systems (BMS) (Ueda et al., 1997)

In their paper the authors provide a good overview of the aforementioned applications on AI aspects in manufacturing environments.

Recapitulating, the application of AI in automated systems has proven to be a promising approach for different objectives. However, the reviewed literature discussed specific problems in isolated fields of research in the manufacturing area. Automated production systems are still not able to cope with unexpected events and situations adequately. To achieve future technical systems with such degrees of adaptability and flexibility, these systems have to be equipped with artificial cognitive capabilities. The following section describes these Cognitive Technical Systems.

20.3 Cognitive Technical Systems

The development and application of Cognitive Technical Systems (CTS) aims at an integrated approach for the planning and execution, as well as the continuous learning and adaptation of processes in technical systems under unpredictable circumstances. The realization of technical systems with cognitive capabilities that are comparable to humans is therefore a promising approach to significantly increase the flexibility and efficiency of industrial systems.

Cognition is the object of investigation of several scientific disciplines like cognitive psychology, cognitive sciences, cognitive engineering, cognitive ergonomics and cognitive systems engineering (Hollnagel and Cacciabue, 1999; Rasmussen et al., 1994). While robots have learnt to walk, navigate, communicate, divide tasks, behave socially and play robot soccer (Mataric, 1998), only a few examples for the application of artificial cognition in manufacturing exist. Therefore, several authors emphasize the need for cognitive systems to overcome deficiencies in automation (Putzer and Onken, 2003), human-robot-interaction (Hoc, 2001) and planning (Shalin, 2005).

Cognitive Technical Systems are hereby equipped with artificial sensors and actuators, are integrated and embedded into physical systems and act in a physical world. They differ from other technical systems in that they perform cognitive control and have cognitive capabilities (Beetz et al., 2007; Zaeh et al., 2007). Cognitive control comprises reflexive and habitual behavior in accordance with long-term intentions. Cognitive capabilities such as perception, reasoning, learning and planning (see Fig. 20.3) turn technical systems into ones that "know what they are doing". More specifically, a CTS is a technical system that can reason using substantial amounts of appropriately represented knowledge, learns from its experience so that it performs better tomorrow than it does today, explains itself and can be told what to do, is aware of its own capabilities, reflects on its own behavior, and responds robustly to surprise (Brachman, 2002). Technical systems being cognitive in this sense will be much easier to interact and cooperate with and they will be more robust, flexible and efficient.

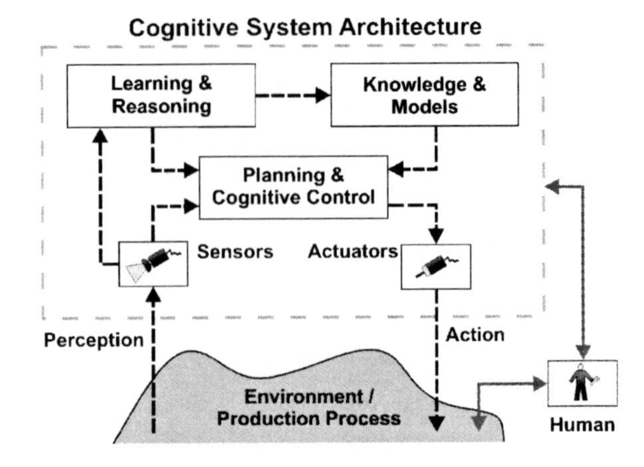

Fig. 20.3 The cognitive system architecture with the closed perception-action loop

A factory environment as a specific occurrence of a technical system forms a superior field of application for artificial cognition. Therefore, the Cognitive Factory as a specific Cognitive Technical System will be described in the following section.

20.4 The Cognitive Factory

As a factory represents a Multi-Agent System with multiple sensors and actuators it is, regarding technical reasons, an object of investigation of outstanding interest. One of the key components of the Cognitive Factory is the cognitive perception-action loop. While performing their tasks, the Cognitive Factory acquires models of production processes, of machine capabilities, work pieces and their properties, and the relevant contexts of production processes. These models that are continuously updated to adapt to changes in the environment are then used for optimized action selection and parametrization (Stulp et al., 2006). Control systems that use predictive models of actions and capabilities achieve reliable operation and high performance (Stulp and Beetz, 2005). To this end, cognitive factories must be equipped with comprehensive perception, learning, reasoning, and plan management capabilities. These mechanisms will be explained in the remainder of this article.

20.4.1 Vision and Goals

The Cognitive Factory as a factory environment consists of different manufacturing resources like production cells, robots and storages, as well as of processes for production planning and control. The paradigm "cognition" in terms of the factory denotes that machines and processes are equipped with cognitive capabilities. In technical terms this comprises sensors and actuators, in order to enable them to assess and increase their scope of operation autonomously (Zaeh et al., 2007). So-called "cognitive sensor networks" allow for real-time acquisition of production and process data and a suitable feedback to the process control. Models for knowledge and learning equip the factory with information about its capabilities and help to expand the abilities of the machines and processes. Continuous information about the status of certain actions and production steps support the improvement of machine parameters without human intervention and enable an autonomous maintenance based on prediction. Confronting the Cognitive Factory with the task of producing a new product, its elements "discuss" this task and decide whether the product can be produced with the available production resources and abilities or not, thus recognizing its own limitations. By storing previously defined strategies within a database in combination with mechanisms of information processing and retrieval, the system is able to derive further decisions. Modular, self-adapting and optimizing plan-based controllers enable the system to create and optimize production procedures during run-time.

The planning intelligence in the Cognitive Factory is therefore shifted to and subsequently embodied in the components of the real factory. Hence, unequaled levels of flexibility, reliability, adaptability and efficiency are reached by providing machine controllers, automated production resources, planning processes and whole factory environments with artificial cognitive capabilities. The Cognitive Factory combines the advantages of automated systems (e.g. low costs, high quality, high efficiency and low manufacturing times) with the flexibility, adaptability and reactivity of common human workshops (Fig. 20.4).

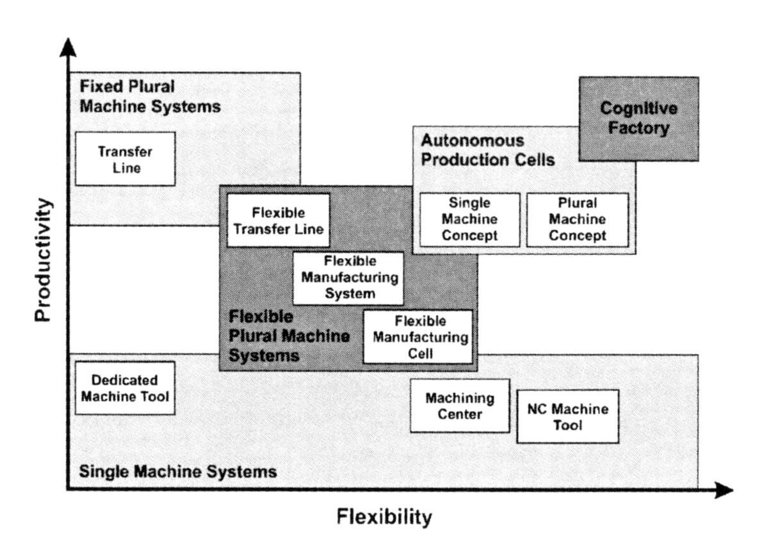

Fig. 20.4 Classification of manufacturing systems and the incorporation of the Cognitive Factory

20.4.2 Core Aspects to Achieve the Cognitive Factory

To realize the vision of the Cognitive Factory, research work needs to be intensified and integrated in different areas. The Cognitive Factory will require knowledge models, autonomous planning capabilities, perception and control mechanisms and a cognitive perception-action loop (Fig. 20.3). The components of this architecture will be described in detail in this subsection.

Perception and Control

Scholz-Reiter and Freitag (2007) recently underlined that future methods for production planning and control must provide on-line, reactive and opportunistic scheduling of multiple products simultaneously. Essential elements to achieve the shift from off-line planning systems to on-line control systems (i.e. the dynamic modification of the manufacturing process after the dispatching of the released production order, e.g. due to machine failures) are distributed control units underneath the central ERP (Enterprise Resource Planning) system. For this purpose autonomous resources and products that are able to identify and locate themselves, to perceive their environment and to communicate with other resources and products within the production system are necessary. As a basic requirement for autonomous resources, intelligent sensor technologies (e.g. RFID, camera, laser scanner) and actual state information from e.g. PLCs (Programmable Logic Controller) or RCs (Robot Controller) are essential to collect information in real-time from the production facility and its subsystems. Hence, a real-time, operation-synchronous monitoring of the production process, the system and the environment with actual sensors is possible and a suitable feedback loop of actual state information to the respective controllers and the process planning level can be realized.

An enabling technology to realize autonomous products that significantly contributes to closed-loop control systems is accomplished by Auto-ID technologies, such as RFID (Radio Frequency Identification) (Scholz-Reiter and Freitag, 2007). There has been little research related to the so-called "intelligent product" – a manufactured item that is equipped with the ability to monitor, assess and reason about its current and future state. It is therefore able to actively influence its own production, distribution, storage and retail. In the established approaches, however, only the unique product identity, as described by the Electronic Product Code (EPC), is used to locate detailed information about the specific product in a specific database (McFarlane, 2002; Wong et al., 2002). The paradigm of the "intelligent product" with the two focuses of a product centered data management (i.e. production, product and quality information) and a product centered production control (i.e. process information) will be developed. Hence, necessary information can be stored directly on the product and manufacturing information can be submitted wirelessly from the product to the machines and vice versa.

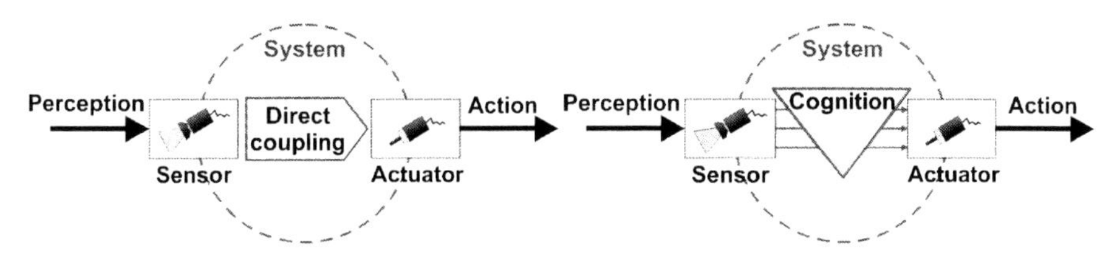

Fig. 20.5 Non-cognitive control mechanism (*left*) and cognitive control mechanism (*right*) (Strube 1998)

Intelligent sensors and sensor networks with an integrated monitoring of the manufacturing environment are barely used in production processes and are so far not considered in intelligent and adaptive production control aspects. Making data and functions accessible, to provide the technical way to perception and control, is the basic principle for the proposed decentralized and distributed intelligence. Intelligent components of a production plant need to be enabled to communicate and share information. A suitable data model and, therefore, a level of semantic information on top of the production system's data is a key enabling technology for cognition purposes to ensure a consistent data processing of the gathered sensor input, a semantic retrieval (i.e. the relevance of the information) and the feedback to the production planning and control process. Semantic information on top of raw data will provide a mechanism for machines and software algorithms to understand, interpret or change that data, which is a precondition to continuously adapting the manufacturing system towards the required state and functionality. Figure 20.5 illustrates the aforementioned aspect and shows the characteristic of a non-cognitive and a cognitive control system. The non-cognitive control mechanism is characterized by a fixed and non-adaptive coupling between the sensor and the actuator, which means that a sensor input generates always the same reaction of the actuators. In the latter case, the sensor input does not generate a direct fixed output. The action of the system is planned according to the actual situation in the production facility and the manufacturing process. Hence, the system is able to adapt its behavior and to increase its scope of operation autonomously.

Embedding Knowledge in the Cognitive Factory

In manufacturing environments, the use of ontologies and explicit semantics allow many unique capabilities, as discussed by Shea et al. (2008). First, ontologies enable logical reasoning to infer sufficient knowledge on the classification of processes that machines are capable of carrying out. Following from ontologies, in the context of the semantic web, a service is a web site that does not only provide static information but also has the possibility to affect actions and change the state of the world (Martin et al., 2005). Second, through knowledge-based planning methods, services can be composed autonomously, using AI planning algorithms, to carry out production sequencing and synchronizing of service execution (Lastra and Delamer, 2006).

As an advancement of the holonic and agent-based architectures developed in the past, the concept of semantic web services is used to build semantic manufacturing services. The devices on the shop floor are able to publish how they exchange information and thus collaborate autonomously. A process planning agent under development will carry out web service composition planning. For the ontology language OWL-S, where a service profile tells what the service does, a service model tells how the service works and a service grounding specifies the way a service can be accessed, i.e. find a sequence of service profiles, considering their groundings, to dynamically create high-level process plans. A current state representation of the shop floor is provided by adding, removing, adapting and learning atomic service descriptions as needed using the hierarchical structure and service-based models of processes. Services, both in relation to the hardware and computational processes, i.e. planning and reasoning, can be physically distributed across machines yet accessible in a unified way.

With this approach, the current state of the Cognitive Factory is directly reflected by the service model and the plan library. If new machines and tools are integrated in the Cognitive Factory, the system can assess their machining capabilities from the machine controller, kinematics and tool descriptions. Through this, the system is aware of its capabilities at any point and can communicate these capabilities, both among machines and between machines and people, to respond to changes in the system and new product requests to the system in a highly flexible manner.

Machining Planning

Computer Aided Process Planning (CAPP) provides the bridge between design and fabrication by assisting humans to make the required translation between the design view of a product and the manufacturing view. This process involves the systematic determination of the detailed methods by which parts can be manufactured, from raw material to a finished product, including, among other steps, selection of machining operations and machine tools as well as tool path planning and generation of NC part programs. CAPP commonly uses feature models, where a feature is a generic sub-shape of a part, with associated attributes useful for reasoning about it. While research has been underway in CAPP for over 25 years, significant research challenges remain for realizing full automation due to the highly knowledge-intensive nature of the task (Corney et al., 2005). A specific example is the internet-based CAD/CAM system Cybercut, which starts through definition of a 2.5-D part in a web-based CAD system that is then passed to three levels of automated planners, a macro planner, micro planner and tool path planner that produces code for direct execution on a 3-axis CNC milling machine (Ahn et al., 2001). Further, beyond pure automation, the majority of CAPP approaches focus on off-line planning rather than benefiting from feedback and experience gained from the shop floor.

A new approach to CAPP is under development that will take advantage of the power of shape grammars as well as move machining planning to the shop-floor (Shea et al., 2008). Shape grammars (Stiny, 1980) are a powerful method for representing valid machining process steps and generation of machining plans. Early theoretical work on a shape grammar that describes a formal language for a single lathe machine has illustrated the possibility, but this was never brought down to the machine level and tested (Brown et al., 1995). Through definition of finite sets of shapes, labels and shape production rules, labeled shapes are generated from an initial labeled shape through iterative application of rules. They provide for a concise and flexible representation of a set of valid shapes, e.g. all shapes producible by a set of machines, along with the process to generate them. Machining a part involves the removal of volume from a piece of stock material such that the geometry of the desired part is produced. The process is described as a sequence of shape transformations, each representing steps in the machining process, providing a mapping between machine process steps and their resulting removed volumes. An additional advantage of a grammatical approach is the use of the grammar for validation of parts to determine, if it is within the set of valid producible shapes. Further, in particular for shape grammars, there is strong potential for producing and recognizing emergent shapes, or learned features, that in turn result in representation of discovered machine capabilities.

For machining planning, the grammars are applied to generate the process steps required to match a desired removal volume, resulting in a complete machining plan with all necessary instructions for execution. The plans are abstracted to create a plan library that can also be used, on-demand, to generate portions of machining plans. This feature can be used for advanced higher level planning and control of the machining devices in the Cognitive Factory. The set of valid manufacturing features is used to create up-to-date descriptions of machines and tools as service models to make known and learned capabilities accessible to the system. This supports high level reasoning and purpose reaction to unforeseen changes in the system.

Assembly Planning

Traditionally, assembly planning, comprising the generation of assembly plans, resource scheduling and system planning, is performed by experienced production engineers (Zha & Lim, 2000). In order to overcome the increasing complexity in assembly operations, several authors have recognized the increasing need for the application of AI techniques in assembly and manufacturing planning (Zha and Lim, 2000; Teti and Kumara, 1997; Cao and Sanderson, 1995). Numerous research approaches address the problem of generating assembly plans, using genetic algorithms (Pan et al., 2006; Dini et al., 1999; Bonneville et al., 1995), graph-based solutions (Yuan and Gu, 1999; Laperrière and ElMaraghy, 1996; Homem de Mello and Sanderson, 1991) or Petri-Nets (Zha and Lim, 2000; Moore and Gupta, 1996). As these approaches for the generation of assembly plans are isolated from underlying production systems, further planning steps are required for producing a product. This comprises the matching of assembly operations to resources and workstations, identifying necessary product and material transports, defining material provision and providing necessary control code.

Future application of AI methods in assembly planning will not be restricted to isolated planning fields like assembly plan generation, control program generation or scheduling, but will focus on integrated approaches to enable the autonomous manufacturing of new products. For this purpose, new methods for the integrated planning and execution of assembly operations need to be developed, that allow production systems to react to unpredicted scenarios and events autonomously. Cognitive planning methods will have to combine information on products as well as on resources of a production system with sensor input data to perform production operations autonomously.

To reach this goal, consistent assembly planning methods on the production system level as well as on the resource level are required. On the system level, this comprises the allocation of assembly operations to resources, the definition and control of necessary material and product transports, and the design of appropriate material provision positions. Furthermore, planning on the system level will have to provide plans, which are directly executable on the workstation level, according to its current state derived from sensor information and user input. Therefore, assembly plans need to be composed of atomic operations that may be executed by the system dynamically in an un-predetermined manner. In addition, resources need information on what operations they are able to perform.

This requires modeling products as a set of operations with given sequence constraints, which are necessary for its production. Matching these product specifications with machine capabilities will allow an autonomous assignment of production operations to resources. On the resource level, not only the necessary assembly operations and sequence constraints are required for an autonomous control, but also local information on both the product to be assembled (e.g. geometries, gripping areas) and the resource (e.g. gripper, material provision positions, buffer positions) need to be combined. This data has to be provided by the assembly planning on the system level. AI methods have to be applied for the execution of assembly operations on the resource level. Genetic algorithms and neuronal networks are strong candidates for this task. Allowing workstations to re-plan their actions according to current sensor and user input will help to make production systems more flexible and changeable.

Cognitive Perception-Action Loop

The Cognitive Perception-Action Loop in the Cognitive Factory is characterized by a multitude of inter-weaved processes. While controlling and monitoring the current production process, the system improves production plan schemes based on experience as well as by acquiring and continually adapting internal system models. Thus, with experience the Cognitive Factory builds production plan schemes tailored to specific production circumstances and learns when to use which production plan. In addition, learned predictive models enable the system to anticipate execution failures and to parametrize sub-processes in order to optimize the overall performance. Hence, the Cognitive Factory is capable of improving its performance by tailoring its operation based on experience and it is capable of automatically adapting its control processes to changes in the circumstances.

Apart from domain and machine setup specific adaptation, the system is able to learn general capabilities such as "how to plan" and even plan "how to learn". An example of learning "how to plan" is the formation of planning knowledge through the analysis of experience collected during plan execution and simulation. Starting from a detailed simulation, execution traces are analyzed for failures like idle times of critical resources. With experience, the planner improves its predictive capabilities and can then correct such inefficiencies without time-intensive simulation.

Planning "how to learn" on the other hand comprises decisions on the sequence of learning activities as well as sensible control of the system complexity to which the learning module is exposed in each stage. Babies, for instance, can only learn complex capabilities efficiently and effectively because they decompose the overall learning task into a carefully laid out plan of learning tasks.

A similar concept governs the Cognitive Technical System in the Cognitive Factory. Starting from a simplified view on the production system, more and more capabilities are learned with an increasingly complex production system model. In early stages, explorative actions and sensing operations are introduced to reduce the model uncertainty, while the focus is set on plan space exploration and plan optimization in later stages. The learned model of the production system is essential for the efficiency and impact of the plan optimization, since the projection of plan variants strongly depends on it. Its conformance to the real system must therefore always be monitored and maintained. Initially general default plans, which might be sub-optimal, are used as a starting point. After collecting enough experience and building informative predictive models from experience, the control system can use the model in order to tailor and optimize the system operation. Otherwise, plan variants are produced based on more general planning knowledge that is not yet adapted to the specific system setup and domain. These plan variants run through a "predict-criticize-revise" loop for further improvement (Fig. 20.6) (Mueller et al., 2007). In the prediction phase, plans are projected through simulation, logical reasoning, or the application of probabilistic models. In the criticize phase, the intended behavior and effects are compared in order to detect discrepancies that might hint at ways of improving the process plans. Finally, in the revision phase the plans are modified through predefined plan transformation rules that are indexed by the shortcomings recognized in the criticize phase. Whenever a new plan variant promises higher performance, the executed plan is instantly exchanged during run-time. Automatic or manual reconfigurations of the set-up and tools are modeled as operations in the planning language, and are proposed by the planner given that the change offers a significant reduction of the current job's processing time.

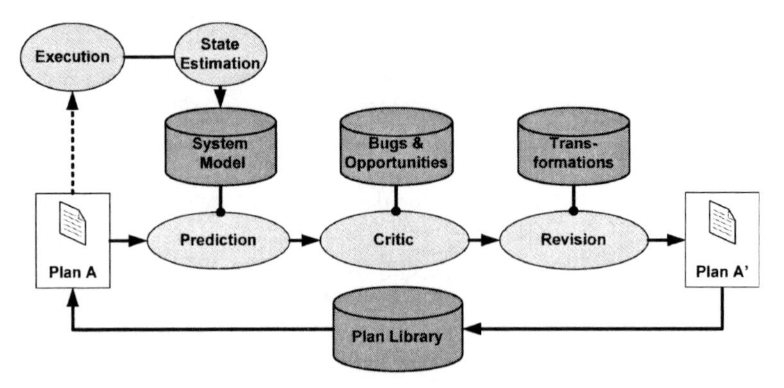

Fig. 20.6 Predict-criticize-revise loop

20.5 Summary and Outlook

Increasing market turbulences and customer demands have lead to the development of changeable and reconfigurable manufacturing systems. However, the highest degree of changeability is still reached by human workshops with skilled workers and their cognitive capabilities, which enable them to react to changes, perceive their environment, plan their next actions and know what they are doing. To reach production systems with a similar level of autonomy and changeability, new paths in production planning and automation need to be struck. Therefore, a new paradigm to reach the next level of changeable production environments, the Cognitive Factory, was proposed.

The Cognitive Factory, as a specific form of Cognitive Technical Systems (CTS), will hereby require methods to enable factory environments to react flexibly and autonomously to changes, similar to human operated facilities. To reach this goal, a cognitive architecture for production systems was introduced. This architecture comprises knowledge models, methods for perception and control, methods for planning, and a cognitive perception-action loop.

Future research areas in production engineering will focus strongly on increasing the autonomy of production systems to enhance their changeability. For this purpose, techniques of Artificial Intelligence (AI) and Cognitive Engineering (CE) need to be applied and improved for an integrated approach. Therefore, high interdisciplinary research efforts can be expected in this area in the next years.

Acknowledgements The authors would like to thank all researchers that are contributing to the concept of the Cognitive Factory. Furthermore, we would like to acknowledge the funding support of the Deutsche Forschungsgemeinschaft (DFG) within the scope of the German cluster of excellence "Cognition for Technical Systems" (CoTeSys) (see also www.cotesys.org).

References

Ahn S., Sundararajan V., Smith C., Kannan B., D'Souza R., Sun G., Mohole A., Wright P., Kim J., McMains S., Smith J., Sequin C., 2001, CyberCut: An Internet Based CAD/CAM System. Journal of Computing and Information Science in Engineering 1/1:52–9

Beetz M., Buss M. and Wollherr D., 2007, Cognitive Technical Systems – What Is the Role of Artificial Intelligence? In: Proceedings of the 30th German Conference on Artificial Intelligence (KI-2007)

Bongaerts L., Monostori L., McFarlane D., Kadar B., 2000, Hierarchy in distributed shop floor control. Computers in Industry 43/2:123–37

Bonneville F., Perrard C. and Henrioud J.M., 1995, A genetic algorithm to generate and evaluate assembly plans, In: Proceedings of the Symposium on Emerging Technologies and Factory Automation

Brachman R., 2002, Systems That Know What They're Doing. IEE Intelligent Systems 17/6:67–71

Brown K.N., McMahon C.A., Sims Williams J.H., 1995, Features, aka the semantics of a formal language of manufacturing. Research in Engineering Design 7/3:151–72

Cao T., Sanderson A.C., 1995, Task sequence planning using fuzzy Petri nets. Systems, Man and Cybernetics, IEEE Transactions on 25/5:755–68

Corney J., Hayes C., Sundararajan V., Wright P., 2005, The CAD/CAM Interface: A 25-Year Retrospective. Journal of Computing and Information Science in Engineering 5/3:188–97

Dini G., Failli F., Lazzerini B., Marcelloni F., 1999, Generation of Optimized Assembly Sequences Using Genetic Algorithms. Annals of the CIRP 48/1:17–20

Feldmann K. and Rottbauer H., 1999, Electronically Network Assembly Systems for Global Manufacturing, In: Proceedings of the 15th International Conference on Computer-Aided Production Engineering

Gou L., Hasegawa T., Luh P.B., Tamura S. and Oblak J.M., 1994, Holonic Planning and Scheduling for a Robotic Assembly Testbed, In: Proceedings of the Fourth International Conference on Computer Integrated Manufacturing and Automation Technology

Hoc J.-M., 2001, Towards a cognitive approach to human-machine cooperation in dynamic situations. International Journal of Human-Computer Studies 54/4:509–40

Hollnagel E. and Cacciabue P.C., 1999, Cognition, Technology & Work: An Introduction, Cognition, Technology & Work, No. 1, pp 1–6

Homem de Mello L.S., Sanderson A.C., 1991, A correct and complete algorithm for the generation of mechanical assembly sequences. Robotics and Automation, IEEE Transactions on 7/2:228–40

Kádár B., Monostori L., Szelke E., 1998, An object-oriented framework for developing distributed manufacturing architectures. Journal of Intelligent Manufacturing 9/2:173–9

Koestler A., 1967, The ghost in the machine. Hutchinson, London

Koren Y., Heisel U., Jovane F., Moriwaki T., Pritschow G., Ulsoy G., Van Brussel H., 1999, Reconfigurable Manufacturing Systems. Annals of the CIRP 48/2:527–40

Laperrière L., ElMaraghy H.A., 1996, GAPP: A Generative Assembly Process Planner. Journal of Manufacturing Systems 15/4:282–93

Lastra J.L.M., Delamer M., 2006, Semantic web services in factory automation: fundamental insights and research roadmap. Industrial Informatics, IEEE Transactions on 2/1:1–11

Leitao P., Restivo F., 2006, ADACOR: A holonic architecture for agile and adaptive manufacturing control. Computers in Industry 57/2:121–30

Márkus A., Kis Váncza T., Monostori L., 1996, A Market Approach to Holonic Manufacturing. Annals of the CIRP 45/1:433–6

Martin D., Paolucci M., McIrlaith S., Burstein M., McDermott D., McGuiness D., Parsia B., Payne T., Sabou M., Solanski M., Srinivasan N., Sycara K., 2005, Bringing Semantics to Web Services: The OWL-S Approach. In: Cordoso J., Sheth A. (eds) Semantic Web Services and Web Process Composition. Springer, Berlin

Mataric M.J., 1998, Behavior-based robotics as tool for synthesis of artificial behavior and analysis of natural behavior. Trends in Cognitive Sciences 2/3:82–7

McFarlane D.C., 2002, Auto ID Based Control Systems – An Overview, In: Proceedings of the 2002 IEEE International Conference on Systems, Man and Cybernetics

Monostori L., Váncza J., Kumara S.R.T., 2006, Agent-Based Systems for Manufacturing. Annals of the CIRP 55/2:697–719

Moore K.E., Gupta S.M., 1996, Petri net models of flexible and automated manufacturing systems: a survey. International Journal of Production Research 34/11:3001

Mueller A., Kirsch A. and Beetz M., 2007, Transformational Planning for Everyday Activity, In: Proceedings of the 17th International Conference on Automated Planning and Scheduling (ICAPS'07)

Pan C., Smith S.S., Smith G.C., 2006, Automatic assembly sequence planning from STEP CAD files. International Journal of Computer Integrated Manufacturing 19/8:775–83

Putzer H., Onken R., 2003, COSA – A generic cognitive system architecture based on a cognitive model of human behavior. Cognition, Technology & Work 5/2:140–51

Rasmussen J., Pejtersen A.M., Goodstein L.P., 1994, Cognitive Systems Engineering. Wiley Series in Systems Engineering. Wiley, New York

Scholz-Reiter B., Freitag M., 2007, Autonomous Processes in Assembly Systems. Annals of the CIRP 56/2:712–29

Shalin V.L., 2005, The roles of humans and computers in distributed planning for dynamic domains. Cognition, Technology & Work 7/3:198–211

Shea K., Engelhard M., Ertelt C. and Hoisl F., 2008, Bridging the gap between design and manufacturing through cognitive technical systems, In: Proceedings of the 7th International Symposium on Tools and Methods of Competitive Engineering

Stiny G., 1980, Introduction to shape and shape grammars. Environment and Planning B 7/3:343–51

Strube G., 1998, Modelling Motivation and Action Control in Cognitive Systems. In: Schmid U., Krems J., Wysocki F. (eds) Mind Modelling: A Cognitive Science Approach to Reasoning, Learning and Discovery. Pabst Science Publishers, Berlin, pp 89–108

Stulp F. and Beetz M., 2005, Optimized Execution of Action Chains Using Learned Performance Models of Abstract Actions, In: Proceedings of the Nineteenth International Joint Conference on Artificial Intelligence (IJCAI)

Stulp F., Isik M. and Beetz M., 2006, Implicit Coordination in Robotic Teams using Learned Prediction Models, In: Proceedings of the IEEE International Conference on Robotics and Automation (ICRA)

Teti R., Kumara S.R.T., 1997, Intelligent Computing Methods for Manufacturing Systems. Annals of the CIRP 46/2:629–52

Tseng M.M., Lei M., Su C., 1997, A Collaborative Control System for Mass Customization Manufacturing. Annals of the CIRP 46/1:373–6

Ueda K., Hatono I., Fujii N., Vaario J., 2000, Reinforcement Learning Approaches to Biological Manufacturing Systems. Annals of the CIRP 49/1:343–6

Ueda K., Hatono I., Fujii N., Vaario J., 2001, Line-less Production System Using Self-Organization: A Case Study for BMS. Annals of the CIRP 50/1:319–22

Ueda K., Kito T., Fujii N., 2006, Modeling Biological Manufacturing Systems with Bounded-Rational Agents. Annals of the CIRP 55/1:469–72

Ueda K., Vaario J., Ohkura K., 1997, Modeling of Biological Manufacturing Systems for Dynamic Reconfiguration. Annals of the CIRP 46/1:343–6

Valckenaers P., 2001, Editorial of the Special Issue on Holonic Manufacturing Systems. Computers in Industry 43/3:233–4

Valckenaers P., Bonneville F., Van Brussel H., Bongaerts L. and Wyns J., 1994, Results of the holonic control system benchmark at KU Leuven, In: Proceedings of the Fourth International Conference on Computer Integrated Manufacturing and Automation Technology

Valckenaers P., Van Brussel H., 2005, Holonic Manufacturing Execution Systems. Annals of the CIRP 54/1:427–30

Van Brussel H., 1990, Planning and Scheduling of Assembly Systems. Annals of the CIRP 39/2:637–44

Van Brussel H., Wyns J., Valckenaers P., Bongaerts L., Peeters P., 1998, Reference architecture for holonic manufacturing systems: PROSA. Computers in Industry 37/3:255–74

Wiendahl H.-P., Ahrens V., 1997, Agent-Based Control of Self-Organized Production Systems. Annals of the CIRP 46/1:365–8

Wiendahl H.-P., ElMaraghy H.A., Nyhuis P., Zäh M.F., Wiendahl H.-H., Duffie N. and Kolakowski M., 2007, Changeable Manufacturing: Classification, Design, Operation, Annals of the CIRP, 56/2

Wong C.Y., McFarlane D., Ahmad Zaharudin A. and Agarwal V., 2002, The Intelligent Product Driven Supply Chain, In: Proceedings of 2002 IEEE International Conference on Systems Man and Cybernetics

Yuan X. and Gu Y. 1999, An integration of robot programming and sequence planning, In: Proceedings of the 1999 IEEE International Conference on Robotics and Automation

Zaeh M.F., Lau C., Wiesbeck M., Ostgathe M. and Vogl W., 2007, Towards the Cognitive Factory, In: Proceedings of the 2nd International Conference on Changeable, Agile, Reconfigurable and Virtual Production (CARV 2007)

Zha X.F., Lim S.Y.E., 2000, Assembly/disassembly task planning and simulation using expert Petri nets. International Journal of Production Research 38/15:3639–76

Chapter 21
Migration Manufacturing – A New Concept for Automotive Body Production

T.P. Meichsner[1]

Abstract The increasing individualization in the automotive industry characterized by so called 'model offensives' together with pressures due to high costs, shorter product life cycle, rising diversity of models and quantity volatility, demand new concepts in the production of vehicles. This is particularly true for 'niche cars'. These difficulties are a motivation to develop a new concept for the body-in-white production, which is highly flexible with respect to models and variants that will be denoted as the "Migration Concept". The concept can be applied to other product categories and industries and can therefore be named as Migration Manufacturing. Its essential characteristics are the manufacturing of different body work models and their variants on one production line, as well as the ability to extend the basic layout along the "migration path". The specific changeability of a body-in-white production with regards to the integration and removal of new models and versions is called "Migration". The production volumes proportion of the specific models can vary in great range and the investments are flexible according to the required volume. The paper describes the basic concept, its components and production phases, as well as its comparison with the conventional transfer lines.

Keywords Migration, Changeability, Body-in-White Production, Real Options, Automotive Manufacture, Niche Vehicles

21.1 Initial Situation

Increasing customization in the car manufacturing sector, combined with high cost pressures and shorter market cycles, have led to a wider range of models and variants, along with unit output rates that are subject to higher volatility. This becomes particularly evident in the fast growing market for the so called "niche" vehicles

[1] Wilhelm Karmann GmbH, Osnabrück, Germany (Specialty Cars)

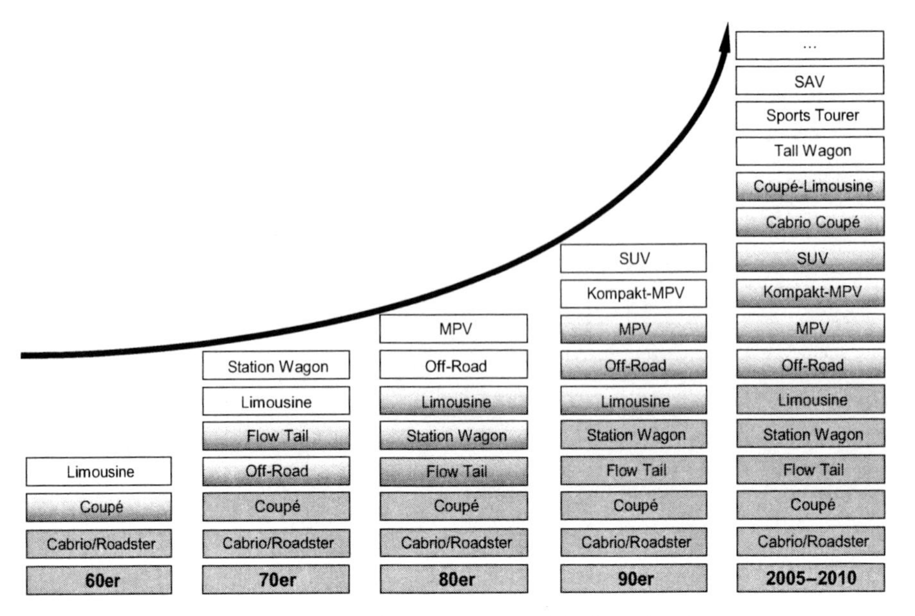

Fig. 21.1 Market development in the niche vehicle sector (Polk Study, 2006)

(Fig. 21.1). These vehicles are targeted at specific customer groups with clearly identifiable lifestyles, and are subject to short-term, unpredictable product life-cycles and widely-varying unit output rates. Thus, there is an unavoidable need to combine a process of permanent change in production with greatly increased flexibility.

This chapter examines the manufacture of such niche vehicles from the point of view of the Contract Manufacturer, a supplier of complete vehicles to OEM's (Original Equipment Manufacturer). Problems arise both when coping with variants and when dealing with models from different manufacturers at the same time.

The vehicle production process can generally be broken down into the main stages of body manufacturing, painting and assembly. The body shop is supplied with pressed parts usually from the own press shop or from external suppliers. At the assembly line the painted car body is completed with the power train, motor, electronics, interior etc.

Figure 21.2 gives an insight into the four main stages of a body shop. Starting with the sub-components of the underbody, the side frames are added in the second stage. Adding the roof stabilizes the body in stage 3, whereas in stage 4 the fenders and movable parts like doors and hoods complete the car body to be ready for painting. Variants of the body pertain typically the side frames, fenders, doors and hoods. Altogether a car body consists of 400 to 500 single press parts, fixed with 4000 to 5000 welding spots, depending on the size and stiffness of the car. The body line encompasses 30 to 40 stations with 300 to 400 welding guns, depending on the length of the cycle time and the complexity of the car body.

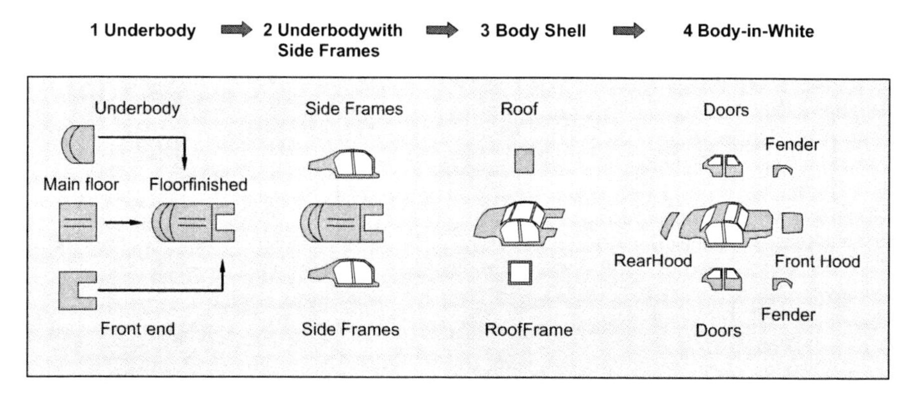

Fig. 21.2 Manufacturing stages of a vehicle body (sedan car example)

The area with the lowest level of flexibility with regards to model and variant, the highest level of value-added and the largest investment for a new product, is the body shop. Therefore, this contribution focuses on the ability of body production, as a major element in competitive production, to change its properties in order to accommodate the needs of low volume and niche vehicles.

The sharp fluctuations in demand for niche vehicles mentioned above often lead to unused body-shop capacity averaging 50% in Europe, which represents over-investment from a financial point of view (Wemhöner, 2006). OEM suppliers attempt to compensate for this effect by producing greater numbers of variants in the form of roadsters, SUVs (Sport Utility Vehicles), convertibles, off-road vehicles and so on, along with variable car body configurations.

The contract manufacturers, with their "single-item" approach to production, lack the ability to meet this need for various vehicle models. To date, this has made it difficult to realize any significant potential for reductions in costs and investment. On the other hand, once OEM suppliers have made the basic investment in flexible body production, little spare investment capacity remains for further vehicles based on the same platform or modular production method.

This set of problems is what has created the need for a new structural concept for car body production. It is based on high flexibility with respect to models and variants in the niche-vehicle sector beyond known borders; a concept which will be called "Migration Manufacturing"*. In addition, this approach makes it possible to choose between manual, semi-automated and highly-automated production. Plant facilities can thus be partly or fully extended, reconfigured, relocated or downscaled to cope with the corresponding model life-cycle.

* Migration Manufacturing is a registered trademark of the author

21.2 Development of the Basic Concept

The newly-coined term "Migratability" refers to the specific ability for changes in vehicle manufacturing, as explained above. In this connection the "migration path" describes the technological route along which the production facilities can move in relation to changes in variants and models as time passes, with the unit output rising and falling accordingly.

Our concept started with an examination of the aspects of changeability (retooling and reconfiguration capacity, flexibility, agility and ability to adjust) that are documented in CIRP keynote papers (Koren, 1999) and (Wiendahl, 2007) as well as in the extensive work at the Institute for Production Systems and Logistics (IFA) at the Leibniz University in Hanover Germany (Wiendahl, 2002, 2000 and 1999).

Vehicle body production involves first the application of flexibility types in terms of the body types, volume (output in units), body variants, processes, components and re-usability, shown in Fig. 21.3 (right). Secondly, the "changeability enablers", likewise developed by IFA (universality, mobility, scalability, modularity and compatibility) could then be applied to the body shop objects. The currently well-known body-production concepts, platform strategies and the various fabrication and joining technologies were analyzed, as was the use of various materials of the car bodies, in order to ensure the practical application of the Migration Concept.

The main characteristics of the Migration Concept encompass the following objectives: incremental investments, short reaction times and robustness. The development of the basic concept for body production was centered on a system model with three levels (Fig. 21.3 – middle).

The first level, which encompasses manufacturing principles, describes the well-known single-line principle. Second, it allows the production of a model family with

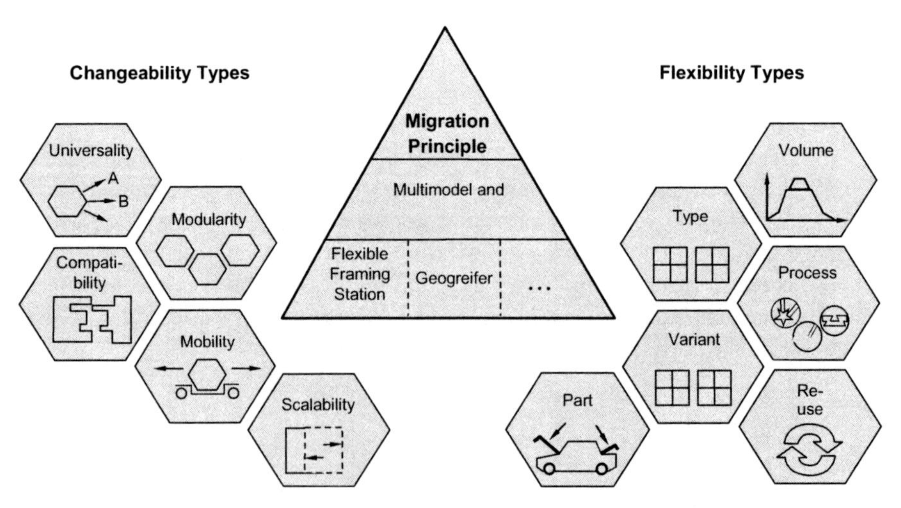

Fig. 21.3 Requirements and objectives of the Migration Concept in vehicle body production

a kind of flexibility that allows bodies to be produced for different variants based on a single platform. The new third principle of vehicle body fabrication involves the implementation of the migration principle, which allows the production of various different models and their variants from different OEMs.

One special characteristic of the Migration Concept, in contrast to other manufacturing concepts, is that there are virtually no limits on the modifications that can be made to vehicle technology (Meichsner, 2006a). These developments can be seen in motive technology (e.g. hybrid drives), in lightweight body manufacture (using such materials as aluminum, carbon and steel) or at the final assembly stage (e.g. with large assembly modules) (Meichsner, 2003a, 2003b and 2002). They are technically and commercially compatible with the Migration Concept.

The unit output characteristic is the most important defining parameter of the Migration Concept. This involves establishing the number of models, the permitted range of their unit output fluctuations and degree of overlap with regards to unit output curves (Meichsner, 2006c). The various manufacturing principles are shown in Fig. 21.4, along with their main characteristics.

The second level of concept development involves the creation of the structural concept that describes the configuration principle of the secondary modules on which the layout of a vehicle body is based. The requirements of this level include: multiple-model compatibility, incremental investment, product-cost reductions and a shortening of the production and model retooling processes, along with an increased range of processes for body production (e.g. multi-joining techniques).

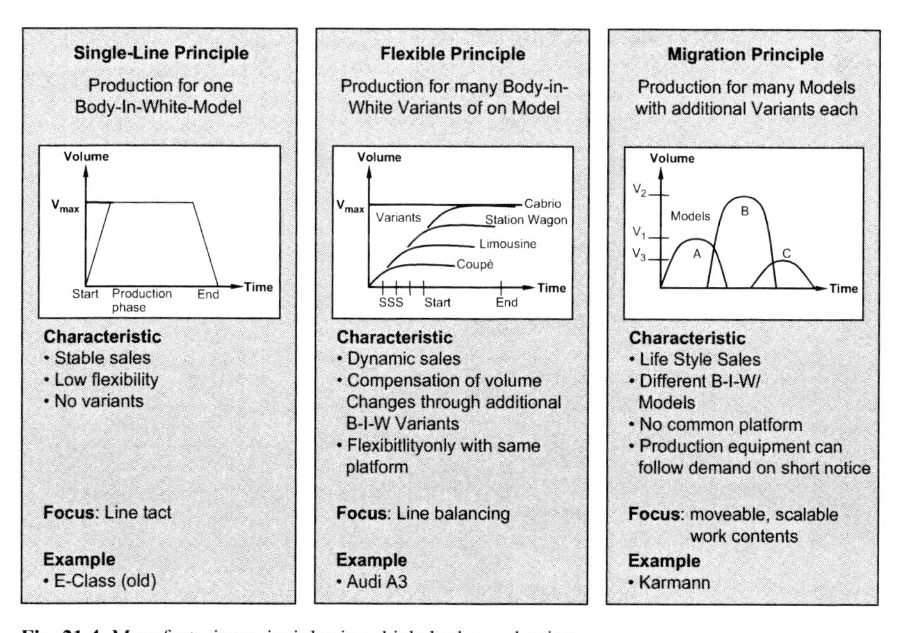

Fig. 21.4 Manufacturing principles in vehicle body production

This involved the creation and evaluation of various structural concepts, along with their corresponding layouts. The "Meandering" and "Tetris" concepts emerged as the preferred candidates.

The meandering structure consists of a basic level during the initial usage phase. This allows the body and its modules to be fully finished, as it includes all the jigs and fixtures that determine geometry. If unit output is increased, the main line is expanded by the addition of a first and, if required, second re-spot welding line. As these re-spot welding lines are reminiscent of bends in a river, the concept is referred to as "meandering". As with a fast-flowing river, a vehicle body can only run through the system in one direction (Fig. 21.5).

The main feature of the Tetris structure is the division into manufacturing cells (e.g. geometry determining, re-spot welding and special cells) and conveyor sections. The term "Tetris" is used because of the similarity with the well-known computer game of the same name (Fig. 21.6). In contrast to the meandering concept, the vehicle body is free to flow in any direction, which needs a more complicated control of the system.

At level three of the Migration Concept, the subsystems that make up the structural concepts are defined and classified into categories (Fig. 21.7). These include: component handling, clamping and fixturing systems (FCS), joining techniques, forming and processing, transport and storage systems, the control systems and the peripheral elements.

The interaction of components is described by means of the already mentioned "migration path". This involves adapting systems to the unit output factors of each individual model at the various phases of production, while applying different technical solutions and keeping control of the correspondingly adapted, phased investments.

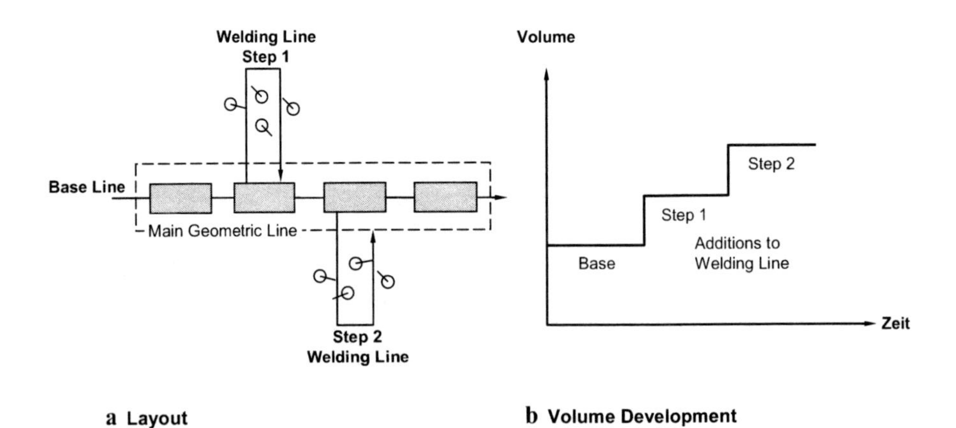

Fig. 21.5a,b Meandering layout concept

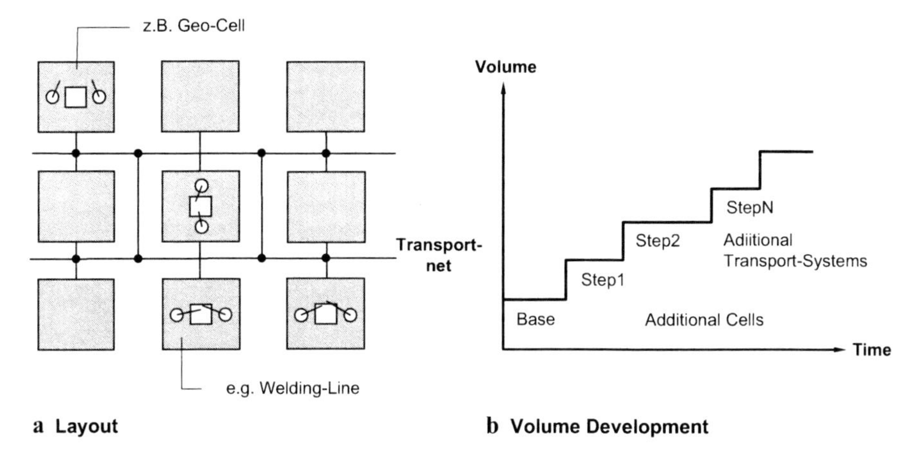

Fig. 21.6a,b Tetris layout concept [IPH]

Handling	Location and Clamp Systems	Joining Technology	Forming and Processing	Transport and Buffer Systems	Control Systems	Peripheral Systems
Examples • Grippers - Geometrical gripper - Multi-functional gripper - Sorting gripper - Modular gripper • Robot used as a manipulator - stationary - mobile	• Clamp systems (LCS) - flexible LCS - adaptive LCS - modular LCS - dynamic LCS	• Welding - Spot welding - MIG welding - MAG welding - WIG welding • Laser welding • Riveting • Bolting • Adhesive	• Soft touch form and pierce • Hemming - Roller hemming - Table top hemmer • Milling • Drilling • Piercing • Parting	• Portal systems • Mobile Systems • Conveyor systems • Stationary buffers • Combined Buffer / Conveyor systems • Loading station for material • Unloading station - manual - automatic	• Networks - LAN - WAN • Control • Regulate • Measure	• Safety and protective systems • Media • Identification systems (RFID)
Type-Specifity low	high	low -medium	high	low	medium	low
Re-usability good	medium	good	medium	medium	good	good

Fig. 21.7 Subsystems relating to vehicle body production, with examples

It is possible to show that based on its special layout structure, the migration principle permits the expansion, model-based adaptability and capacity reduction of plant facilities with a high degree of re-integration for the individual equipment components used in the next generation system. Figure 21.8 shows example scenarios for "Vehicle A" with a shrinking unit output rate and reduced production, and for "Vehicle B" where unit output is rising and the production facilities are being extended in phases. By applying investment in accordance with the required unit production output at each point along the migration path, the economic viability of production is maintained with respect to high fixed costs and variable production costs.

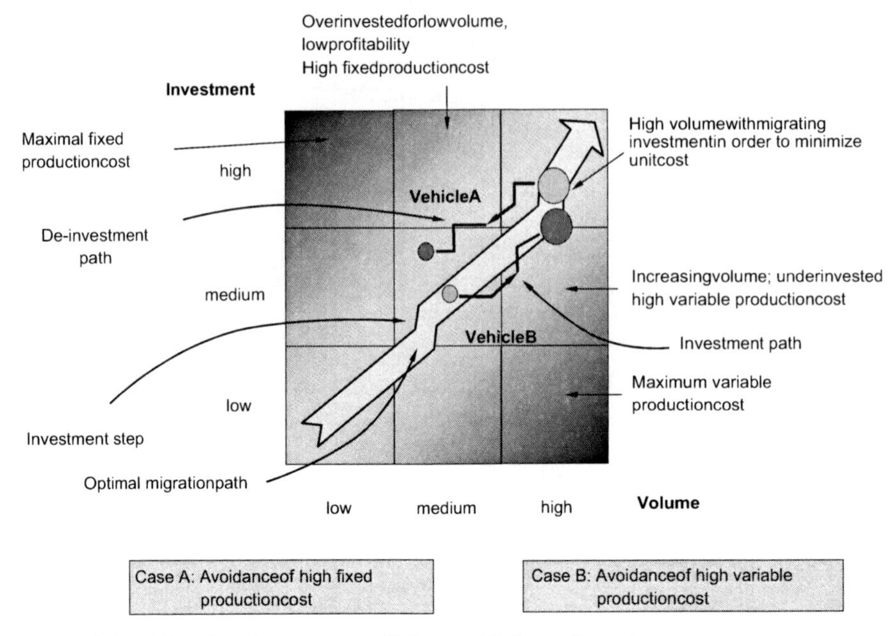

Fig. 21.8 The Migration Concept: cost-efficiency with fluctuating unit output

The next phase in the application of the Migration Concept involves the definition of further requirements regarding plant operation. These are: faster setup, retooling and dismantling of plant systems, type-independent subsystems, high levels of re-usability, simple expansion capacity and the ability to test plant equipment before production begins. The most important operating subsystems (e.g. robots, welding tongs) were defined against the background of these requirements.

21.3 Operating Phases of the Migration Concept

In order to closely define the migration principle, the layout structures for four scenarios with different unit output patterns were planned from initial start-up to the integration of a third model. The required control strategies for material flow were also drawn up for taking different models of body through the system. Evaluation of the layout structures revealed a slight advantage to the meandering structure.

The Migration Concept calls for new skills of the operating staff. This makes it necessary to create a specific human resources concept, in order to provide a reinforced basis for expertise and skills on the part of technical staff and for the use, for example, of a specialist in migration techniques.

The conclusion of the setup phase resulted in guidelines that can be applied to the creation of a migration compatible infrastructure.

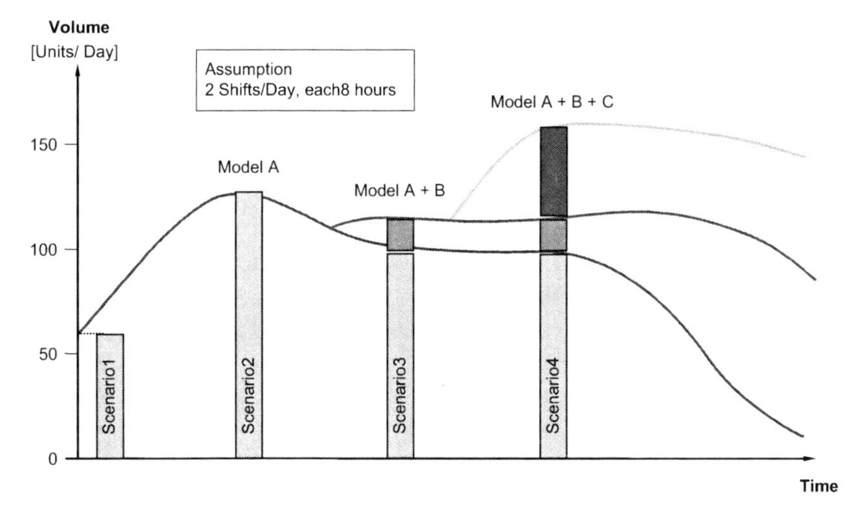

Fig. 21.9 Unit output scenarios for three models from different OEMs

In order to ensure correct functioning, all concepts are tested with structural layout modeling and a simulation involving three car models with various unit output rates and time scales. Four scenarios from each were then chosen for explanation and are shown in Fig. 21.9.

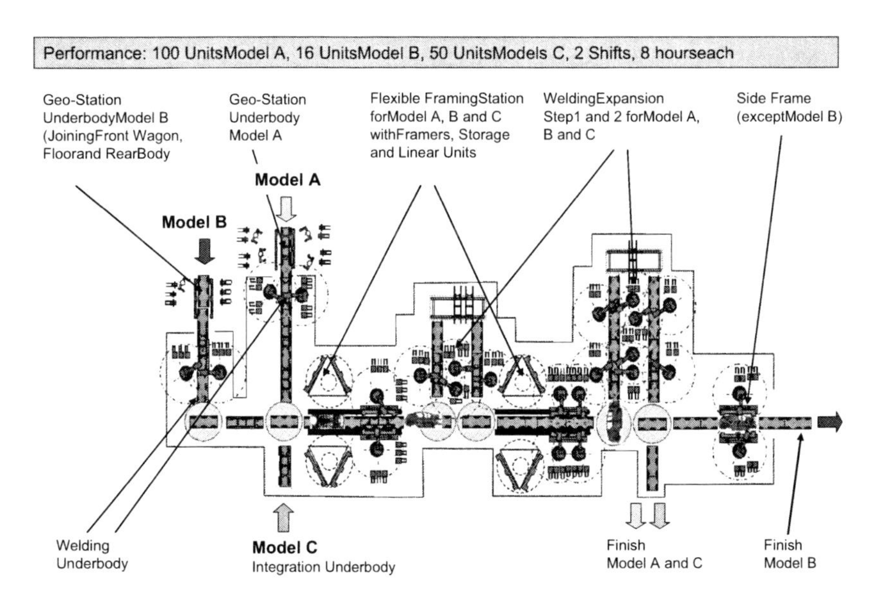

Fig. 21.10 Meandering layout scenario no. 4, with three models of vehicle body

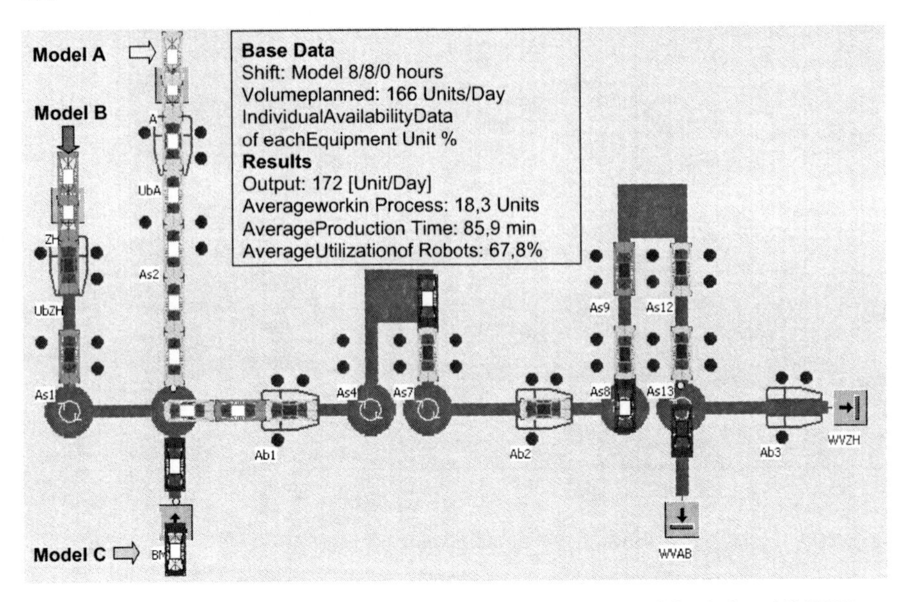

Fig. 21.11 Simulation-model scenario no. 4: partial load with three models, A, B and C [IPH]

Figure 21.10 shows the example of a meandering layout in scenario 4 with three models of body, while Fig. 21.11 shows the corresponding simulation model.

The simulation included start-up (scenario 1), peak production (scenario 2), a model changeover (scenario 3) and modification with up to three models of body (scenario 4) for various platforms. This showed that the dynamic approach also delivers clear advantages over the classic concepts. A short migration time for the plant system in the various operating phases is made possible by modular configuration, compatibility, scalability and the ability to pre-test extensions to the system. The meandering structure shows slight advantages over the Tetris structure. This is due to a higher output rate, somewhat lower investment cost for handling, conveying and a simpler control system.

21.4 Practical Evaluation and Implementation

This phase generally revealed that the meandering and Tetris concepts fulfill all requirements regarding the Migration Manufacturing for flexible body production, in terms of models and their variants, migration and model compatibility and flexibility with respect to unit output rates.

A commercial evaluation was then made of the meandering concept in order to compare it with the single-line structure commonly used to date. After discussion of the classical cost justification methods, an approach based on the cash-flow per-

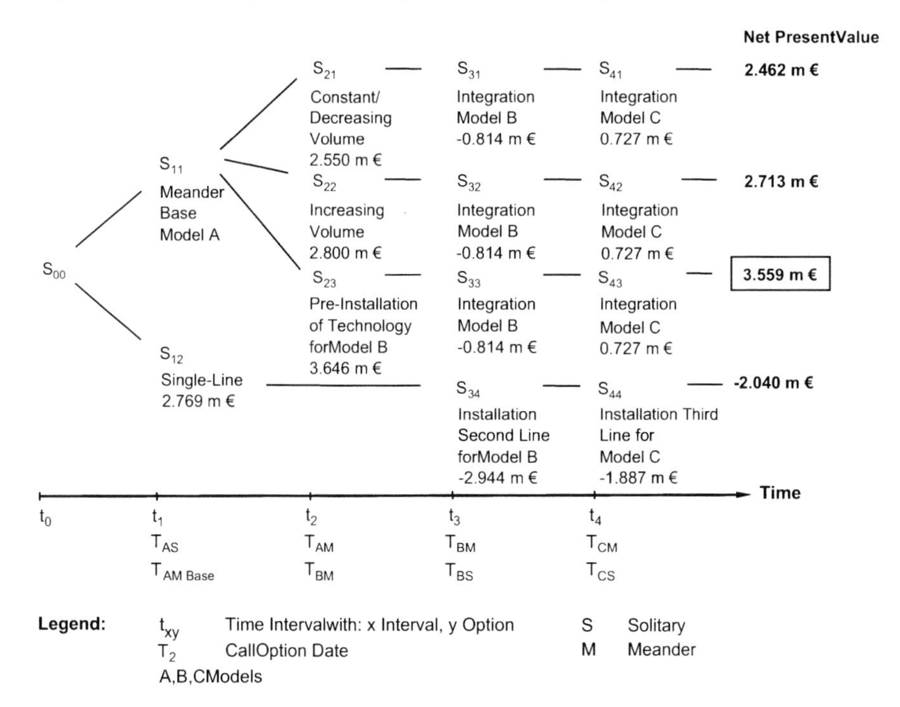

Fig. 21.12 Schematic outline of investment evaluation

formance of investments for the structural concept and its expansion options was adopted. In each case, a simplified set of real options adapted to suit these special tasks was applied (Hommel, 1999; Zäh, 2003). Figure 21.12 shows the set of real options applicable to the Migration Concept, along with alternative single-line approaches to production. The highest positive cash value is reached with the decision path $S_{11} - S_{43}$, whereas the Net Present Value for the single line with the subsequent investments is even negative.

The results clearly show that the meandering structure with an initial basic investment and an option to expand the plant is more likely, compared with a single-line structure, to increase unit output for model A and favor subsequent phased investment for models B and C.

The advantage of a phased, migrating investment is made particularly clear by the faster return on investment. Ongoing production creates the investment resources for the expansion phase of the first model A, and also finances the plant investment for subsequent products (Fig. 21.13). At six months, the concept shows a clearly shorter lead-in time for planning and implementation for initial start-up than the usual ten months for a single-line concept.

The evaluation of investments with real options clarifies the expanded decision-making possibilities available to management. Investment costs are no longer re-

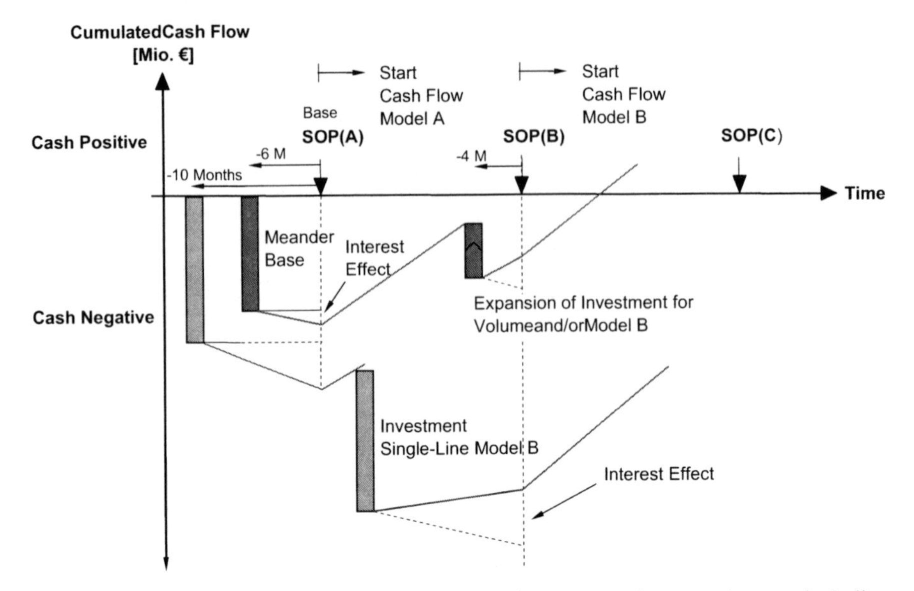

Fig. 21.13 Average cash flow of a migrating production structure in comparison to single-line production (principle process)

garded as fixed items, unable to be changed once the budget is approved. Due to dynamic market requirements with increased risk, value-orientated management requires technical concepts with structures that can be adapted over the lifetime of the project.

The latest research on Migration Manufacturing shows not only a significant advantage on the investment or cash flow side, but also a significant reduction on product cost. These items are for example maintenance, supply chain, space, energy, management, quality and most important up-time improvement. Related on the complexity of the production the variable production cost can be reduced between 5% and 14% compared to separate lines. Installed production equipment designed for Migration Manufacturing can be adopted more easily for a new model, than completely new installed lines. The efforts for planning, engineering, installation and production start-up can be reduced between 50% and 80% compared to individual production lines.

The migration-based approach furthermore leads to an expansion of the corresponding manufacturing methods, with previously-fixed production concepts made compatible with change. This includes a permanent inflow of risk, market-related and technological factors into the migration of structural concepts.

The concept has been already realized and practically tested in sections of the body-in-white shop. The risk to install such a new concept for a complied production line with different car models in one step is high, mainly because of the necessary fast ramp up of each new model. However, this risk can be compensated by an experienced engineering team, intensive teamwork with the line builder and a detailed virtual production simulation. The requirement for management to apply this innovative manufacturing concept becomes increasingly important due to the economic pressure, the volatile market demand and the shorter product life cycle.

21.5 Conclusion and Outlook

Where international mobility and willingness to facilitate change are concerned, the Migration Concept for a flexible approach to models and variants in the production of vehicle bodies is not just applicable to the car industry, but is a characteristic of a whole new trend.

Migration (from the Latin for "to move from one place to another") refers in this context to the technological modification or transformation of all or part of a production system (e.g. of vehicle body lines) at a manual, semi-automated or highly-automated level, with the possibilities that this offers, in terms of the growth, division, partial or complete consolidation of capacity, and/or reduction of a plant system.

The Migration concept is distinguished from flexible and reconfigurable manufacturing systems in the following aspects:

Flexible production systems for a vehicle body shop require a known production program to justify a 10 to 30% higher investment level compared to a single line approach. In the nineties, the large car manufacturing companies invested into Flexible Manufacturing Systems (FMS) for the body-in-white production with the expectation of a broad re-use of the line for its next models. With a continuously changing car design for safer and stiffer car bodies, the use of new materials and joining technologies and a shortening product life cycle, the FMS did not fulfill the financial and technical requirements for model flexibility.

Thereafter automotive production lines became less flexible, less sophisticated and, therefore, less expensive, but could only be used for one model cycle. Flexible Production Systems were on the decline, except for the manufacturing of a defined product family based on the same body platform.

Migration Manufacturing Systems (MMS) consist of a less complex base line with a planned migration track for additional production equipment when needed. The ability for migration does not require additional investment. The new not yet decided product can be integrated into the existing line by using virtual planning and integrated digital engineering tools.

Reconfigurable Manufacturing Systems for car body production become popular in times of high financial pressure. In theory, used robots or other non-product specific equipment such as turn tables or welding guns can be re-integrated into new production lines. In practice, such equipment does hardly find its way into new product lines. The reasons are the availability of the equipment for the new line, while the old product is still running and the state of technology with its influence in line utilization, integration ability and maintenance. Migration Manufacturing consists of an open, state of the art base system. The meanders or other production segments can consist of reconfigurable and older systems. The re-integrated equipment suits the temporary scalable extension of the body shop line or specialty applications, such as end of line measurement equipment.

The new approach of Migration Manufacturing Systems focuses on an "expansion-on-demand" approach for production that avoids any investment into flexibility, which may be needed for future products.

The difficulties to realize Migration Manufacturing originates from manufacturing engineering. Planning and layouts must be based on virtual and digital engineering process. The manufacturing engineers must learn to work with scenario techniques instead of using precise model forecasts from design and marketing. In addition, the production layout and required space will be variable during the complete life cycle of the car. New engineering tools and manufacturing concepts must be used consequently in order to achieve the lowest investment level for automotive production.

The engineering methods and manufacturing technologies required for implementing the Migration Concept in vehicle body construction are largely already available and practically tried-and-tested, although certain individual elements and factors still need to be adapted to the specific requirements of the Migration Manufacturing. These include such items as the further development of migration-compatible layout planning, different plant system components, floating work content in the welding cells and transport control logic for the production units used in vehicle body manufacture.

It can be affirmed that a company that gets in early with the principles of migration manufacturing can count on gaining a clear commercial advantage over its competitors. The basic investment in the new production infrastructure pays for itself in a relatively short period of time. The migration of the production structure for the integration of new products is then carried out at a cost that is always lower and more quickly implemented than that of new investment in a single-line concept. The entry barriers to competitors are thus increased. An investment level is also established for additional models, which corresponds to proportional investment in a flexible, variant-compatible approach to body production. However, the main differentiating concept characteristic of Migration Manufacturing lies in the ability to use new technologies to integrate highly different models of vehicle bodies from various OEM customers into a single production process.

The concept of Migration Manufacturing is well applicable for larger production installations outside of the automotive sector, such as electronics, food and beverage or even the pharmaceutical and chemical industries. Also in these industries a reduced investment in production equipments with scalable production lines enables management to improve the financial side of the business case as well as to mitigate market risks.

References

Hommel U., Müller J. 8/1999, Realoptionsbasierte Investitionsbewertung, (Real Option Based Investment Evaluation), Finanz Betrieb, pp 177–188

Karmann, der Spezialist für exklusive Aufträge (Karmann, the Specialist for Exclusive Cars), Interview Karmann Post, Magazine 183, Spring 2006, pp 4–8

Koren Y., Jovane F., Heisel U., 1999, Reconfigurable Manufacturing Systems. A keynote paper. Annals of the CIRP 48/2:527–540

Meichsner T.P., 2003, Das veränderte Anforderungsprofil in der Prozesskette Kleinserienkarosserie aus der Sicht des Systemlieferanten am Beispiel Maybach und Lamborghini Gallardo (The Changing Requirements in the Process Chain for Low-Volume Car Body from the Perspective of a System Supplier with Examples of Maybach and Lamborghini Gallardo). 1st European Strategy Conference: Future Production Systems Car Body Process Chain, International Expert Group on Car Body Design, 1st/2nd July Bad Nauheim, Germany, Proceedings pp 249–274

Meichsner T.P., 2003, Engineering Loop im Karosseriebau aus Sicht des Zulieferers (Engineering Loop for the Car Body Production from the Perspective of a Supplier). International Car Body Expert Circle, 16. Conference, Process Chain Automotive Body, Car Body Manufacturing and Joining Technology Today and in the Future, Proceedings, Esslingen, Germany, 7.-9. May

Meichsner T.P., 2002, Engineering Loop: Best-Practice-Beispiele bei der Realisierung einer Al-Leichtbau-Karosserie (Best-Practice-Examples for the Realization of an Aluminium Light-Weight Body-in-White), The Aluminium Automobile Process Chain, Automotive Circle International Conference, 2nd European Al-Automobile Conference, Bad Nauheim, Germany, 7./8. November, Proceedings pp 175–204

Meichsner T.P., 2006, Fertigungskonzepte im Karosseriebau für die Spezialserie (Manufacturing Concepts for the Car Body Production of Specialty Cars); Proceedings; Conference Car Body; Ulm, Germany; 9th und 10th October

Meichsner, T. P.: Migrationskonzept für einen Modell- und Variantenflexiblen Karosseriebau (Migration Concept for a Model and Variants Flexible Automotive Body Manufacturing), PhD Thesis University of Hanover, Germany, 2007. Publisher: PZH Produktionstechnisches Zentrum GmbH, Garbsen Germany. ISBN: 978-3-939026-63-1. ISSN: 1865-5513

N.N., 2006, Kompakte SUV überholen die großen Offroader (Compact SUV overtake large Offroader), Autobild, 31/2006, p 11

Polk R.L. 2006, Marketing Services, Marktentwicklung Nischenfahrzeuge, Datenerhebung; Studie; Essen (Marketing Services, Market Development, Data Survey Report), Essen, Germany

Wemhöner N., 2006, Flexibilitätsoptimierung zur Auslastungssteigerung im Automobilbau, (Optimization of Flexibility to Increase Utilization in the Automotive Production); WZL RWTH Aachen, Germany, PhD Thesis. University of Aachen. Shaker Verlag, Report 12/2006, p 192

Westphal J.R., 2001, Komplexität in der Produktionslogistik. Ein Ansatz zu flussorientierten Gestaltung und Lenkung heterogener Produktionssysteme, p 8. (Complexity in the Production Logistics. An Approach for a Flow-Oriented Design and Control of Heterogeneous Production Systems, p 8), issued by Wissenschaft, E., Wiesbaden: Gabler

Wiendahl, H.-P., Hernández, R., 1999, Bausteine der Wandlungsfähigkeit zur Planung wettbewerbsfähiger Fabrikstrukturen (Modules of the Changeability for Planning Competitive Plant Structures); Proceedings 2nd German Conference for Factory Planning 2000+, Stuttgart, Germany, 26nd/27th October

Wiendahl H.-P., 2002, Wandlungsfähigkeit: Schlüsselbegriff der zukunftsfähigen Fabrik. wt Werkstattstechnik online 92/4:122–127, Changeability: the Key for Future Factories

Wiendahl H.-P., ElMaraghy H.A., Nyhuis P., Zäh M.F., Wiendahl H.-H., Duffie N., Brieke M., 2007, Changeable Manufacturing – Classification, Design and Operation. Annals of the CIRP 56/2:783–809, A keynote paper

Wiendahl H.-P., Hernández R., 2000, Wandlungsfähigkeit – ein neues Zielfeld der Fabrikplanung. Industrie Management 16/5:37–41, Changeability – A new Target Field for Factory Planning

Zäh M.F., Sudhoff W., Rosenberger H., 2003, Bewertung mobiler Produktionsszenarien mit Hilfe des Realoptionsansatzes. ZWF (Journal of Economic Factory Operation) 98/12:646–65, Evaluation of mobile Production Scenarios with an real Option Approach

Chapter 22
Changeable Factory Buildings – An Architectural View

J. Reichardt[1] **and H-P. Wiendahl**[2]

Abstract The construction of factories is an extensive and complex single-piece production in our economy. Only a well-balanced consideration of all planning criteria can ensure a project's success in the long run. A factory's design cannot be derived from production requirements only but it also grows out of the context of location, climate, society and human beings within an extremely creative process. Over and above its purely functional suitability the sensible structure of a building can give a positive impulse to future changeability aspects as well as motivation and communication. This chapter presents construction relevant design fields and their elements as they arise from the planning of the manufacturing processes. The versatile network of buildings will be analytically classified according the design aspects of buildings structure as well as the future changeability of manufacturing processes.

Keywords Design fields, Changeability enablers, Architects View

Our cities and rural areas are increasingly marred by inhospitable industrial estates. Widespread confusion over economic goals such as 'inexpensive' and 'economically efficient' are used to justify anonymity, banality and ugliness. Architectural critic Christoph Hackelsberger once ironically called these areas 'industrial steppes'. Nobody would willingly stay there for longer than the contracted hours of work.

When industrial enterprises wish to present themselves to the public with their buildings, they do so mostly with their headquarters. Their factories, however, are rarely presentable. Minimal budgets, tight construction schedules and hierarchical instead of co-operative planning procedures prevent the most natural way of building, i.e. of developing good architectural solutions for a construction project. It is, after all, in the field of industrial building that architecture has retained a measure of

[1] Architects BDA, Germany

[2] University of Hannover, Hannover, Germany

freedom removed from the stylistic fashions to which the classical types of building are subjected.

Industrial architecture is open to new technologies, construction systems and materials; and the quest for new building concepts is an absolutely thrilling affair. The results of joint endeavors in this direction can be extremely cost-efficient production plants, well-proportioned interior spaces, fascinating structures and pleasant working environments.

What must be done to make people aware of this and to promote a new era of 'industrial culture'? The joint efforts of all those involved in industrial building, and a growing mutual understanding of the others' aims are the only promising path to long-term success.

The buildings and the intermediate residual spaces are generally accepted without complaint as a social no-man's-land. Every single Euro invested in excess of the absolute minimum of the 'economic' building, and every additional day spent in planning or constructing is regarded as undue extravagance.

Findings from basic research into the decay of the aesthetic quality of buildings for business, trade and industry reveal a concentration of different irregularities in our society. In the rat race for market-economical competitiveness, urban and rural authorities sacrifice their architectural identity. Vast areas, highly subsidized or simply given away for nothing, are thus removed from the influence of urban or regional planners committed to quality design. The entrepreneurs themselves literally spoil their own future by short-sighted, strategically unwise decisions to build. Already the next change in production will often force companies to move to another location, mostly leaving behind them permanently denaturised wastelands in cities or rural areas.

22.1.1 The Factory Planners View

The aim of the whole factory must be oriented to serving the market with products at the desired quality and with the lowest possible cost. The main scope of factory design remains the planning of the facilities, the organization and the employees, Fig. 22.1 (Wiendahl and Hernández, 2006).

Facilities are comprised of the equipment for manufacturing and assembly, as well as supporting equipment for logistics and information technology. The factory organization includes the organizational principles, their processes within the factory and the external interfaces. Aspects associated with labor such as the working environment, the payment system and the working time model are major tasks of the employees related design field. The business culture and the increasing importance of sustainability frame these three tasks whereas the factory location and its buildings are the foundation. The flow of material, information, personnel, work, energy, media and capital are the foundation for the manufacturing processes.

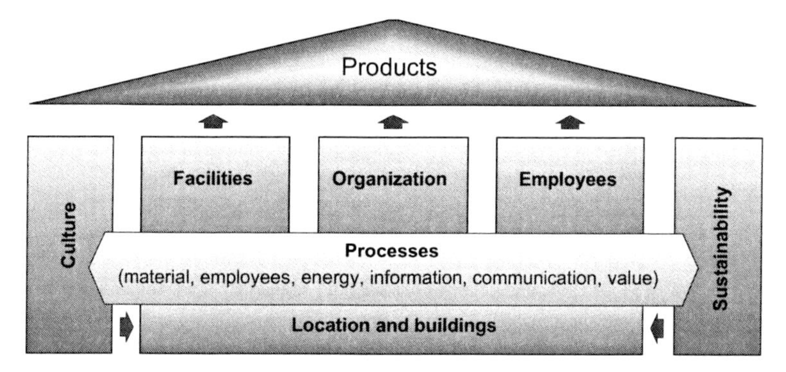

Fig. 22.1 Design fields of factories

The life cycle of the products and equipment often was in the same range of the life cycle of the technical building equipment; however, this is no longer the case. Therefore, the buildings must be able to follow more or less fast changes of the product, the processes and the facilities. For this reason, it is especially important to synchronize the design of factories location and buildings to be well timed with the three main design fields.

In addition producing companies have to react to the turbulence in the manufacturing environment by changing their factories in ever decreasing intervals. The extent of those necessary changes today often exceeds the mere possibilities concerning an individual technical system (e.g. a single machine) but has to include related areas as well as the whole site. This leads to the concept of the transformable factory (Wiendahl, 2002; Wiendahl and Hernández, 2006; Dashchenko, 2006; Westkämper, 2006; Wiendahl et al., 2007).

For the purpose of deriving clearly defined objects that are looked at for changeability, they can be classified into means, organization and space (Nyhuis et al., 2005) and allocated according to four structure levels of the factory. The lowest level is represented by a workstation where a single process is performed e.g. turning or assembly. The next level embraces a system or cell arranged in a working area: here a part family would be manufactured completely. The next higher level is a segment located in a building, in which a group of products is produced. The highest level is the site, on which several buildings and facilities are arranged according to a master plan. Here not only the actual production for a portfolio of products is performed but also the design department and support functions like sales, engineering, human recourses, information systems etc. are located.

Based on the work of Hernández (2003), Wiendahl et al. (2005), Nyhuis et al. (2005) and Heger (2007), all together 261 objects have been identified and aggregated into 25 categories, which are depicted in Fig. 22.2.

In this context reconfigurable manufacturing systems and reconfigurable assembly systems can be seen as a basis for a transformable factory. Examples for such

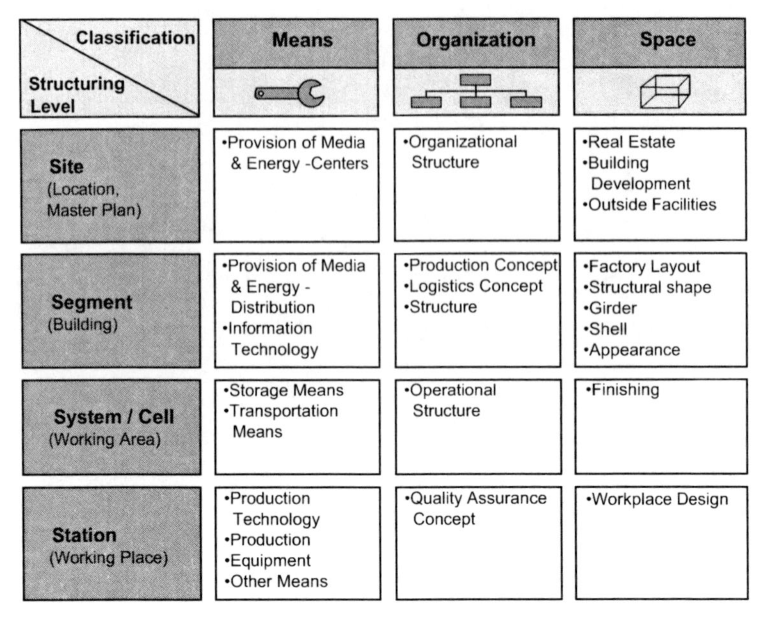

Fig. 22.2 Objects of a factory

change processes are the extension of buildings, the adaptation of the company organization or the relocation of a sub-factory to a low wage country.

22.1.2 The Challenge: Multi-User, Changeable and Scalable Buildings

Design Level: Location

With regard to the current global exchange of goods between company networks the choice of location is affected by a sum of global, regional and local factors. At local level, the strategic positioning of the location within the logistic network as well as considering the infrastructure capacities of the road, rail, air and sea transportation systems is of primary importance. In order to ensure trouble-free movement of goods in the long run access to motorways and/ or high ways should avoid areas susceptible to congestions. Traffic routing should also avoid cross-town links and bridges that impose restrictions to height, width and weight. Especially for export-oriented industries it is of increasing importance to be well situated with regard to airports, as this is beneficial to reducing reaction time. As to transport by railway attention should be paid to being nearby main tracks and it should be analyzed if it makes sense to maintain ones own rail link. Being located nearby distribution centers for goods as well as container terminals is advantageous. Regional shipping on rivers and canals is profitable only for either very bulky and heavy goods or cheap raw materials. When making a decision on choice of location the planned development of transport over a period of 15 to 30 years should be scrutinized.

A valuation system for alternative locations takes into account the aspects of infrastructure, supply, disposal, site, labor market, environment, expandability, planning and building laws as well as purchasing price and communal promotion. The various site/location evaluation criteria should be assessed and rated according to their relevance to a project's objective.

Design Level: Master Plan

The current and future efficiency of a concept for urban development is specified on the design level of general development sometimes called master plan. The general development's characteristics can either be stimulating or restraining for future factory changeability. The types of requirements that have to be met determine the choice of shapes as well as the criteria regarding protection of property. A general master plan combines guidelines for the layout of buildings and zoning of public thoroughfares and open spaces within potential construction stages. The deliberate choice and combination of structural shapes is crucial in achieving the highest possible degree of changeability.

Infrastructure	Supply, Disp. Media	Site	Environment	Laws and Conditions	Location Valuation
• road • rail • air • sea	• electricity • water • gas • hot water, heating steam • drainage • waste • data networks	• geometric shape / land register • soil conditions • obstacles existing buildings	• weather data • ventilation • plantation	• zoning plans • legally binding land-use plans • design ordinances • state laws • special regulations • masterplan	• preparation of land for building • supply, disposal • site • labour market • environment • expandability • planning and building laws • purchasing price • promotions

Fig. 22.3 Overview of design fields and elements of a master plan

22.2 Performance and Constituent Components of Factory Buildings

According to the four levels of design in factory planning explained in Fig. 22.2 it is highly advantageous, also with regard to the spatial considerations, to focus step by step on the corresponding levels of design. This enables joint and results-oriented rough to detailed planning. The special quality of this kind of synergetic operation occurs on each level in the form of integrative compilation of process aspects and architectural aspects. The special value of synergetic design is displayed in addition in a stronger joint focus concerning changeability in the process-related as well as in spatial approaches.

22.2.1 Form Follows Performance

At the end of the 19th century, the slogan 'form follows function', coined by the American architect and theoretician Louis Sullivan, saw functional necessity as the key factor for developing formal architectural solutions. In the heyday of the Bauhaus (a famous movement in German architecture), functionalist architects adhered to it to free architecture from the fetters of eclectic styles. In the second half of the 20th century many architects, reacting to the plain 'box', hoped for more variety and formal eminence of their designs by adhering to the motto 'function follows form'.

When it comes to adaptable factory buildings, these two strategies do little to achieve this goal, because it addresses only one aspect of the complex interaction between nature, man and the function and form of architecture. A frequent question arising in a building project is whether the present function and form will be viable in the long-term. Temporary programmes or short-lived aesthetic fashions are ill suited to foster robust, solid forms.

This is why holistic solutions, holistic in terms of process (function) and space (form), are in demand. What matters is to carefully and deliberately combine a number of significant elements with partial, ideally complementary solutions to complex problems. The term 'performance' is meant to denote the process of finding such solutions, from which the conviction of 'form follows performance' is deduced (Reichardt et al., 2004).

Hence, the specific formal design is not fixed beforehand, but the result of any spatial solution which develops out of the required functional performance. Based on these performance criteria new construction technologies should be used; energy consumption should be optimized and ecological parameters be meshed. The degree of flexibility deemed necessary is to be secured by defining the adaptability criteria for every area and level of the building. Interior design and furnishing should facilitate communication from person to person. On the whole, industrial architecture should help to foster corporate culture and identity and facilitate changeability.

22.2.2 Building Components

Architectural design of a single part of the master plan, namely the building, embodies the five shaping design fields regarding the components of a built structure: load bearing structure, outer shell, media, finishing and graceful appearance (Fig. 22.4).

As already explained it is almost impossible to predict future production processes and the areas they will require, or the structural changes an industrial enterprise will undergo. The greatest possible adaptability of a factory building to future changes is a touchstone of a company's operational flexibility.

This entails defining the adaptability/convertibility of every architectural section and the adequate efficiency of every supply and disposal system. This mainly concerns span widths, clear heights; floors load carrying capacity and structural provisions for horizontal and vertical extensions to the building. The supply and distribution systems must allow for different interior air-conditioning methods. This makes it necessary to conceive exposed, easily accessible installations, independent of other systems in the building so that they can be altered without disturbing ongoing production.

Load Bearing Structure

A load bearing structure consists of plane and columnar structural elements, bracings and foundations all of which are important for a building's stability. On-site fabricated or prefabricated parts made of steel, reinforced concrete, wood or light metal can be used.

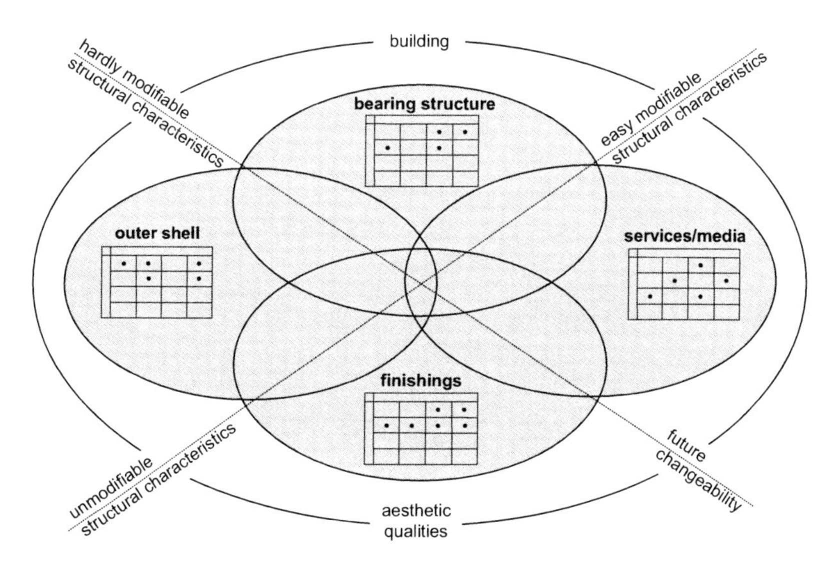

Fig. 22.4 Buildings design fields

The load bearing structure is the most durable part of a building's structure and, therefore, is the most difficult to change. It is normally designed to last for the entire service life of the building. The choice of the load bearing structure has great influence on the long-term usability as well as on the architectural interior and exterior shapes. A formidable architectural design sensibly integrates the requirements concerning the structural set-up and cost (Reichardt, 1999).

Unfortunately in practice the aspect of the required profitability concerning the load bearing structure is commonly mistaken for the search of the least expensive construction – often a momentous mistake.

Outer Shell

The outer shell draws the outline of the interior as an independent climatic space protected from the exterior. It consists of fixed closed or transparent elements for facades and roofs as well as of mobile elements like gates, doors, windows or smoke outlets.

Services and Media

The term media denotes the entirety of all control centers, routings and connections that are essential for production processes, user comfort and building security.

Finishings

Finishings include stairs, core spaces, special fixtures as well as all elements irrelevant to stability. Non-load bearing walls should be easy to change; flexible office walls allow for spatial changes within a few hours.

Aesthetic Qualities

These functional and construction relevant criteria are complemented by the focus on the more subjective aesthetic qualities of a building within the design field of impression and graceful appearance.

Our powers of aesthetic perception receive countless bits of information in the field of tension between monotony and chaos. Yet unity and diversity are interdependent; they are the necessary extremes the balance of which has to be readjusted for every building project. If and when the scales tip towards regularity and monotony, the immediate effect is boredom, while agitated multiformity is perceived as chaotic.

The ideal solution is to formulate permanently valid, sustainable design frameworks for both architectural and urban structures. Building heights and a material canon fixed over long terms, but leaving scope for creative designs will make original solutions possible also in the future.

22.3 Synergetic Planning of Processes, Logistics and Buildings

The questions of how to deal with the design parameters for industrial buildings outlined above, and how to structure the project-determining contract specifications leads on to the subject of space-planning methodology.

A critical review of current practices in industrial building reveals serious differences between the state-of-the-art work-flow of the automobile industry and the planning and construction method applied to a factory complex. Pioneer architectural thinkers like Richard Buckminster Fuller of the 1930s used the term 'cultural lag' to describe the traditional backwardness of the construction industry as against other more progressive branches of industry such as the automobile or aircraft industry which were said to be twenty years ahead.

A closer look at the usual practice of industrial architecture reveals that it is substantially different from the 'digital' working methods of progressive industries. As a rule, the separate definition of project sections is carried out in a linear, sequential way, while the automobile industry has adopted the time-saving method of 'simultaneous engineering'.

The new quality of co-operative factory planning, from both process and spatial perspectives, results from an early integration of spatially defined project sections (process, site, building and services).

In this approach the three-dimensional structure is continuously refined from rough draft (assumptions) to final design (fixations), with decisions being constantly evaluated in joint discussions of alternative solutions. The targets for every project section, such as adaptability or defined functional efficiency/performance should be clearly laid down in the respective contract specifications, translated into three-dimensional models and tested with the overall design. Figure 22.5 depicts this twofold yet integrated approach with the relevant fields of design on the process and the building side.

The capacities of currently available CAD/CAM database technology are suited to develop and constantly up-date consistent integrated three-dimensional digital models in order to optimize the project in its entirety and cyclically to control overall spatial quality.

In addition, the 3D synergetic factory model should be optimized by identifying and eliminating potential 'collisions' between different contracting firms and by constant quality checks. This ensures that conflicts between process, structural and services design – which would influence costs, construction time and quality negatively – are recognized early on and do not have to be dealt with and eliminated on the construction site afterwards.

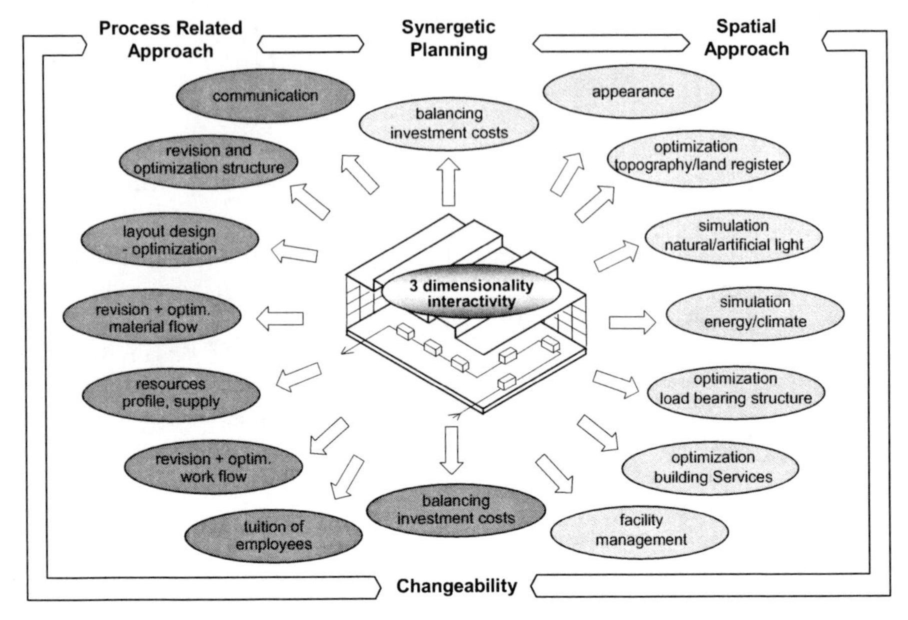

Fig. 22.5 Synergetic data model approach

22.4 Industrial Example of a Transformable Factory

As far back as around 1900, the Modine Company produced the radiators for Henry Ford's legendary 'Tin Lizzy'. Today, almost 10,000 employees around the world produce progressive motor-cooling systems. In Wackersdorf, Modine Montage Ltd. in Germany exclusively produces radiators for all new 3-series and X-series BMWs.

When the company received a five-year-contract with BMW of about 500 Million Euro the original building was too small for installing the required capacity. Therefore, a green-field factory planning project was initiated. The requirements were an expected growth of production volume and number of variants by 110% over the next five years. The future after those five years was uncertain. Therefore, the new factory was planned on the principles of synergetic factory planning (Nyhuis and Reichardt, 2005) and with a special focus on transformability. The cost was estimated with 20 Million Euro.

The special requirement of both architectural planners and production process engineers co-operating from the first sketch through to completion made joint workshops necessary from early on to ensure that the architectural optimally supports the lean production process. The sustainability, flexibility and communicativeness of the building are now the result of the efficient interaction of the factory structure, outer shell, services and interior finishing work (Reichardt and Gottswinter, 2004).

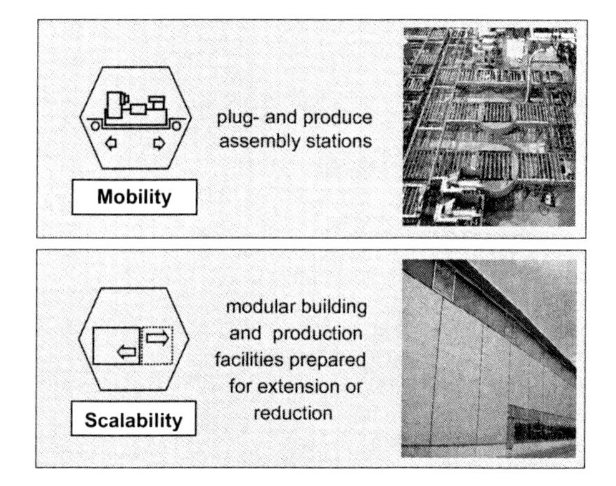

Fig. 22.6 Examples for transformable factory elements

The resulting factory is able to react to changes in the environment in many different ways. Figure 22.6 illustrates two examples. The assembly stations are mobile and can be relocated within hours where they are needed. The modular building and the layout characteristics enable a growth of the different production departments as needed.

This is achieved by means of a modular building (e.g. panels, offices, piping, and cladding) and modular processes (e.g. autonomous teams, plug-and-produce equipment). The media distribution covers all areas of the factory via the ceiling and therefore imposes no restrictions on the layout and during relocation of equipment or structural partitions.

In addition to transformability, the support of the personal communication was a special concern. The whole management of the factory is located in a gallery housed in bureaus with glass walls having a direct view of the running manufacturing process. If any disturbance occurs the responsible manager reaches the floor within minutes or they meet in a special room positioned directly above the production lines in the ceiling of the building.

Figure 22.7 gives an impression of the building structure. It consists of two modules each of which has two sub-modules with a span width of 36 respectively 18 meters. Each module is autonomous with regard to media and energy supply. The building can be expanded using two more modules if necessary without interrupting the production.

The high grid width supports the transformability in the building. Besides the technical basis the workforce had a strong will to adapt and change. All these aspects lead to a highly transformable factory (Busch, 2005) that received the "Best Assembly Factory" award in Germany 2006.

This example does not mark the end of the factory evolution. In some cases, even temporary factories have been installed, e.g. to assemble streetcars, in only 6 months.

Fig. 22.7 Layout of the transformable factory

22.5 Conclusion

Factory buildings are usually treated as the last step in designing a factory and are often viewed as a mere protection against the weather influences, and their cost should be minimized. Thereby, three important aspects are traditionally neglected. First a cheap building may cost through out its lifetime much more than the initial investment, because of bad insulation, inefficient energy systems and rigid technical installations. Secondly the appearance of a cheap building gives the wrong impression to the customers since it does not coincide with the desired product/company image. And third the ability of the building to react to changes in the production with respect to the products design and volume as well as to changes in production technology and equipment is completely neglected.

Three principles can help to avoid these traps. Fist the approach of "form follows performance" aims for a design which sees the building as part of the performance of the whole factory fostering especially personal communication, the aesthetic appearance as part of the corporate identity and the motivation for the employees. Secondly a synergetic planning method bringing together factory planners and architects from the very beginning of the project avoids wrong interpretation of the main objectives of the factory and brings up innovative solutions. And finally the early consideration of transformability characteristics leads to constructing factories, which are able to follow the fast changes in the production conditions through a modular and scalable design with sometimes even mobile components.

References

Busch M., 2005, Synergetisches Fabrikplanungsprojekt am Beispiel der Automobilzulieferindustrie (Synergetic factory planning project with an example of the automotive supplier industry), 6. Deutsche Fachkonferenz Fabrikplanung, Fabriken für den globalen Wettbewerb (6th german symposium factory planning, factories for the global competition) Ludwigsburg, 08.–09.11.2005

Dashchenko O., 2006, Analysis of Modern Factory Structures and Their Transformability. In: Dashhchenko A.I. (ed) Reconfigurable Manufacturing Systems and Transformable Factories, pp 395–422

Heger C., 2007, Bewertung der Wandlungsfähigkeit von Fabrikobjekten (Evaluation of the transformability of factory objects). PZH-Verlag, Garbsen, PhD Thesis, Univ. Hannover

Hernández R., 2003, Systematik der Wandlungsfähigkeit in der Fabrikplanung (Systematics of transformabiliyt in factory planning). VDI Verlag, Düsseldorf, PhD thesis

Nyhuis P., Reichardt J., Elscher A., 2005, Synergetische Fabrikplanung (Synergetic factory planning), 6. Deutsche Fachkonferenz Fabrikplanung, Fabriken für den globalen Wettbewerb (6th german symposium factory planning, factories for the global competition), Ludwigsburg, 08.–09.11.2005

Nyhuis P., Kolakowski M., Heger C.L., 2005, Evaluation of Factory Transformability, 3rd International CIRP Conference on Reconfigurable Manufacturing, Ann Arbor, USA, 11.–12.05.2005

Reichardt J., Wandlungsfähige Gebäudestrukturen 1999 (Transformable building structures) Proceedings 2nd Deutsche Fachkonferenz Fabrikplanung – Fabrik 2000+, Stuttgart

Reichardt J., Eckert A., Baum T., 2004, Form follows performance: Entwicklung des modularen, weit gespannten Tragwerks für ein neues Montagewerk (Form follows performance: Development of a wide span load bearing structure for a new assembly factory) Stahlbau, 7/2004, p 475ff

Reichardt J., Gottswinter C., 2004, Synergetische Fabrikplanung – Montagewerk mit den Planungstechniken aus dem Automobilbau realisiert, (Synegetic Factory Planning – Assembly factory realized with planning methods from the automotive industry) industrieBAU, 3/2004, p 52ff

Westkämper E., 2006, Factory Transformability: Adapting the Structures of Manufacturing In: Dashhchenko A.I. (ed) Reconfigurable Manufacturing Systems and Transformable Factories, pp 371–381

Wiendahl H.-P., 2002, Wandlungsfähigkeit: Schlüsselbegriff der zukunftsfähigen Fabrik (Transformability: key concept of a future robust factory). wt Werkstattstechnik online 92/4:122–127, http://www.werkstattstechnik.de/wt/2002/04

Wiendahl H.-P., Nofen D., Klußmann J.H., 2005, Planung modularer Fabriken – Vorgehen und Beispiele aus der Praxis (Planning of modular factories – approach and practical examples). Carl Hanser Verlag, München/Wien

Wiendahl H.-P., Hernández R, 2006, The Transformable Factory – Strategies, Methods and Examples. In: Dashhchenko A.I. (ed) Reconfigurable Manufacturing Systems and Transformable Factories, pp 383–393

Wiendahl H.-P., 2006, Global Supply Chains – A New Challenge for Product and Process Design, 3rd International CIRP Conference on Digital Enterprise Technology, Setúbal, Portugal 18.–20.9.2006

Wiendahl H.-P., ElMaraghy H.A., Nyhuis P., Zäh M.F., Wiendahl H.-H., Duffie N., Brieke M., 2007, Changeable Manufacturing – Classification, Design and Operation. Annals of the CIRP 56/2:783–809, A keynote paper

Index

A

B

C

N

O

P

Q

R

S

T

U

V